UB MAGDEBURG   MA9
002 494 957

Titles in This Series

14 **Yu. S. Il'yashenko, editor,** Nonlinear Stokes phenomena, 1992

13 **V. P. Maslov and S. N. Samborskiĭ, editors,** Idempotent analysis, 1992

12 **R. Z. Khasminskiĭ, editor,** Topics in nonparametric estimation, 1992

11 **B. Ya. Levin, editor,** Entire and subharmonic functions, 1992

10 **A. V. Babin and M. I. Vishik, editors,** Properties of global attractors of partial differential equations, 1992

9 **A. M. Vershik, editor,** Representation theory and dynamical systems, 1992

8 **E. B. Vinberg, editor,** Lie groups, their discrete subgroups, and invariant theory, 1992

7 **M. Sh. Birman, editor,** Estimates and asymptotics for discrete spectra of integral and differential equations, 1991

6 **A. T. Fomenko, editor,** Topological classification of integrable systems, 1991

5 **R. A. Minlos, editor,** Many-particle Hamiltonians: spectra and scattering, 1991

4 **A. A. Suslin, editor,** Algebraic K-theory, 1991

3 **Ya. G. Sinaĭ, editor,** Dynamical systems and statistical mechanics, 1991

2 **A. A. Kirillov, editor,** Topics in representation theory, 1991

1 **V. I. Arnold, editor,** Theory of singularities and its applications, 1990

Advances in SOVIET MATHEMATICS

Volume 14

Nonlinear Stokes Phenomena

Yu. S. Il'yashenko
Editor

American Mathematical Society
Providence, Rhode Island

ADVANCES IN SOVIET MATHEMATICS

EDITORIAL COMMITTEE

V. I. ARNOLD
S. G. GINDIKIN
V. P. MASLOV

Translation edited by A. B. SOSSINSKY

1991 *Mathematics Subject Classification.* Primary 32G34, 34A20, 35C35, 58D27; Secondary 34A30, 43C05.

Library of Congress Catalog Card Number: 91-640741

ISBN 0-8218-4112-2;   ISSN 1051-8037

COPYING AND REPRINTING. Individual readers of this publication, and nonprofit libraries acting for them, are permitted to make fair use of the material, such as to copy an article for use in teaching or research. Permission is granted to quote brief passages from this publication in reviews, provided the customary acknowledgment of the source is given.

Republication, systematic copying, or multiple reproduction of any material in this publication (including abstracts) is permitted only under license from the American Mathematical Society. Requests for such permission should be addressed to the Manager of Editorial Services, American Mathematical Society, P.O. Box 6248, Providence, Rhode Island 02940-6248.

The appearance of the code on the first page of an article in this book indicates the copyright owner's consent for copying beyond that permitted by Sections 107 or 108 of the U.S. Copyright Law, provided that the fee of $1.00 plus $.25 per page for each copy be paid directly to the Copyright Clearance Center, Inc., 27 Congress Street, Salem, Massachusetts 01970. This consent does not extend to other kinds of copying, such as copying for general distribution, for advertising or promotional purposes, for creating new collective works, or for resale.

Copyright © 1993 by the American Mathematical Society. All rights reserved.
Printed in the United States of America.
The American Mathematical Society retains all rights
except those granted to the United States Government.
The paper used in this book is acid-free and falls within the guidelines
established to ensure permanence and durability. ∞
This publication was typeset using $\mathcal{AMS}$-TeX,
the American Mathematical Society's TeX macro system.

10 9 8 7 6 5 4 3 2 1   98 97 96 95 94 93

*To the memory of Jean Martinet,
master mathematician
and admirable human being*

Contents

Foreword	ix
Nonlinear Stokes Phenomena YU. S. IL'YASHENKO	1
Finitely Generated Groups of Germs of One-Dimensional Conformal Mappings, and Invariants for Complex Singular Points of Analytic Foliations of the Complex Plane P. M. ELIZAROV, YU. S. IL'YASHENKO, A. A. SHCHERBAKOV, and S. M. VORONIN	57
Tangents to Moduli Maps P. M. ELIZAROV	107
The Darboux-Whitney Theorem and Related Questions S. M. VORONIN	139
Nonlinear Stokes Phenomena in Smooth Classification Problems YU. S. IL'YASHENKO and S. YU. YAKOVENKO	235

Foreword

The nonlinear Stokes phenomenon occurs in the local theory of differential equations, in short—in *local dynamics*, and finds its application in singularity theory. Local dynamics dates back to Poincaré's work related to the famous classification of phase portraits of autonomous systems near nondegenerate singular points. For analytic systems, these portraits can be of only one of the four types: saddle, node, focus, or center. These four have become so firmly associated with the new "qualitative" theory that on the front cover of the recently published biography of H. Poincaré they appear just below the portrait of the great mathematician.

The main efforts in analytic local dynamics till about the mid-sixties were focused on the problem of analytic equivalence of a vector field or a holomorphism to its linearization in some neighborhood of the singular point. Necessary conditions for such an equivalence, close to sufficient ones, were obtained by Bryuno and Siegel. Recently Yoccoz found a necessary and sufficient condition for an analytic one-dimensional map to be analytically equivalent to its linear part.

Starting from the mid-sixties, the main interest in local dynamics switched to normal form theory of resonance maps and fields. Resonances appear when there are certain integer identities between the eigenvalues of the linearization. After the work of Bryuno (1967, 1971) and Il'yashenko (1981) it became clear that the formal classification in the resonance case usually differs from the analytic one: the normalizing series diverge, as a rule. Nevertheless, the geometric reason for such a divergence remained unrevealed.

"*The existing proofs of the divergence,*" wrote Arnold in 1972, "*are based on computations of the growth of coefficients and do not explain its nature (in the same sense that computation of the coefficients of the series* arctan z *does not explain the divergence of this series for* $|z| > 1$, *although it proves divergence).*"

The first resonance problem for which the cause of divergence was discovered turned out to be the problem of analytic classification of germs of holomorphisms $(\mathbb{C}, 0) \to (\mathbb{C}, 0)$ with identity linear part. In 1975 Ecalle discovered that the Borel transforms of the normalizing series in this problem have singularities. It is their presence which explains the divergence, since Borel transforms of converging series are entire functions on the complex plane. Such an approach could be termed "analytic"; it developed into the theory of *fonctions résurgentes* by Ecalle.

Another, geometric, approach consists in the following. The normalizing series turns out to be asymptotic for a holomorphic change of variables conjugating the initial germ with its formal normal form. But this change (the normalizing transformation) is defined in a sector (having its vertex in the fixed point) rather than in an entire neighborhood of the point. A punctured neighborhood of the singularity can be covered by a finite number (an atlas) of such sectors, each of them being endowed with a normalizing transformation (chart). *These transformations do not coincide on the intersections of the sectors*, and this is what is called the *nonlinear Stokes phenomenon*. The transition functions between the charts, defined in the intersections of the sectors, are (up to inessential technical details) the invariants of analytic classification for line holomorphisms tangent to the identity.

In solving different classification problems, the *realization* problem for moduli arises. More exactly, with each germ under consideration one can associate a complete set of invariants (the modulus of the classification). These moduli belong to a certain functional space, the *moduli space*. A map from the space of germs to the moduli space, called the *moduli map*, arises. An inverse image of an arbitrary point under this map is an equivalence class of germs. The realization problem is whether the moduli map is epimorphic. In other words, is it true that each element of the moduli space can be realized as the modulus for some germ?

For germs $(\mathbb{C}, 0) \to (\mathbb{C}, 0)$ tangent to the identity, the answer is positive. In order to construct a germ with the prescribed modulus, an almost complex structure is used. Such a geometric approach was developed independently by Voronin and Malgrange (1981). The functional moduli thus obtained are usually called the *Ecalle-Voronin moduli*.

The most important singular points of vector fields from the point of view of local dynamics are the so-called elementary singular points, i.e., those having at least one nonzero eigenvalue. Any isolated singularity of an analytic vector field on the complex plane after finite number of applications of the so-called σ-process, or *blow-up*, can be split into a finite number of elementary ones. This is the assertion of the famous desingularization theorem by Bendikson-Seidenberg-Dumortier. Therefore the analytic classification of elementary singularities is of extreme importance for the local theory of planar analytic differential equations. This classification was obtained by Martinet and Ramis (1982, 1983); a certain contribution is also due to Voronin, Elizarov, and Il'yashenko (1982, 1983).

The survey of this activity, starting with Stokes' pioneering work on linear theory and the above-mentioned Voronin theory, is given in Paper I (the paper number n of this volume is simply referred to as Paper n).

In the theory of polynomial planar vector fields, finitely generated groups of germs $(\mathbb{C}, 0) \to (\mathbb{C}, 0)$ naturally appear as monodromy groups (holonomy is another term) for solutions at infinity. After desingularization of nonelementary singularities, spherical fibers with rich monodromies also arise. At the same time, the analytic classification of these groups yields invariants of the analytic classification of holomorphic flows with complex singularities. This is treated in Paper II. Moreover, this paper contains the analytic classification of the simplest nonelementary singularities, namely, those having nilpotent Jordan cell linearization. An exposition of articles by Moussu (1985) and Cerveau-Moussu (1989) is included in it together with some original results.

Formally equivalent vector fields and maps can have different formal invariants represented in the form of Fourier series. Not a single coefficient of these series can be calculated from a finite segment of the Taylor expansion of the initial germ. This causes difficulties when attempting exact computations of the functional invariants. In this situation an approximate computation turns out to be fruitful. In order to carry it out, a germ equivalent to its own formal normal form is taken, and the tangent to the moduli map is computed at this point. Such computations yield numerous examples of pairwise analytically nonequivalent resonance vector fields. Apparently, rich information about the monodromy of a polynomial vector field at infinity can be obtained in this way. Computation of derivatives of the moduli maps is the subject of Paper III.

March 24, 1991, Arnold wrote a letter to Il'yashenko (see Figure 1 on next page) which, among other things, contained the following:

" ... *instantly I became keenly interested in the normal form for a germ $(\mathbb{C}, 0) \to (\mathbb{C}, 0)$ of the form $t \mapsto t+t^2+\cdots$. The formal normal form is $t+t^2$. Is it true that divergence is a rule? (...) The question arises in singularity theory (normalization of the semicubic parabola^1 by transformations of the form $(X(x), Y(y))$).*

The C^∞-reducibility to $(x+y)^3 = (x-y)^2$ takes place here, the analytic one is unknown, and it seems that it does not exist at all, due to the above reason (the map of the semicubics into itself is the superposition of the two involutions σ_x, σ_y). If possible, leave the answer here."

The answer was obtained by Voronin and appeared later in the form of two papers (Voronin 1981, 1982). In the first the analytic classification of one-dimensional holomorphisms tangent to the identity was obtained; the other was concerned with analytic classification of certain diverging diagrams, including classification of pairs of involutions, semicubic singularities, and envelopes of families of planar curves. In the same way as in Arnold's letter, in all the problems an indicator appears, namely the germ of a map $(\mathbb{C}, 0) \to (\mathbb{C}, 0)$ with identity linear part. The functional modulus of this indicator

---
^{1} $x(t) = t^2 - t^3$, $y(t) = t^2 + t^3 + \cdots$.

24.3.79

Ю. С. Ильяшенко

Юлия, меня вдруг стал остро интересовать вопрос о нормальной форме ростка $(\mathbb{C},0) \to (\mathbb{C},0)$ вида $t \mapsto t + t^2 + \ldots$.
Формальная н.ф. есть $t + t^2$.
Верно ли, что тут обычная расходимость?
И как с топологической приводимостью к н.ф.?

Я забыл совершенно — но Вы, по моему, занимались (с Найшулем?) почти этим (или этим?).

Вопрос возникший в теории особенностей (приведение полукубической $+\ldots$ особенности преобразованиями $(X(x), Y(y))$.

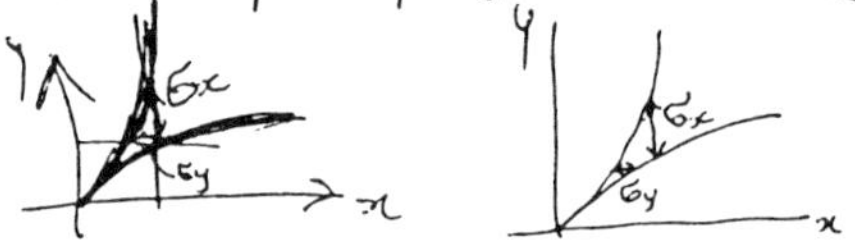

[figures]

Здесь есть C^∞ приводимость к $(x+y)^2 = (x-y)^3$, аналитическая же, видимо, неизвестна — и, кажется отсутствует по изложенной выше причине (отображение полукубики в себя есть произведение двух инволюций, δx и δy). Если можно, ставьте мне ответ тут.

FIGURE 1. Copy of a letter from V. I. Arnold to Yu. S. Il′yashenko.

forms a complete system of analytic invariants for the initial problem from singularity theory. Recently Voronin has discovered that analogous although more complicated phenomena occur in the problem of bypassing an obstacle, previously studied by Arnold and Melrose. Paper IV deals with all these matters.

Paper V stands somewhat aside from the topics described above. It is concerned with the smooth classification of local families of diffeomorphisms and vector fields in real phase space. From a roughly heuristic point of view these families are of three types: deformations of generic and "weakly degenerate" germs; deformations of "moderately" and, at last, "strongly degenerate" germs. The first class consists of deformations of hyperbolic nonresonance and single-resonance germs of vector fields and diffeomorphisms and deformations of multidimensional saddle-node vector fields (exactly one zero eigenvalue, the others being rationally independent). For this class there exist polynomial integrable normal forms (Il'yashenko and Yakovenko, 1991). These normal forms make it possible to give a transparent exposition of numerous known results from bifurcation theory, and obtain some new results at the same time. This activity goes beyond the scope of the present volume.

"Moderately degenerate germs" are real line diffeomorphisms with modulus one multiplier at the fixed point, and planar vector fields with a pair of purely imaginary eigenvalues. Smooth classification of generic one-parameter deformations of such germs involves functional moduli whose origins and nature are similar to those described in Paper I. This classification along with some applications to nonlocal bifurcation theory constitutes the body of Paper V.

Finally, "strongly degenerate germs" are germs of real plane diffeomorphisms with a pair of complex conjugate multipliers on the unit circle together with all more degenerate objects. Apparently, for this case neither smooth nor even topological classifications can be obtained.

Concluding this foreword, I would like to point out that one of the main applications of the Ecalle-Voronin functional moduli has become the solution of the Dulac problem ("a polynomial vector field on the real plane can have but a finite number of limit cycles"), obtained independently by Ecalle and the author of these lines.

It is my pleasure to express my cordial gratitude to S. Yu. Yakovenko and A. D. Vaĭnshteĭn, who translated the original Russian text, sometimes cumbersome, and to A. B. Sossinsky, who strongly improved the linguistic appearance of this book.

* * *

I would like to stress the fact that investigations of nonlinear Stokes phenomena began in the late seventies simultaneously in Moscow and in France. The French studies, pioneered by Malgrange and Ecalle, were then extended and considerably developed by Martinet and Ramis. Fortunately, the Russian and the French research more often completed than repeated each other. With animation we exchanged reprints and news about recent achievements. It was (and continues to be) cooperation rather than competition. I got acquainted with Martinet and Ramis in Lumini in September 1989, and we became friends. We hoped for mutual and fruitful joint work. But on July 2, 1990, after ten months of distressing illness, professor Jean Martinet passed away.

We dedicate this book to his memory.

Belyaevo-Bogorodskoe, May 11, 1991   Yulii S. Il'yashenko

Translated by A. VAĪNSHTEĪN

Nonlinear Stokes Phenomena

YU. S. IL'YASHENKO

This paper is devoted to the exposition of research in the geometric theory of normal forms, originating in the pioneering works of S. Voronin [37], J. Ecalle [10] and B. Malgrange [24]. At the end of the first section the classical Stokes phenomenon for linear equations, discovered in the 19th century [36], is discussed.

§1. What is a Stokes phenomenon?

As we mentioned in the Foreword, local dynamics near a fixed point determines a canonical chart (defined up to a finite number of parameters) in the typical nonresonant case, and an atlas of charts in the resonant one. The Stokes phenomenon consists in nontriviality of the transition functions for this atlas. Let us proceed with a more detailed exposition, starting with the nonresonant case.

1.1. Linearizing chart in the nonresonant case. The germ of a vector field at its singular point on an n-dimensional manifold is called *resonant*, if its eigenvalues satisfy a certain arithmetical condition of the form

$$\lambda_j = (\lambda, k).$$

Here $\lambda = (\lambda_1, \ldots, \lambda_n)$ is the string of the eigenvalues, $k = (k_1, \ldots, k_n) \in \mathbb{Z}_+^n$ is an integer vector such that

$$|k| \stackrel{\text{def}}{=} k_1 + \cdots + k_n \geq 2.$$

SIEGEL THEOREM [2, 35]. *For almost all (in the sense of the Lebesgue measure) vectors $\lambda \in \mathbb{C}^n$ the germ of a holomorphic vector field with spectrum λ is biholomorphically equivalent to a linear germ. The transformation taking the initial germ to linear form (called the normalizing transformation or chart) is uniquely determined by the 1-jet of this transformation.*

REMARKS. 1. The exceptional set of zero measure contains all resonant hyperplanes as a proper subset.

1991 *Mathematics Subject Classification.* Primary 32C34, 34A20; Secondary 34A30.

2. It is the normalizing chart which is canonically defined by the local dynamics. The constants determining it are the coefficients of its 1-jet.

There exists an analog of this theorem for germs of maps. We are mostly interested in the one-dimensional case.

SCHRÖDER THEOREM. *The germ of a conformal mapping of the complex line whose multiplier (i.e., the derivative at the origin) has modulus different from 1 is analytically equivalent to the linearization of the germ. The normalizing chart is determined uniquely up to a scalar factor.*

SIEGEL THEOREM FOR MAPS [35]. *The assertion of the Schröder theorem is valid for the case of maps with multiplier equal to $e^{\pi i \varphi}$ for almost all real irrational φ.*

Recently J.-C. Yoccoz [39] has given a complete description of the set of all φ such that each germ with multiplier $e^{\pi i \varphi}$ is analytically equivalent to its linearization independently of nonlinear terms.

Here we end the description of charts determined by nonresonant local dynamics, and pass to that of atlases occurring in the resonant case.

1.2. Normalizing atlas. In the resonant case it is impossible to linearize vector fields even in the formal category: in general the resonance condition $\lambda_j = (\lambda, k)$ between the eigenvalues forms an obstruction to removing the monomial

$$z^k \partial/\partial z_j, \quad \text{where } z^k = z_1^{k_1} \cdots z_n^{k_n}.$$

In the resonant case, the formal normal form contains only resonant monomials of the above type, while the normalizing series bringing the germ to its formal normal form diverge in general [6, 17].

Nevertheless, in the case of small dimensions (less than 3), there exists a geometric object corresponding to those series. A punctured neighborhood of the fixed point is covered by a finite number of sector-like domains containing the fixed point on their boundaries. In each domain the field (respectively, the map) is analytically equivalent to its formal normal form. The normalizing series is the asymptotic expansion for the transformation normalizing the field or the map in each of these domains, when the argument tends to the singular (fixed) point. Thus the "geometric normal form" for a resonant map or a vector field in small dimensions is the triple: (the formal normal form; the covering of a neighborhood by the sectors; the transition functions between the normalizing charts defined in these sectors). These transition functions contain all information on geometric properties of the germ.

In the following sections we describe in detail the corresponding constructions for complex line holomorphisms and planar vector fields. Here we proceed with the description of the classical Stokes phenomenon. Let us start with the formal theory.

1.3. Formal normal forms for linear systems with irregular singularities.

Consider a linear nonautonomous system of the form

$$t^{r+1}\dot{z} = A(t)z, \tag{1.1}$$

where $A(0) \neq 0$. The construction given below is aimed at answering several questions, among them the following one: when are any two given systems of the form (1.1) linearly equivalent? Linear equivalence means a transformation of the form

$$z = H(t)w, \tag{1.2}$$

taking one system into the other; the operator-valued function H is assumed to be holomorphic and nondegenerate in an entire neighborhood of the origin in the z-plane.

V. I. Arnold has pointed out that it is convenient to regard linear nonautonomous systems as nonlinear autonomous ones by adding another equation and changing the time variable:

$$z' = A(t)z, \qquad t' = t^{r+1}. \tag{1.3}$$

The eigenvalues of the system thus obtained are equal to $\lambda_1, \ldots, \lambda_n, \lambda_{n+1}$, where $\lambda_1, \ldots, \lambda_n$ is the spectrum of $A(0)$, and $\lambda_{n+1} = 0$. After this procedure, equation (1.3) can be investigated by using methods of nonlinear normal form theory.

This program is carried through as follows. A transformation linear in z takes the system (1.1) to a system of the same form with another matrix $A(t)$. So only linear terms are involved in the computations. These terms have the form $z_i t^m \partial/\partial z_j$ and may be resonant only if $\lambda_j = \lambda_i + m\lambda_{n+1}$, that is, $\lambda_i = \lambda_j$.

DEFINITION. The equation (1.1) is called *nonresonant*, if all the eigenvalues of the operator $A(0)$ are distinct; otherwise it is called *resonant*.

REMARK. From the point of view of nonlinear theory, equation (1.3) is always resonant, since $\lambda_{n+1} = 0$. The preceding definition means that there are no extra resonances besides those implied by the latter equality.

For simplicity we restrict ourselves to the nonresonant case.

POINCARÉ-DULAC THEOREM [2]. *By formal linear transformations of the form* (1.2), *any system of the form* (1.1) *can be put in a form containing only resonant monomials*:

$$\dot{\tilde{w}}_i = t^{-(r+1)}\tilde{b}_i(t)\tilde{w}_i. \tag{1.4}$$

Here the symbols $\tilde{b}_i$ *stand for formal series.*

The formal normal form (1.4) for the nonresonant system (1.1) is diagonal, hence integrable. Now we carry out further simplifications, reducing (1.4) to polynomial form.

Let
$$\widetilde{B}(t) = \text{diag}(\widetilde{b}_1(t), \ldots, \widetilde{b}_n(t)),$$
$$\widetilde{B}(t) = B(t) + t^{r+1} B_1(t), \qquad B = \text{diag}(b_1(t), \ldots, b_n(t)).$$

Here B is a matrix polynomial of degree $\leq r$. In order to compute the fundamental matrix of solutions $\widetilde{W}$ for the system (1.4), let us represent the primitive $\int B(t) t^{-(r+1)}$ in the form
$$\int B(t) t^{-(r+1)} = D(t^{-1}) + C \ln t,$$
where C is a constant diagonal matrix, and $D(t^{-1})$ is a diagonal matrix with polynomials of degree $\leq r$ in t^{-1} on the diagonal. Then
$$\widetilde{W} = t^C \exp D(t^{-1}) \exp B_2(t),$$
where $B_2 = \int B_1$. Finally we substitute $w = \exp(-B_2(t))\widetilde{w}$, so that the system (1.4) is transformed into a *normalized* one, i.e., a system of the form
$$\dot{w}_i = t^{-(r+1)} b_i(t) w_i, \tag{1.5}$$
the functions b_i being polynomials of degree $\leq r$. The fundamental matrix of solutions for this system has the form
$$W = t^C \exp D(t^{-1}) \tag{1.6}$$
with C, D the same as above.

The initial equation becomes a normalized one via the transformation $z = \widehat{H}(t)w$, where $\widehat{H}(t)$ is a formal Taylor series, called the *normalizing series*. The initial system has the formal solution
$$Z = \widehat{H}(t)W = \widehat{H}(t) t^C \exp D(t^{-1}).$$

REMARK. The normalizing series $\widehat{H}$ conjugating the initial system with its formal normal form is *not* uniquely determined.

1.4. Sectorial normalization. As shown at the end of this subsection, normalizing series generically diverge. Nevertheless they serve as asymptotic expansions for transformations which are defined in some sectors and conjugate the initial system with its normal form in these sectors. This fact was discovered by H. Poincaré: it turns out that such transformations in fact exist in sectors with vertex zero and any given direction as bisector, provided that these sectors are sufficiently narrow. Only recently [33] the maximal width of a sector in which the normalizing transformation is holomorphic has become known. This width depends on the position of the sector, that is, on the direction of its central radius. In order to describe such "maximal domains of the normalizing maps", we need the following

DEFINITION. A *dividing ray* corresponding to a pair of complex numbers λ, μ is any of the $2r$ rays defined by the equality

$$\operatorname{Re}(\lambda - \mu)t^{-r} = 0.$$

SECTORIAL NORMALIZATION THEOREM [33]. *Let the system* (1.1) *be nonresonant, i.e., let all eigenvalues of $A(0)$ be different. Then for an arbitrary sector S not containing any two dividing rays that correspond to any pair of eigenvalues, there exists a holomorphic transformation $z = H_S(t)w$ conjugating* (1.1) *with its normal form* (1.5).

For each normalizing series $\widehat{H}$ the transformation H_S may be chosen so that $\widehat{H}$ is the asymptotic expansion for H_S. If the angle at the vertex of the sector is greater than π/r, then such a normalizing transformation is unique.

Cover the punctured disk $D^* = \{0 < |t| < \rho\}$ by a finite number of sectors S_j with vertical angles less than π/r, enumerated in their natural counterclockwise order. Then the domain $D^* \times \mathbb{C}^n$ for ρ sufficiently small turns out to be endowed with the atlas of charts

$$(w_j, t) = (H_j^{-1}z, t), \qquad H_j \stackrel{\text{def}}{=} H_{S_j},$$

normalizing the initial system in the "extended sectors" $\widetilde{S}_j = S_j \times \mathbb{C}^n$. The transition functions $H_{j+1} \circ H_j^{-1}$ differ from the identity by corrections decreasing more rapidly than any finite power of $|t|$ as $t \to 0$, $t \in S_j \cap S_{j+1}$, since their asymptotic expansions coincide:

$$\|H_{j+1}^{-1} \circ H_j(t) - E\| = o(|t|^{-N}), \quad \forall N \in \mathbb{N}. \tag{1.7}$$

It turns out that these transition functions are characterized by linear Stokes operators $C_j \colon \mathbb{C}^n \to \mathbb{C}^n$ and their corrections are exponentially flat at the origin, as will be shown below.

1.5. Stokes operators and normalizing cochains. Each normalizing map generates an isomorphism between the spaces of solutions of the initial system and its normal form. The superposition ratio of any two such maps in the intersection of their domains defines an automorphism of the space of solutions of the normalized system into itself. This automorphism is called the Stokes operator.

The formal definition is as follows. Let W be the fundamental matrix of solutions of the normalized system (1.6), and let H_j, H_{j+1} be the normalizing linear maps corresponding to adjacent sectors S_j, S_{j+1} having nonempty intersection. Then

$$Z_j = H_j(t)W, \qquad Z_{j+1} = H_{j+1}(t)W$$

are the fundamental matrices for the initial system in the sectors S_j, S_{j+1} respectively. On the intersection of the sectors, they are related by a linear operator C_j:

$$Z_{j+1}(t) = Z_j(t)C_j.$$

This is the *Stokes operator*. It is connected with the transition function by the following relation:
$$H_j^{-1} \circ H_{j+1} = W C_j W^{-1}.$$

From the "asymptotic identity" condition (1.7) imposed on the transition functions and the formulas for solution of the normalized system, it follows that *the corrections of the transition functions are exponentially flat, and the Stokes operators are unipotent*.

Indeed, denote by $a_{ik}(t)$ and c_{ik} the matrix elements of WC_jW^{-1} and C_j respectively (j being fixed). Then, since $W = \operatorname{diag}(w_1(t), \ldots, w_n(t))$ and
$$w_i(t) = \exp(-r^{-1}t^{-r}(\lambda_i + o(1))),$$
we have
$$a_{ik}(t) = c_{ik}\exp(r^{-1}t^{-r}(\lambda_k - \lambda_i + o(1))).$$
As $t \to 0$, $a_{ik}(t)$ tends to δ_{ik}. Hence $c_{ii} = 1$; if $i \neq k$ and the function $\operatorname{Re}(t^{-r}(\lambda_k - \lambda_i))$ does not tend to $-\infty$ as $t \to 0$ in the intersection $S_j \cap S_{j+1}$, then $c_{ik} = 0$. Therefore only one of the two elements c_{ik} and c_{ki} may be nonzero. This implies that the operator C_j is unipotent. If the function $\operatorname{Re}(t^{-r}(\lambda_k - \lambda_i))$ tends to $-\infty$ in $S_j \cap S_{j+1}$, then the function $a_{ik}(t)$ is exponentially decreasing. □

The set of normalizing maps H_j forms a "normalizing cochain" with coboundary (i.e., the set of "iterative ratios" $H_j^{-1} \circ H_{j+1}$) very close to the identical one: the corrections of the ratios decrease more rapidly than any holomorphic function in the sector with angle $> \pi/r$ possibly can. This cochain may be considered as a kind of sum of the series $\widehat{H}$.

Matrix elements of the normalizing cochains together with germs of holomorphic functions generate a differential algebra of "functional cochains", whose germs at the origin are uniquely determined by their Taylor series. This circumstance permits to prove that the Stokes operators for the initial system belong to its Galois group [29, 30, 22]. All necessary definitions from Galois theory can be found in [23]: a detailed exposition here would lead us too far away from our main topic.

The set of the Stokes operators completely determines the equivalence class of nonresonant equations (1.1) with respect to the equivalence relation (1.2) (see [35]). The normalizing series converges if and only if all the Stokes operators are trivial: $C_j = \operatorname{id}$. On the other hand, any collection of linear operators satisfying the above conditions can be realized as the set of Stokes operators for an appropriate system (Birkhoff, [4]). This concluding note implies that the normalizing series diverges for a generic system, as was mentioned above.

§2. One-dimensional mappings tangent to the identity

In this section we construct a normalizing atlas for conformal mappings $f: (\mathbb{C}, 0) \to (\mathbb{C}, 0)$ with identity linear part at the origin: $j_0^1 f = \operatorname{id}$.

The analytic classification of such mappings is given; this classification is determined by the so-called *Ecalle-Voronin functional moduli*. Geometric properties of the mappings are investigated as they depend on the moduli. A linear approximation formula is obtained for computing these moduli.

From now on we suppose that $f \neq \text{id}$; let the Taylor expansion for $f - z$ begin with the term az^p, $a \neq 0$. We denote by $\mathscr{A}_p$ the set of all such germs.

Let $f \in \mathscr{A}_p$, $f(z) = z + az^p + \cdots$. By a linear transformation $z \mapsto \alpha z$ one can get $a = 1$. In order to avoid some irrelevant details, we consider only the problem of analytic classification of germs of the form

$$z \mapsto z + z^p + \cdots \tag{2.1}$$

with respect to transformations having identity linear parts only. For the sake of simplicity we shall refer to this very problem as the *analytic classification problem*.

EXAMPLES. 1. Consider the holomorphic vector field $v(z) = z^p + \cdots$, the dots denoting higher order terms. The time 1 map for the flow of this vector field is a germ belonging to the class $\mathscr{A}_p$ and will be denoted by g_v^1. Any germ which can be represented in this form is called *embeddable*.

2. Any (nonzero) iterative power $f^{[n]} = f \circ f \circ \cdots \circ f$ (n times) of the germ $f \in \mathscr{A}_p$ belongs to $\mathscr{A}_p$.

REMARK. The iterative power will be denoted by square brackets in order to distinguish it from the multiplicative one.

3. Any germ analytically equivalent to a germ of class $\mathscr{A}_p$ is also in the same class.

2.1. Questions and problems on one-dimension mappings. There are three main questions to be answered in this section.

1. THE EMBEDDING PROBLEM. Is any germ of class $\mathscr{A}_p$ embeddable or not?

2. THE ROOT EXTRACTION PROBLEM. For what germs f from $\mathscr{A}_p$ and for what natural n is the following equation

$$g^{[n]} = f, \qquad g \in \mathscr{A}_p$$

solvable (with respect to g)?

3. THE ANALYTIC CLASSIFICATION PROBLEM. For what germs $f, g \in \mathscr{A}_p$ does there exist an analytic transformation $h: (\mathbb{C}, 0) \to (\mathbb{C}, 0)$ conjugating these germs, i.e.,

$$h^{-1} \circ g \circ h = f,$$

and tangent to the identity?

All these questions make sense if instead of germs of class $\mathscr{A}_p$ one deals with formal power series of the form $f = z + z^p + \cdots$. The class of all such series will be denoted by $\widehat{\mathscr{A}_p}$. Denote the Taylor series of the germ f by

$\hat{f}$. The formal time 1 map for the formal vector field $\hat{v} \in \widehat{\mathscr{A}_p}\partial/\partial z$ is well defined in the class $\widehat{\mathscr{A}_p}$ and will be denoted by $g_{\hat{v}}^1$.

Replacing germs by power (Taylor) series (possibly divergent) in the previous equations, one obtains their formal counterparts:

$$g_{\hat{v}}^1 = \hat{f}, \qquad \hat{g}^{[n]} = \hat{f}, \qquad \hat{h}^{-1} \circ \hat{g} \circ \hat{h} = \hat{f}.$$

In the first two equations the series $\hat{f}$ is given, and $\hat{v}$, $\hat{g}$ are to be found; in the third f, g are given, and the problem is to find $\hat{h}$.

It turns out that there are no obstructions to the formal solution of the first two problems, and almost none to the solution of the third problem. More precisely, the first two problems always possess a formal solution, while the third one is formally solvable if the $(2p+1)$-jets of the series $\hat{f}$, $\hat{g}$ coincide (this is a sufficient condition which is far from necessary: in fact, there is only one independent relation between coefficients of $(2p+1)$-jets of the series $\hat{f}$, $\hat{g}$ that must be satisfied).

Nevertheless all these problems hardly ever have analytic solutions. To understand this phenomenon, one should first investigate the (third) problem of analytic classification of germs. This classification admits functional invariants. The construction described below allows us to "see" them without any calculations.

We shall conclude this section by stating the formal classification theorem for germs of class $\mathscr{A}_p$.

THEOREM 1. *Any germ of class $\mathscr{A}_{p+1}$ is formally equivalent to the unique germ $f_{p,\lambda} = g^1_{w_{p,\lambda}}$ of a time 1 map for the flow of the vector field*

$$w_{p,\lambda} = z^{p+1}(1 + \lambda z^p)^{-1}\partial/\partial z. \tag{2.2}$$

The set of all germs of class $\mathscr{A}_{p+1}$ formally equivalent to the germ $f_{p,\lambda}$ is denoted by $\mathscr{A}_{p+1,\lambda}$.

2.2. Topological classification. The orbit space. Before stating the analytic classification results, let us first state the topological ones.

THEOREM 2 [8, 31]. *Any germ of class $\mathscr{A}_{p+1}$ is topologically equivalent to the time 1 map for the flow of the vector field*

$$v(z) = z^{p+1}\partial/\partial z.$$

An important illustration of this theorem is provided by the following set of examples.

EXAMPLE 1. The phase portrait of the field $v(z)$ is given in Figure 1. The punctured neighborhood of the origin is covered by $2p$ sectors of the form

$$S_j = \{z : |\arg z - \pi j/p| < \alpha\}, \qquad \alpha \in (\pi/p, \pi/2p), \ j = 1, \ldots, 2p.$$

The sector S_{2p} includes a segment of the positive semiaxis, while the remaining sectors are enumerated in their natural counterclockwise order. In what

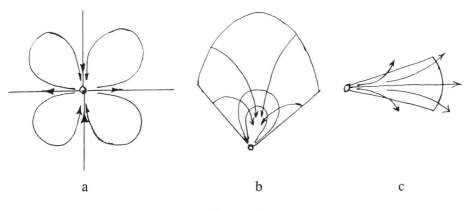

FIGURE 1

follows this covering will be referred to as the *nice p-covering*. Sectors with even numbers are weakly attracting. This means that all positive semitrajectories starting in any such sector close enough to the origin entirely belong to this sector and tend to the origin as time increases. By replacing the circular arc bounding the sector by an appropriate curve, one can obtain a sector-like domain which is attracting in the classical sense.

The odd sectors are weakly repellent, that is, they become weakly attracting when time is reversed.

EXAMPLE 2. The phase portrait of the vector field (2.2) in some small neighborhood of the origin is topologically equivalent to that of the field v. This means that there exists a homeomorphism of the neighborhood preserving the origin and taking phase trajectories of the first field to those of the second. The topological equivalence is due to the fact that if the attracting/repelling sectors constructed above are sufficiently small, then the addition of the perturbation $w_{p,\lambda} - v$ (small with respect to v) does not destroy the attracting/repelling properties.

Now consider the quotient space of the punctured neighborhood of the origin by the action of a representative of a germ $f \in \mathscr{A}_{p+1}$. Points of this space are identified with orbits of f. The topology is inherited from the natural projection. As is shown below, the quotient space turns out to be a non-Hausdorff Riemann surface. Let us start with examples.

EXAMPLE 3. The quotient space of the complex plane $\mathbb{C}$ by the unit shift $t \mapsto t+1$ is the Riemann sphere minus two points $0, \infty$:

$$\mathbb{C}/(\mathrm{id}+1) \sim \mathbb{C}^* \stackrel{\mathrm{def}}{=} \widehat{\mathbb{C}} \setminus \{0, \infty\}.$$

The natural projection is given by the mapping $t \mapsto \exp 2\pi i t$.

EXAMPLE 4. The quotient space M of the exterior of the circle

$$\{z : |z| > R, R > 1\}$$

on the complex plane by the same action is a pair of doubly punctured spheres

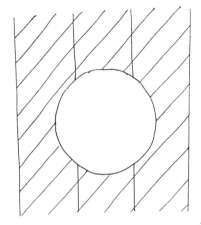 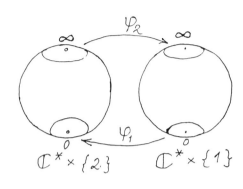

FIGURE 2

$\mathbb{C}^*$ with identified neighborhoods of the removed points (Figure 2).

Indeed, any orbit of the unit shift entirely belonging to M corresponds to a unique point of the quotient space. On the other hand, each orbit having nonempty intersection with the circle corresponds to two distinct points. The space thus obtained possesses nonseparable points belonging to the boundaries of identified neighborhoods.

EXAMPLE 5. The quotient space of the punctured neighborhood of the origin on the complex plane $\mathbb{C}$ by the action of time 1 map for the flow of the field $v(z) = z^2 \partial/\partial z$ is the same as (i.e., homeomorphic to) that of Example 4.

Indeed, the substitution $z \mapsto t = -1/z$ straightens the field v, i.e., transforms it into the field $\partial/\partial z$, so the action becomes the standard unit shift on the image $M = \{t \in \mathbb{C} : |t| > R = \varepsilon^{-1}\}$ of the initial neighborhood $|z| < \varepsilon$.

EXAMPLE 6. For any germ of the class $\mathscr{A}_2$ there exists a punctured neighborhood U of the origin in the plane $\mathbb{C}$ such that the corresponding quotient space by the action of f is the same as in Example 5. This follows immediately from Theorem 1.

EXAMPLE 7. For any natural p there exists a punctured neighborhood U of the origin such that the quotient space by the time 1 map for the field

$$w_{p,0} = z^{p+1} \partial/\partial z$$

is a collection of $2p$ doubly punctured spheres with cyclically identified neighborhoods of the removed points (see Figure 3).

EXAMPLE 8. The statement of Example 7 holds also for the field $w_{p,\lambda}$ with $\lambda \neq 0$ (see (2.2)).

EXAMPLE 9. The statement of Example 7 holds not only for the embeddable germs g_v^1, $v = z^{p+1} \partial/\partial z$, but also for any germ from the class $\mathscr{A}_{p+1}$. This follows from Theorem 1.

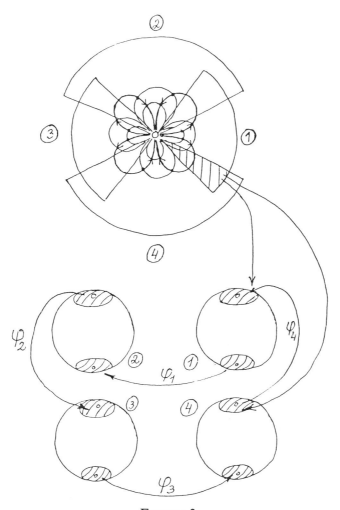

FIGURE 3

2.3. Functional invariants. Denote the class of germs formally equivalent to the time 1 map $f_{p,\lambda} = g_w^1$, $w = w_{p,\lambda}$ (see (2.2)) as $\mathscr{A}_{p+1,\lambda}$. For germs of the class $\mathscr{A}_{p+1}$ to be analytically equivalent, it is necessary that they be formally equivalent, which means that they must belong to the same class $\mathscr{A}_{p+1,\lambda}$.

The quotient space of the action of the germ f inherits the complex structure from the source space of the natural projection mapping, thus being a non-Hausdorff Riemann surface (i.e., a real 2-manifold endowed with the complex structure but without the separation axiom, see Figures 2, 3). Their identified components may be of one of the following three conformal types: doubly punctured sphere $\mathbb{C}^*$, punctured disk $\{0 < |z| < 1\}$ and an annulus $\{a < |z| < b\}$. It turns out that only the first possibility always occurs. This is proved below; now we shall use this fact to "see" the functional invariants of the analytic classification for germs of class $\mathscr{A}_{p+1}$ in Figure 3.

The Riemann sphere has the following important property: its biholomorphic map onto itself is uniquely determined provided that the coordinates of three points are given. The doubly punctured sphere $\mathbb{C}^*$ is obtained from $\widehat{\mathbb{C}}$ by removing two points 0 and ∞; after fixing these two points, the global chart on $\mathbb{C}^*$ is determined uniquely up to multiplication by a scalar.

The identifying maps on Figure 3 are conformal and can be extended to the removed point by the removable singularity theorem. The quotient space of the punctured neighborhood U of the origin by the action of a germ of class $\mathscr{A}_{p+1}$ can be represented as the union $\bigcup_{j=1}^{2p} \mathbb{C}^* \times \{j\}$ with identified neighborhoods of the removed points. After ascribing the coordinate values 0 and ∞ to the pair of removed points on the sphere $\mathbb{C}^* \times \{1\}$, the remaining spheres can be endowed with coordinate systems in such a way that the identifying maps take the form

$$\begin{aligned}\varphi_j: (\widehat{\mathbb{C}}, 0) \times \{j\} &\to (\widehat{\mathbb{C}}, 0) \times \{j+1\} \quad \text{for } j \text{ even}, \\ \varphi_j: (\widehat{\mathbb{C}}, \infty) \times \{j\} &\to (\widehat{\mathbb{C}}, \infty) \times \{j+1\} \quad \text{for } j \text{ odd},\end{aligned} \quad (2.3)$$

(the enumeration is mod $2p$, so $\varphi_{2p+1} = \varphi_1$). The choice of charts transforms the maps φ_j into functions. The charts are determined uniquely up to scalar multiplication as global charts on $\mathbb{C}^* \times \{j\}$. The following normalizing condition excludes arbitrariness in the choice of all constants except one: let us require that

$$\text{all functions } \varphi_j, \; j = 1, \ldots, 2p-1, \text{ are tangent} \atop \text{to the identity either at the origin or at infinity.} \quad (2.3')$$

This condition determines our choice of the charts on all doubly punctured spheres except the first one, after the chart z on the first sphere is chosen. The linear part νz of the function $\varphi_{2p}(z) = \nu z + o(z^{-1})$ at infinity is now uniquely determined and does not depend on the choice of z. Indeed, if the chart z is replaced by another chart Cz, $C \in \mathbb{C}$, then, according to the normalizing condition, all the remaining maps will be multiplied by C. The map φ_{2p} is replaced by the map $C \circ \varphi_{2p} \circ C^{-1}$ with the same linear part νz. This construction motivates the following definition.

DEFINITION. Two collections φ, $\widetilde{\varphi}$ of $2p$ homeomorphic functions (2.3), all of them except the last one being tangent to the identity either at the origin (for j odd) or at infinity (for even j), are called *equivalent*, if the first collection can be transformed into the second by the linear transformation $z \mapsto Cz$ both in the source and in the target spaces:

$$\varphi \sim \widetilde{\varphi} \iff \exists C \in \mathbb{C} : \varphi_i \circ C = C \widetilde{\varphi}_i, \; i = 1, \ldots, 2p. \quad (2.4)$$

The set of all equivalence classes will be denoted by M^*_{p+1}. Thus every germ f of the class $\mathscr{A}_{p+1}$ is associated with a unique equivalence class (2.3), (2.4) denoted by μ^*_f. This class is constructed independently of the choice of the chart in which the initial germ is given; hence it is the same for all germs analytically equivalent to f; that is, μ_f^* is a functional invariant of analytic classification of germs from the class $\mathscr{A}_{p+1}$. Thus we have proved the first statement of the following classification theorem [33, 10, 24, 18, 3]:

THEOREM. 1. *Invariance. To every germ* $f \in \mathscr{A}_{p+1,\lambda}$, $f = z + z^p + \cdots$, *there corresponds a unique class* $\mu_f^* \in M_{p+1}^*$, *which is an invariant of the analytic classification.*

2. *Equivalence and equimodality. The germs* $f, g \in \mathscr{A}_{p+1,\lambda}$ *with coinciding invariants* $\mu_f^* = \mu_g^*$ *are analytically equivalent with respect to the above group action.*

3. *Realization. Every class* $\varphi \in M_p^*$ *can be realized as an invariant* μ_f^* *for some germ* $f \in \mathscr{A}_{p+1}$

4. *Analytic dependence on parameters. If the germ* f *analytically depends on a finite-dimensional complex parameter* ε, *then* μ_f^* *also analytically depends on it.* (*Analytic dependence of an equivalence class means it is possible to choose an analytic family of representatives.*)

An analog of this theorem is proved below. In order to investigate the quotient space depicted in Figure 3, in particular, to make sure that the conformal type of its identified components is indeed that of the doubly punctured sphere, we use the notions of quasicomplex structure and quasiconformal mapping.

2.4. The problem of conformal type and quasiconformal mappings.

What remains after the complex structure (CS) on a manifold is lost? More precisely, let M be an n-dimensional complex manifold, considered at the same time as a $2n$-dimensional real manifold $^{\mathbb{R}}M^{2n}$. The holomorphic atlas on the manifold is replaced by a C^1-smooth one, so the C^1-structure on M still remains. The lost complex structure leaves its trace in the form of an *almost complex structure* (ACS). By this expression we mean the decomposition of the complexified cotangent bundle into the direct sum of two subbundles consisting of forms which were $\mathbb{C}$-linear (of type $(1, 0)$) and $\mathbb{C}$-antilinear (of type $(0, 1)$) in the sense of the initial complex structure. It is sufficient to define only the first subbundle, because the second is obtained as the set of adjoints. Both subbundles are $\mathbb{C}$-linear.

EXAMPLE 1. An almost complex structure on $\mathbb{C}$ can be defined by the form
$$\omega = dz + \mu\, d\bar{z}$$
as the subbundle $\mathbb{C}$-spanned by this form. For the sake of convenience, we call it a μ-*almost complex structure*, μ being a smooth function on $\mathbb{C}$. The choice of another holomorphic chart $w = \varphi(z)$ leads to another expression for the same almost complex structure: $\tilde{\omega} = d\tilde{w} + \tilde{\mu}\, d\overline{\tilde{w}}$, where
$$\tilde{\mu} \circ \varphi(z) = \mu(z)\overline{\varphi}_z/\varphi_z, \qquad \varphi_z \stackrel{\text{def}}{=} \partial\varphi/\partial z$$

(recall that $\varphi_{\bar{z}} = 0$). Thus it is not the function which is associated with the μ-almost complex structure on our one-dimensional complex manifold, but rather the ratio of two 1-forms, the so-called *Beltrami differential* $\mu \, d\bar{z}/dz$. Note that $|\mu|$ is well defined by the corresponding μ-almost complex structure.

The almost complex structure obtained from the lost complex structure is called *integrable*. Not all almost complex structures defined as decompositions of the above kind are integrable: in §3 sufficient integrability conditions are given for the case $n > 1$. The case $n = 1$ is covered by the following

THEOREM [28]. *For the integrability of a μ-almost complex structure defined on a one-dimensional complex manifold by the corresponding Beltrami differential $\mu \, d\bar{z}/dz$, it is sufficient that μ be continuous and the inequality $|\mu| \leqslant k < 1$ be satisfied.*

REMARK. In the classical theory of functions, μ is only assumed to be measurable with charts holomorphic in the (Sobolev) generalized sense [1]. But for our purposes it is sufficient to consider only C^1-smooth charts and continuous Beltrami differentials.

EXAMPLE 2. Consider a C^1-smooth mapping f of the punctured sphere $\mathbb{C}^*$ onto the Riemann surface S homeomorphic to an annulus. Let z and w be global holomorphic charts on $\mathbb{C}^*$ and S respectively, $w = f(z)$. Then the complex structure on $\mathbb{C}^*$ induces the almost complex structure on S by means of 1-form

$$df = (\partial f/\partial z) \, dz + (\partial f/\partial \bar{z}) \, d\bar{z},$$

or, equivalently, by the form

$$\omega = dz + \mu \, d\bar{z}, \qquad \mu = \frac{\partial f/\partial \bar{z}}{\partial f/\partial z}.$$

The mapping f is called *regular* if μ can be extended to the points 0, ∞ by the zero value while remaining continuous.

COROLLARY. *Let $f: \mathbb{C}^* \to S$ be a regular mapping (see the preceding example). Then there exists a conformal mapping $\mathbb{C}^* \to S$; this means that S has the conformal type of $\mathbb{C}^*$ rather than that of an annulus or a punctured disk.*

PROOF. The continuation of μ on $\widehat{\mathbb{C}}$ by zero values in 0, ∞ defines an almost complex structure on the Riemann sphere. This structure is integrable, so there is a global chart u on $\widehat{\mathbb{C}}$ such that $u(0) = 0$, $u(\infty) = \infty$ and du is $\mathbb{C}$-linear in the sense of the μ-almost complex structure (as in Example 1). Therefore du is proportional to df on $\mathbb{C}^*$; hence $u \circ f^{-1}: S \to \mathbb{C}^*$ is C^1-smooth and satisfies the Cauchy-Riemann equations, thus being the holomorphic mapping with the desired property. □

Let us return to the investigation of the quotient space by the action of the germ $f \in \mathscr{A}_{p+1}$. According to the formal classification theorem, this germ

can be put in the form
$$f(z) = f_{p,\lambda} + o(z^N), \qquad f_{p,\lambda} = g^1_{w_{p,\lambda}}, \qquad w_{p,\lambda} = z^{p+1}(1+\lambda z^p)^{-1}\partial/\partial z$$
for any $N < \infty$ by means of a holomorphic change of variable.

The transformation
$$z \mapsto t(z) = -p^{-1}z^{-p} + \lambda \ln z \tag{2.5}$$
maps every sector S_j of the nice p-covering of U onto the domain $\widetilde{S}_j$ containing the complement of the sector $|\arg t| \leqslant \pi/2 + \varepsilon$ to the big circle $|t| \leqslant R$ for sufficiently small ε if j is odd (if j is even, then the set is of the form $|\arg t - \pi| \leqslant \pi/2 + \varepsilon$). The field $w = w_{p,\lambda}$ is transformed into the constant field $\partial/\partial t$, the corresponding time 1 map being simply the unit shift $\mathrm{id} + 1 : t \mapsto t + 1$. Hence f in the coordinate t takes the form
$$\widetilde{f} : t \mapsto t + 1 + R(t), \qquad R(t) = O(|t|^{-M}).$$
By the choice of an appropriate N, the number M can be made as large as we wish.

REMARK. From now on in S_1 we consider the main branch of the logarithmic function which takes real values on the positive semiaxis, while in the remaining sectors S_j we take the natural continuation of the logarithm along an arc of the unit circle in the positive (counterclockwise) direction.

The quotient space of the domain $\widetilde{S}_j = t(S_j)$ by the action of $\widetilde{f}$ can be described as follows. Assume that j is even and consider the curvilinear semistrip bounded by the line $\operatorname{Re} t = C$ and its image $\{\widetilde{f}(C + i\tau) : \tau \in \mathbb{R}\}$ for sufficiently large C; identify the points $C + i\tau$ and $\widetilde{f}(C + i\tau) = C + i\tau + O(|C + i\tau|^{-M})$ thus obtaining an abstract Riemann surface S.

Let us represent the punctured sphere $\mathbb{C}^*$ as the quotient space of the strip bounded by the lines $\operatorname{Re} t = C$ and $\operatorname{Re} t = C + 1$ whose boundary points $C + i\tau$, $C + 1 + i\tau$ with the same ordinates are identified, and define the mapping F of the sphere thus realized to the Riemann surface S (constructed above) as the linear interpolation on horizontal segments:
$$F : C + \theta + i\tau \mapsto C + \theta + i\tau + \theta R(t), \qquad \theta \in [0, 1],$$
$$R(t) = \widetilde{f}(t) - (t + 1) = O(|t|^{-M}), \qquad t = C + \theta + i\tau.$$
Evidently for such a mapping one has
$$F_t = 1 + R(t)/2 + \theta R'(t), \qquad F_{\bar{t}} = R(t)/2 \quad \text{since } 0 - \operatorname{Re} t - C,$$
$$\lambda_t = \lambda_{\bar{t}} = 1/2, \qquad \mu = F_{\bar{t}}/F_t = o(|t|^{-M}).$$
The mapping F can be modified so as to become smooth, μ also being smooth on $\mathbb{C}^*$ and continuous on $\widehat{\mathbb{C}}$, $\mu(0) = \mu(\infty) = 0$. After such a modification, the preceding corollary can be applied to prove that all components of the quotient space are conformally equivalent to the doubly punctured sphere $\mathbb{C}^*$.

2.5. The sectorial normalization theorem. In this section the normalizing atlas for germs of the class $\mathscr{A}_{p+1}$ is constructed precisely as described above. Let S_j, $j = 1, \ldots, 2p$, be the sectors of the nice covering (see 2.2).

NORMALIZATION THEOREM FOR A SINGLE SECTOR. *In every sector S_j of the nice covering there exists a unique holomorphic mapping H_j of the form*

$$z \mapsto z + h_j(z), \qquad h_j = o(z^{p+1}),$$

conjugating f with its formal normal form

$$f_{p,\lambda} = g_w^1, \qquad w = w_{p,\lambda} = z^{p+1}(1 + \lambda z^p)^{-1} \partial/\partial z.$$

PROOF. *Existence.* Let j be even. In the straightening chart t in S_j given by (2.5), f takes the form

$$\widetilde{f}_j : t \mapsto t + 1 + R_j(t), \qquad R_j = o(t^{-M}),$$

for M as large as we need (see the end of 2.4); $\widetilde{f}_j$ is defined in the sector $|\arg t| \leqslant \pi/2 + \varepsilon$, $\varepsilon > 0$. The mapping $\widetilde{H}_j = t \circ H_j \circ t^{-1}$ conjugates $\widetilde{f}_j$ with the unit shift $\mathrm{id} + 1$. Let $H_j = \mathrm{id} + h_j$. The correction h_j satisfies the following functional equation:

$$h_j = h_j \circ \widetilde{f}_j + R_j. \tag{2.6}$$

After iterating this equation (the subscript j is omitted for simplicity) one obtains

$$h \circ \widetilde{f}^{[n]} = h \circ \widetilde{f}^{[n+1]} + R \circ \widetilde{f}^{[n]}.$$

If both right- and left-hand sides of all these equations are defined for some t, then formal summation yields

$$h(t) = \sum_{n=0}^{\infty} R \circ \widetilde{f}^{[n]} \tag{2.7}$$

(here we take into account the fact that $R \to 0$ as $t \to \infty$). Let us establish this formula and prove the convergence of the series (2.7).

Note that if $|t_0|$ is sufficiently large, then the orbit of the point t_0 belongs to the sector $|\arg(t - t_0)| < \pi/4$ (see Figure 4), in which $|R(t)| < \varepsilon$. Hence $|\widetilde{f}^{[n]}(t_0)| > cn$ for some $c > 0$, and $|R \circ \widetilde{f}^{[n]}(t_0)| < c_1 n^{-M}$. This proves the convergence of the series (2.7) and the existence of the normalizing transformation in the even sectors. The case of j odd is treated analogously.

REMARKS. 1. $h_j(t) \to 0$ as $t \to \infty$, $t \in S_j$.

2. The above reasoning allows us to find, in each sector S_j, the maps τ_j which conjugate the restriction $f|_{S_j}$ with the unit shift. These maps satisfy the condition $\tau_j(z) - t(z) \to 0$.

Uniqueness. It follows immediately from (2.6) and the convergence of h_j to zero at infinity, hence the equality (2.7) must be satisfied.

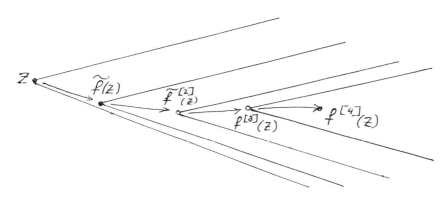

FIGURE 4

The normalizing atlas was constructed by Fatou in 1920. More than 60 years passed until S. M. Voronin [37], J. Ecalle [10] and B. Malgrange [24] began considering the transition functions.

How do the normalizing charts corresponding to different sectors of a nice covering behave on the intersections of these sectors? It turns out that they coincide up to flat functions. In order to state this fact more precisely, we need some estimates.

Let the germ f be of the class $\mathscr{A}_{p+1,\lambda}$, S_j being the sectors of the nice covering, and H_j the maps conjugating f with $f_{p,\lambda}$ in S_j. In the curvilinear sectors $H_j(S_j \cap S_{j+1})$, the transition functions $\Phi_j = H_{j+1} \circ H_j^{-1}$ are defined. The main property of the transition functions is that they commute with the standard mapping $f_{p,\lambda}$. Indeed, both H_{j+1} and H_j map orbits of f into orbits of $f_{p,\lambda}$, hence Φ_j maps orbits of $f_{p,\lambda}$ into themselves:

$$\Phi_j \circ f_{p,\lambda} = f_{p,\lambda} \circ \Phi_j.$$

This fact implies the following important property of the Φ_j: their difference from the identity decreases *exponentially* as $z \to 0$. Let us prove this fact. Write Φ_j in the chart t and denote the functions thus obtained by $\widetilde{\Phi}_j$. The above commutation means that

$$\widetilde{\Phi}_j(t+1) = \widetilde{\Phi}_j(t) + 1,$$

so $\psi_j(t) = \widetilde{\Phi}_j(t) - t$ is a 1-periodic holomorphic function defined in the upper half-plane $\operatorname{Im} t > c$ for some $c > 0$ if j is odd, and in the lower half-plane $\operatorname{Im} t < c'$ for some $c' < 0$ if j is even.

Hence ψ_j, $j = 1, \ldots, 2p - 1$, can be expanded in a Fourier series with zero free terms. Indeed, the corrections of the maps H_j (i.e., their differences from the identical map) tend to 0 as $t \to \infty$.

For j odd (respectively even) the function t maps the intersections $S_j \cap S_{j+1}$ onto domains contained in some upper (respectively lower) half-plane.

Hence
$$\psi_j(t) = \sum_{l=1}^{\infty} c_{j,l} e^{2\pi i l t}, \qquad j = 1, 3, 5, \ldots, 2p-1 \text{ (odd)}, \tag{2.8}$$
and for even $j \neq 2p$ the summation is taken over l satisfying $-1 \geqslant l > -\infty$.

On the intersection $S_1 \cap S_{2p}$, the function $t(z)$ defined by (2.5) is multivalued: its values obtained by continuation over the domains $S_1, S_2, \ldots, S_{2p}$ differ from its initial values in S_1 by the constant $c_0 = 2\pi i \lambda$ (λ being the same as in the formal normal form): $t_{2p}(z) - t_1(z) = c_0$.

From Remark 2 above it now follows that in the intersection $S_1 \cap S_{2p}$ we have $\tau_{2p}(z) - \tau_1(z) \to c_0$ as $z \to 0$. Hence
$$\psi_{2p}(t) = -c_0 + \sum_{l=-1}^{-\infty} c_{2p,l} e^{2\pi i l t}. \tag{2.9}$$

Denote the exponentially decreasing sum $\psi_{2p} + c_0$ by $\tilde{\psi}_{2p}$. The corrections ψ_j of the transition functions decrease as $\exp(-c|t|)$ with $|\operatorname{Im} t| \to \infty$ while t is in the domain of $\tilde{\Phi}_j$ for $j = 1, \ldots, 2p - 1$. In the initial chart z, one may conclude that the corrections of Φ_j decrease as $\exp(-c'/|z|^p)$. For Φ_{2p} the estimate is the same:
$$H_1 \circ H_{2p}^{-1} = t_1^{-1} \circ (\operatorname{id} - c_0 + \tilde{\psi}_{2p}) \circ t_{2p} = t_1^{-1} \circ (t_{2p} - c_0 + \tilde{\psi}_{2p} \circ t_{2p})$$
$$= t_1^{-1} \circ (t_1 + \tilde{\psi}_{2p} \circ t_{2p}) = \operatorname{id} + O(\exp(-c'/|z|^p)).$$

Thus the punctured neighborhood U lies in the union of the sectors S_j forming some nice p-covering of U. In each sector the normalizing mapping H_j is defined, the collection $H = (H_1, \ldots, H_{2p})$ constitutes a string which we call the *functional cochain*. The collection $\Phi = (\Phi_1, \ldots, \Phi_{2p})$, $\Phi_j = H_{j+1} \circ H_j^{-1}$, is called the *iterative coboundary* and denoted by $\delta_\circ H$ (the sign $\circ$ reminds one of the iterative nature of operations). The corrections of the maps Φ_j decrease as $\exp(-c|z|^{-p})$. The functions $H_{j+1} - H_j$ of the *differential coboundary* have the same rate of decrease.

Thus we have proved the first two assertions of the following theorem stated in cohomological terms (the third assertion is proved in 2.6):

SECTORIAL NORMALIZATION THEOREM [33, 10, 24]. 1. *For any germ of class $\mathscr{A}_{p+1}$ there exists a normalizing cochain*
$$H = (H_1, \ldots, H_{2p}).$$
The functions H_j are biholomorphic in the sectors S_j of the nice p-covering and conjugate the germ f with its formal normal form $f_{p,\lambda}$ (2.2).

2. *The coboundary of the cochain $\delta_\circ H = \{H_{j+1} \circ H_j^{-1}\}$ is a collection of maps with the corrections exponentially decreasing as $z \to 0$:*
$$|(\delta_\circ H)_j - \operatorname{id}| < (\exp(-c/|z|^p))$$
for some $c > 0$ depending on f.

3. *The normalizing cochain is uniquely defined up to superposition with a time θ map ($\theta \in \mathbb{C}$) for the flow of $w_{p,\lambda}$: any two normalizing cochains G, H satisfy $G = g^{\theta}_{w_{p,\lambda}} \circ H$ for some (complex) θ.*

It can be proved that all H_j have a common asymptotic Taylor series $\widehat{H}$ that conjugates the series $\widehat{f}$ and $\widehat{f}_{p,\lambda}$; for the case $(p, \lambda) = (1, 0)$ this has been demonstrated in [24].

REMARK 3. The map t (2.5) straightening the field $w = w_{p,\lambda}$ conjugates the time θ map g^{θ}_w with the shift $t \to t + \theta$. The choice of the normalizing transformation in S_j is equivalent to the choice of the chart $\tau_j t \circ H_j$ conjugating f with the unit shift and defined up to the addition of a constant $\tau_j \to \tau_j + c$.

Normalizing cochains are a particular case of so-called *functional cochains*. Namely, the set of all holomorphic collections $\{H_j\}$, $j = 1, \ldots, 2p$, such that each H_j is holomorphic in S_j and the *differential coboundary*

$$\{H_{j+1} - H_j\}, \qquad j = 1, \ldots, 2p \bmod 2p,$$

consisting of functions, exponentially decreasing like $\exp(-c|z|^{-p})$, forms a differential algebra. These algebras of functional cochains associated with nonlinear Stokes phenomena play the key role in the solution of the *Dulac problem* on the finiteness of the number of limit cycles for analytic vector fields on the plane [20, 21].

2.6. An alternative description of functional moduli and the proof of the analytic classification theorem. Let us summarize some material from the previous section. The sectorial normalization theorem associates to each germ $f \in \mathscr{A}_{p+1}$ its normalizing cochain $H = \{H_j \mid j = 1, \ldots, 2p\}$, also called the normalizing atlas for f. A chart H_j of this atlas (for any j) conjugates the initial germ with the time 1 shift $f_{p,\lambda}$ along the phase curves of the vector field $w = w_{p,\lambda}$ for some complex number λ. The map t (2.5) rectifies the field w and conjugates $f_{p,\lambda}$ with the unit shift $t \mapsto t + 1$. Let $t_j = t|_{S_j}$; the details related to the nonunivalence of the logarithm are discussed in 2.5. The transition functions

$$\widetilde{\Phi}_j = \tau_{j+1} \circ \tau_j^{-1} = t_j \circ \Phi_j \circ t_j^{-1}$$

commute with the unit shift. Summarizing formulas (2.8), (2.9), we get

$$\widetilde{\Phi}_j(\tau_j) = \tau_j + \sum_{l=-1}^{-\infty} c_{j,l} \exp 2\pi i l \tau_j \qquad \text{for } j < 2p \text{ even},$$

$$\widetilde{\Phi}_{2p}(\tau_{2p}) = \tau_{2p} + C + \sum_{l=-1}^{-\infty} c_{2p,l} \exp 2\pi i l \tau_{2p}, \qquad (2.10)$$

$$\widetilde{\Phi}_j(\tau_j) = \tau_j + \sum_{l=1}^{\infty} c_{j,l} \exp 2\pi i l \tau_j \qquad \text{for } j \text{ odd}.$$

The formula
$$C = -2\pi i \lambda \qquad (2.10')$$
for the free term in $\widetilde{\Phi}_{2p}$ gives the relationship between formal and analytic classifications. These series do converge in some sufficiently low (respectively, high) half-plane. The normalizing charts H_j are uniquely determined by the condition $H_j - \mathrm{id} = o(z^{p+1})$. The choice of another chart z leads to another normalizing atlas $G = \{G_j\}$. Let us show that $H_j \circ G_j = g^c_{w_{p,\lambda}}$ for some $c \in \mathbb{C}$. This is equivalent to the statement
$$t_j \circ H_j \circ G_j^{-1} \circ t_j^{-1} = \mathrm{id} + c,$$
because the map t_j transforms the field $w_{p,\lambda}$ to the standard one $\partial/\partial \tau_j$. Let us prove the last equality. Its left-hand side commutes with the unit shift in the "large" sectors
$$\widetilde{S}_j = \{|\arg \tau_j| < \pi/2 + \varepsilon, \ |\tau_j| > R\} \times \{j\}$$
for j odd. The function ψ by which this left-hand side differs from the identity is 1-periodic on $\widetilde{S}_j$, so the expression $\psi((\ln w)/2\pi i)$ is a well-defined, holomorphic and bounded function on the cylinder $\mathbb{C}^* = \{w\}$. Such a function is necessarily a constant. The case of even j is similar. □

This proves the third statement of the sectorial normalization theorem and motivates the following

DEFINITION. Two collections ψ, $\widetilde{\psi}$ of the form (2.10) are said to be *equivalent* if one can be obtained from the other by a shift:
$$\psi \sim \widetilde{\psi} \iff \exists c \in \mathbb{C}: \ \psi = \widetilde{\psi} \circ (\mathrm{id} + c). \qquad (2.11)$$

Thus to every germ $f \in \mathscr{A}_{p+1,\lambda}$ corresponds the equivalence class μ_f of collections (2.10). The set of all equivalence classes will be denoted by M_p^+ and referred to as the *moduli space*.

ANALYTIC CLASSIFICATION THEOREM. *The correspondence between $f \in \mathscr{A}_{p+1}$ and $\mu_f \in M_p^+$ described above provides a complete system of invariants for the analytic classification: this means that all four assertions of the classification theorem stated in 2.3 hold for the class M_p^+ instead of M_p^*.*

PROOF. 1. *Invariance of the class μ_f*. Let f, g be two analytically equivalent germs of class $\mathscr{A}_{p+1}$, h being the conjugating holomorphism: $h \circ f = g \circ h$. Let H, G be the cochains normalizing f, g respectively. Then the cochain $G \circ h$ normalizes f. By the third assertion of the sectorial normalization theorem, there exists a $c \in \mathbb{C}$ such that $G \circ h = g^c_w \circ H$, $w = w_{p,\lambda}$ (2.2). Then
$$G_{j+1} \circ G_j^{-1} = (G_{j+1} \circ h) \circ (G_j \circ h)^{-1} = g^c_w \circ H_{j+1} \circ H_j^{-1} \circ g^{-c}_w.$$

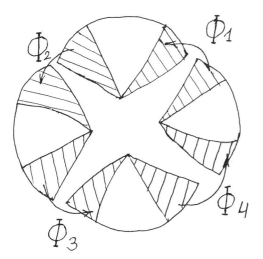 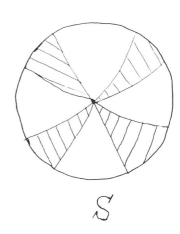

FIGURE 5

Let the germs $G_{j+1} \circ G_j^{-1}$ and $H_{j+1} \circ H_j^{-1}$ in the chart (2.5) have the form $\mathrm{id}+\psi_j$, $\mathrm{id}+\widetilde{\psi}_j$. According to Remark 3, the map g_m^c in this chart is the shift $t \to t+c$. Therefore

$$\mathrm{id}+\psi_j = (\mathrm{id}+c) \circ (\mathrm{id}+\widetilde{\psi}_j) \circ (\mathrm{id}-c) \implies \psi_j = \widetilde{\psi}_j \circ (\mathrm{id}-c)$$
$$\implies \{\psi\} \sim \{\widetilde{\psi}\} \implies \mu_f = \mu_g.$$

2. *Equimodality and equivalence.* Let the collections ψ, $\widetilde{\psi}$ associated with the germs f, $\widetilde{f} \in \mathscr{A}_{p+1}$ be equivalent in the sense of definition (2.11). Then, choosing another normalizing cochain for g, one can obtain the *coincidence* of ψ and $\widetilde{\psi}$. Therefore the coboundaries $\delta_\circ H$, $\delta_\circ \widetilde{H}$ also coincide. In the sectors of the nice p-covering define the map $H_j^{-1} \circ G_j$, which makes sense in the punctured neighborhood of the origin, because on the intersections of the sectors the individual maps do continue each other. Indeed, the relation $H_j^{-1} \circ G_j = H_{j+1}^{-1} \circ G_{j+1}$ is equivalent to $H_{j+1}^{-1} \circ H_j = G_{j+1}^{-1} \circ G_j$, which in turn follows from the equality $\delta_\circ H = \delta_\circ G$. The map $H^{-1} \circ G$ admits an analytic extension to the origin by the Riemann bounded extension (i.e., removable singularity) theorem.

3. *Realization.* The proof of the realization given below uses one of the basic ideas relevant to nonlinear Stokes phenomena.

Consider the sectors S_j of the nice p-covering as lying on different copies of the complex plane: $S_j \subset \mathbb{C} \times \{j\}$. Let $\mu_f \in M_p$ be the given module (the equivalence class). Consider any of its representatives: it is a collection of "transition functions", that is, a set of maps Φ_j, each of them defined and biholomorphic in some sector containing the left boundary radius of S_j in such a way that the image contains the right boundary radius of S_{j+1} (see Figure 5). Moreover, the corrections $\Phi_j - \mathrm{id}$ are exponentially flat at the origin. Denote by z_j the chart on S_j induced by the embedding $S_j \hookrightarrow \mathbb{C} \times \{j\}$.

Consider the disjoint union $\bigcup_{j=1}^{2p} S_j$ and identify the sectors containing S_j using the maps Φ_j (Figure 5). The quotient space is an abstract Riemann surface S homeomorphic to the punctured disk. Let us construct a *conformal* mapping of this surface onto the punctured disk $D^* \subset \mathbb{C}$. To do this, first construct the quasiconformal mapping $H_0 \colon S \to D^*$ as follows. The surface S is covered by the sectors S_j and in each one the chart z_j is defined. Let $\theta = \{\theta_j\}$ be an infinitely smooth partition of unity subordinate to the covering $\{S_j\}$ with the following additional condition: all derivatives of θ_j grow no more rapidly than some (negative, depending on the number of the derivative) power of $|z_j|$.

Then $H_0 = \sum_j \theta_j z_j$ is the chart on S defining the map $S \to D^*$. From the polynomial growth condition and the following identities

$$H_0|_{S_j \cap S_{j+1}} = z_j \theta_j + z_{j+1} \theta_{j+1} = z_j(\theta_j + \theta_{j+1}) + (z_{j+1} - z_j)\theta_{j+1}$$
$$= z_j + (\Phi_j(z_j) - z_j)\theta_{j+1},$$

taking into account the fact that the function in the parentheses decreases exponentially, we conclude that

$$\left.\frac{\partial H_0}{\partial \bar{z}_j}\right|_{S_j \cap S_{j+1}} = (\Phi_j(z_j) - z_j)\frac{\partial \theta_{j+1}}{\partial \bar{z}_j} = o(1), \qquad \frac{\partial H_0}{\partial z} = 1 + o(1).$$

This means that the almost complex structure induced on D^* by H_0^{-1} is defined by the Beltrami differential $\mu \, d\bar{z}/dz$, where μ exponentially converges to zero as $z \to 0$ along with all its derivatives. Thus μ can be smoothly extended to the origin. By the theorem in 2.3, the μ-almost complex structure is integrable and there exists a diffeomorphic mapping $G \colon (\mathbb{C}, 0) \to \mathbb{C}$ holomorphic in the sense of this structure. The superposition $H = G \circ H_0 \colon S \to \mathbb{C}$ is analytic. Thus S is conformally equivalent to the punctured disk D^*.

The standard time 1 map $f_{p,\lambda}$ in S_j correctly defines the mapping $\tilde{f}$ on S, because $f_{p,\lambda}$ commutes with Φ_j. This is the second central step in the argument. In the neighborhood of the origin in D^* we have the mapping f conjugated with $\tilde{f}$ by H and extendable to the origin by the bounded extension theorem. So f can be regarded as a germ $(\mathbb{C}, 0) \to (\mathbb{C}, 0)$.

It remains only to prove that f belongs to the class $\mathscr{A}_{p+1,\lambda}$. First let us check that f is tangent to the identity. In every sector S_j the normalized germ $f_{p,\lambda}$ is conjugated with f by the map H. Complete S_j by adding the vertex 0. The map $H \colon S \cup \{0\} \to \mathbb{C}$ defines a smooth structure on the manifold thus obtained. The theorem from 2.3 asserts that the map G constructed above belongs to the class C^1. But the function μ is also smooth on $S \cup \{0\}$. By the Newlander-Nirenberg theorem [28], the function G can be chosen smooth. Then the linear parts of $f_{p,\lambda}$ and f are conjugated by the linear part of H at the origin. The first one is the identity; therefore so is the second. Hence the germ f belongs to the class $\mathscr{A}_l$ for some $l \in \mathbb{N}$. From the construction of S it follows that f and $f_{p,\lambda}$ are topologically equivalent. But germs of the classes $\mathscr{A}_l$ and $\mathscr{A}_p$ are not equivalent if $p \neq l$, because they have different numbers of "petals" (see Figure 5). Hence $f \in \mathscr{A}_{p+1}$, because the germ f has exactly $2p$ "petals" (their number is equal to the number of sectors S_j). The formal classification parameter λ is obtained from the functional invariant (2.8) as the limit $-(2\pi i)^{-1} \lim_{\operatorname{Im} \tau \to \infty} \psi_{2p}(\tau)$. So $f \in \mathscr{A}_{p+1,\lambda}$.

Finally note that the maps $H_j = z_j \circ G^{-1}$ form a normalizing atlas for f on U. The transition functions for this atlas are equal to $z_{j+1} \circ z_j^{-1}$, that is, they are as prescribed. So the collection $\{\Phi_j\}$ is a functional modulus for the germ f: the realization is over.

4. *Analytic dependence.* If the germ f depends analytically on a finite number of parameters ε, then the normalizing cochain also depends analytically on ε. This follows immediately from the sectorial normalization theorem: if all terms of the series (2.7) depend analytically on ε, then so does the sum of the series.

2.7. Applications. The construction of the Ecalle-Voronin moduli solves the first two problems posed in 2.1; the classification theorems stated in 2.3 solve the third.

1. *The embedding problem.*

THEOREM 1. *The germ of a conformal mapping, formally equivalent to the time 1 map $f_{p,\lambda} = g_w^1$ for the field*

$$w = z^{p+1}(1 + \lambda z^p)^{-1} \partial / \partial z,$$

is analytically equivalent to this map if and only if the corresponding modulus (2.10) *is equal to*

$$\operatorname{id}, \ldots, \operatorname{id}, \operatorname{id} - 2\pi i \lambda. \tag{2.12}$$

PROOF. Indeed, the Ecalle-Voronin modulus for $f_{p,\lambda}$ has (2.12) as one of its representatives; the simultaneous conjugation of all maps in this collection by the shift makes no changes, so (2.12) is the *unique* representative in its equivalence class. Any germ analytically equivalent to $f_{p,\lambda}$ has the same functional modulus.

REMARK. The realization part of the classification theorem asserts that embeddable germs are extremely scarce among all formally equivalent germs.

2. *The root extraction problem.*

THEOREM 2. *Suppose that the germ $f \in \mathscr{A}_{p+1,\lambda}$ admits extraction of the nth root, tangent to the identity:*

$$\exists g: \quad g^{[n]} = f, \ g'(0) = 1.$$

Then the modulus μ_g consists of maps commuting not only with the unit shift, but also with the nth root $\mathrm{id} + n^{-1}$ of this shift.

PROOF. The map g commutes with f, therefore it takes orbits of f onto themselves, thus generating a quotient map on the orbit space. This space is the union of doubly punctured spheres as this is described in 2.3 and shown in Figures 2, 3. Let (2.3) be the corresponding collection of identifying maps, i.e., the module μ_f^*. Denote by π the projection of the phase space onto the space of orbits, by $\pi_* g$ the quotient map. Since g is tangent to the identity, $\pi_* g$ maps every punctured sphere onto itself, and is linear. After n iterations of g (but not earlier!) every point $z \in (\mathbb{C}, 0)$ turns out to be on the same f-orbit. Therefore the map g restricted to each punctured sphere is simply multiplication by a primitive root of unity. Since this map is well defined on the orbit space, all the identifying maps commute with this multiplication. This means that all the φ_j in (2.3) are of the form

$$\varphi_j(\zeta) = \zeta \psi_j(\zeta^n). \tag{2.13}$$

Let us state a similar property for the modulus μ_f constructed in 2.6. For odd j the following diagram

$$\begin{array}{ccc} \tau_j(S_j \cap S_{j+1}) & \xrightarrow{\widetilde{\Phi}_j} & \tau_{j+1}(S_j \cap S_{j+1}) \\ {\scriptstyle \exp \circ 2\pi i} \downarrow & & \downarrow {\scriptstyle \exp \circ 2\pi i} \\ (\mathbb{C}^*, 0) \times \{j\} & \xrightarrow{\varphi_j} & (\mathbb{C}^*, 0) \times \{j+1\} \end{array} \tag{2.14}$$

is commutative, while for even j the symbol 0 must be replaced by ∞, see (2.13). If in this diagram φ_j commutes with multiplication by the primitive root of unity, then Φ_j commutes with the shift $\mathrm{id} + m/n$ (m/n is an irreducible fraction). But the map Φ_j commuting simultaneously with $\mathrm{id} + 1$, $\mathrm{id} + m/n$, commutes also with $\mathrm{id} + 1/n$.

REMARKS. 1. It can be shown that in the above argument g is multiplication by $\exp 2\pi/n$ on each sphere, and $m = 1$.

2. The commutative diagram above explains the relation between the moduli $\{\varphi_j\}$ and $\{\widetilde{\Phi}_j\}$, see (2.3) and (2.10), respectively. Namely, by the definition from 2.3, for $1 \leq j \leq 2p - 1$ we have

$$\varphi_j(\zeta) = \zeta(1 + \widetilde{\varphi}_j(\zeta)), \quad \widetilde{\varphi}_j(\zeta) = o(1) \quad \text{for} \quad \begin{cases} \zeta \to 0, & j \text{ odd}, \\ \zeta \to \infty, & j \text{ even}, \end{cases}$$

$$\varphi_{2p}(\zeta) = \alpha \zeta(1 + \widetilde{\varphi}_{2p}(\zeta)), \quad \widetilde{\varphi}_{2p}(\zeta) = o(1), \quad \zeta \to \infty.$$

The relationship between $\tilde{\Phi}_j$, φ_j, α and λ is the following:

$$\tilde{\Phi}_j(t) = t + \ln(1 + \tilde{\varphi}_j(\exp 2\pi i t)), \qquad j \leq 2p - 1;$$
$$\tilde{\Phi}_{2p}(t) = t + (2\pi i)^{-1} \ln \alpha + \ln(1 + \tilde{\varphi}_{2p}(\exp 2\pi i t)). \qquad (2.15)$$

On the other hand, $\Phi_{2p}(t) = t - 2\pi i \lambda + o(1)$. Therefore

$$\alpha = \exp 4\pi^2 \lambda. \qquad (2.16)$$

Formula (2.13) along with (2.14) gives an alternative proof of the root extraction theorem.

3. *The centralizer.* Let us describe the centralizer of a germ $f \in \mathscr{A}_{p+1,\lambda}$ in the group of conformal germs tangent to the identity, that is, the set of germs $g: (\mathbb{C}, 0) \to (\mathbb{C}, 0)$ commuting with f and having a multiplier equal to 1. The map g takes orbits of f onto themselves, thus defining the quotient map $\pi_* g$ of the space of orbits. Since $g'(0) = 1$, every sphere is mapped by g onto itself. Therefore $\pi_* g$ is multiplication by some constant c_j on the jth sphere. Since it admits the factorization, we have

$$\varphi_j \circ c_j = c_{j+1} \circ \varphi_j. \qquad (2.17)$$

Computation of linear terms at the origin for j odd (respectively, at infinity for j even) yields

$$c_j = c_{j+1} \stackrel{\text{def}}{=} c, \qquad j = 1, 2, \ldots, 2p.$$

Consider three cases:

(1) f is a typical map not admitting any nontrivial root extraction;
(2) f admits root extraction but is not embeddable;
(3) f is embeddable.

In the first case all the identifying maps do not commute simultaneously with multiplication by any primitive root, that is, they do not commute with any linear mapping at all. This means that $c = 1$ and $g = f^{[m]}$ for some $m \in \mathbb{N}$.

Suppose the germ f is not embeddable (that is, there are nonlinear functions among the φ_j), but (2.17) holds for some c. Then c is a root of unity. If it is a primitive one, then (2.13) holds and $(\pi_* g)^{[n]} = \text{id}$ for some n. Therefore $g^{[n]} = f^{[m]}$ for some m. So we conclude that f admits extraction of the nth root, and g must be the root's mth power.

At last suppose that f is embeddable. Then the collection μ_f is of the form (2.12) and $(\pi_* g)$ corresponds to simultaneous multiplication by c. In this case f is analytically equivalent to the time 1 map for the field w (2.2). This field is transformed by the projection π into the field $\pi_* w = 2\pi i \zeta \partial/\partial \zeta$ on each sphere. The projection π is of the form $z \mapsto \exp 2\pi i t(z)$, where t straightens w and is given by (2.5). The multiplication by c on spheres corresponds to a time T shift along the phase curves of the field $\pi_* w$, $T = (\ln c)/2\pi i$. After returning to the initial phase space of f, this map becomes the time T_1 map for the field w with complex $T_1 = T + m$, $m \in \mathbb{Z}$. Conversely, for any complex T_1, the germ $g_w^{T_1}$ commutes with $f_{p,l}$. So we have proved

THEOREM 3. *Consider the germs* $(\mathbb{C}, 0) \to (\mathbb{C}, 0)$ *with multiplier equal to* 1. *Centralizers of typical (i.e., not admitting root extraction) germs consist of their iterates (both positive and negative); if the germ admits root extraction, then its centralizer consists of iterates of the root of maximal possible degree. The centralizer of an embeddable germ* g_w^1 *consists of the flow transformations* g_w^T, $T \in \mathbb{C}$.

2.8. Derivative of the moduli mapping. Consider the class of formally equivalent germs $\mathscr{A}_{p,\lambda}$ and the corresponding moduli space M_p. The mapping which associates the modulus μ_f with every germ $f \in \mathscr{A}_{p,\lambda}$ will be called the *moduli mapping*. Its derivative maps the tangent space to $\mathscr{A}_{p,\lambda}$ onto the tangent space to M_p. This derivative can be explicitly calculated at points corresponding to embeddable germs with trivial Ecalle-Voronin moduli (see (2.12)). The calculation was carried out by P. M. Elizarov [13] using ideas suggested by A. A. Shcherbakov [32]. For the sake of simplicity, consider only the class $\mathscr{A}_{p,\lambda}$ with $(p, \lambda) = (1, 0)$. Germs of this class can be put to the form

$$f(z) = g_w^1 + R(z), \qquad w = z^2 \partial/\partial z, \qquad R = O(z^4).$$

The change of variable $t = -1/z$ straightens the field w and conjugates the germ g_w^1 with the shift $\mathrm{id} + 1$, while the germ f is transformed into the germ

$$\widetilde{f}(t) : t \mapsto t + 1 + \widetilde{R}(t), \qquad \widetilde{R}(t) = O(t^{-2}),$$

the correction $\widetilde{R}(t)$ being analytic at infinity. For a germ with trivial modulus we can put $\widetilde{R} = 0$.

Perturbations of such a germ have the form

$$f_\varepsilon : t \mapsto t + 1 + \varepsilon h(t), \qquad h(t) = O(t^{-2})$$

with h holomorphic at $t = \infty$. Denote by μ_ε the Ecalle-Voronin modulus for f_ε:

$$\mu_\varepsilon = \varphi_\varepsilon^\pm, \qquad \varphi_\varepsilon^\pm(t) = \sum_{\pm 1}^{\pm\infty} c_k(\varepsilon) e^{2\pi i k t}.$$

Our problem is to calculate $(d\mu_\varepsilon/d\varepsilon)|_{\varepsilon=0}$, that is, to find the variations

$$\gamma_k = (dc_k(\varepsilon)/d\varepsilon)|_{\varepsilon=0}.$$

The answer is given in terms of Borel transform. Recall that the Borel transform of an analytic function

$$h = \sum_{k=0}^{\infty} h_k t^{-(k+1)}$$

is the function

$$\mathscr{B}h: \zeta \mapsto \sum_k h_k \zeta^k / k!.$$

REMARKS. 1. The Borel transform of a polynomial of degree $N+1$ in $1/\tau$ is the polynomial of degree N in ζ.

2. The Borel transform can be expressed in integral form

$$(\mathscr{B}h)(\zeta) = -\frac{1}{2\pi i} \int_\Gamma h(t) e^{t\zeta} dt, \qquad (2.18)$$

where Γ is a sufficiently large circle.

3. One can regard the parameters γ_k as coordinates in the tangent space to the "manifold" of Ecalle-Voronin moduli. When considered this way, the formulas defining γ_k are the expressions for the coordinates $((d\mu)_0 \cdot h)_k$ of the image $(d\mu)_0 \cdot h$.

THEOREM [13]. *The derivative $d\mu$ of the moduli map μ at the point corresponding to the time 1 shift in class $\mathscr{A}_{1,0}$ is given by the formula*

$$((d\mu)_0 \cdot h)_k = -2\pi i \mathscr{B}h(-2\pi i k).$$

PROOF. Here we give a proof which serves as the model for the more intricate calculations of Paper III. Use the definition of functional moduli as the transition function for the normalizing atlas. Let $\widetilde{S}_1$, $\widetilde{S}_2$ be sectors forming the nice 2-covering of the neighborhood of infinity as in 2.5, $\widetilde{H}_{1\varepsilon}$, $\widetilde{H}_{2\varepsilon}$ being the corresponding normalizing maps. By (2.7),

$$\widetilde{H}_{1\varepsilon} = t + \varepsilon \sum_0^\infty h \circ f_\varepsilon^{[n]}(t), \qquad \widetilde{H}_{2\varepsilon} = t + \varepsilon \sum_{-1}^{-\infty} h \circ f_\varepsilon^{[n]}(t).$$

By definition, $\widetilde{\Phi}_{j\varepsilon} = \widetilde{H}_{j+1,\varepsilon} \circ \widetilde{H}_{j-1,\varepsilon}$; $\widetilde{\Phi}_{1\varepsilon}$ and $\widetilde{\Phi}_{2\varepsilon}$ are defined in the half-planes $\operatorname{Im} t > c \gg 0$, $\operatorname{Im} t < c \ll 0$, respectively. We now calculate the variation of $\widetilde{\Phi}_{1\varepsilon}$ which will yield the formulas for the coefficients γ_k, $k \geq 1$. The variation of $\widetilde{\Phi}_{2\varepsilon}$ is similar. It is easy to see that

$$\frac{d}{d\varepsilon} \widetilde{\Phi}_{1\varepsilon}(t) = -\sum_{n=-\infty}^{+\infty} h(t+n), \qquad \operatorname{Im} t \gg 0.$$

This is a 1-periodic function of t, and its Fourier coefficients are equal to γ_k, $k \geq 1$. Thus for sufficiently large β we have

$$\gamma_k = -\int_{i\beta}^{i\beta+1} e^{-2\pi i k t} \sum_{n=-\infty}^{+\infty} h(t+n) dt = -\int_{i\beta-\infty}^{i\beta+\infty} e^{-2\pi i k t} h(t) dt. \qquad (2.19)$$

From (2.18) it follows that this is precisely the value of the Borel transform. To prove this, we must verify that (2.18) coincides with (2.19) for $\zeta = -2\pi i k$ up to change of sign. Indeed, let $\Gamma = \Gamma_R$ be the circle of radius R centered at the origin, and $\Gamma^\pm$ be its arcs lying above and below the line $L = \{\operatorname{Im} t = \beta\}$. The integral (2.18), if taken over Γ^- instead of Γ_R, tends to zero as $R \to \infty$ because the exponent is bounded below the line L, and h tends to zero more rapidly than a linear function does. If taken over Γ^+, the integral (2.18) tends to the integral (2.19) by the Cauchy theorem. This finishes the proof. □

COROLLARY. *Let P, Q be polynomials of different degrees without terms of order zero and one. Then the maps*

$$t \mapsto t + 1 + \alpha P(t^{-1}) \quad \text{and} \quad t \mapsto t + 1 + \beta Q(t^{-1})$$

are analytically nonequivalent for all sufficiently small pairs (α, β) and the same holds also for all $(\alpha, \beta) \in \mathbb{C}^2$ that lie outside some analytic codimension one (possibly, empty) subset.

§3. Degenerate elementary singular points

In this section both the orbital analytic and the topological classifications for degenerate singularities of holomorphic planar vector fields are described. The functional moduli giving the first classification are very similar to those for the classification of one-dimensional mappings discussed in §2. This similarity essentially permits to reduce the first problem to the second. Somewhat surprisingly, the topological classification is based on the analytic one and cannot be obtained in terms of jets of any finite order.

3.1. Sectorial normalization theorem. Consider an analytic vector field with an isolated degenerate singularity at the origin. Assume that this singularity is *elementary*, that is, exactly one eigenvalue (say, the first one) is nonzero.

DEFINITION. Two holomorphic vector fields are called *analytically orbitally equivalent*, if the first field can be transformed into the second by a biholomorphic change of variables and subsequent multiplication by a nonvanishing holomorphic function (the last step is a change of the time variable).

In other words, two fields are analytically orbitally equivalent if and only if there exists a holomorphic mapping of the phase space which takes complex phase curves of the first field to those of the second one.

Multiplying the field by a nonzero constant, we can make the first (nonzero) eigenvalue equal to -1. Let us start with the formal classification, which operates with formal power series transformations without free terms and multiplication by power series with nonzero free terms, without any convergence assumptions.

FORMAL CLASSIFICATION THEOREM. *Each holomorphic planar vector field v with a degenerate elementary singularity at the origin is formally orbitally equivalent to the field corresponding to the equation $\underline{\dot{z}} = w_{p,\lambda}(\underline{z})$, $\underline{z} = (z, w)$:*

$$\dot{z} = z^{p+1}(1 + \lambda z^p)^{-1}, \qquad \dot{w} = -w. \tag{3.1}$$

PROOF. First use the formal power series transformation in order to put the field v in resonance normal form [2, 3]. The resonance monomials are of the form $z^k \partial/\partial z$ and $yz^k \partial/\partial z$. So the field v is formally equivalent to the field

$$\widehat{f}(z)\partial/\partial z - y\widehat{g}(z)\partial/\partial y,$$

where $\widehat{f}$, $\widehat{g}$ are formal power series, and $\widehat{g}(0) = 0$. It is easy to prove that if $\widehat{f} \equiv 0$, then the singularity is not isolated: there is a whole curve of singular points tangent to z-axis in the origin. Hence the series $\widehat{f}$ is nonzero. Let

$$\widehat{h}(z) = \widehat{f}(z)/\widehat{g}(z) = az^{p+1}(1 + O(z)), \qquad a \neq 0.$$

By a linear substitution $z \mapsto cz$ with an appropriate c one can get $a = 1$.

The formal vector field $\widehat{h}(z)\partial/\partial z$ is equivalent to the field $w_{p,\lambda}$ for some $\lambda \in \mathbb{C}$ (see §2 and [7]). This proves the theorem. □

The class of germs having (3.1) as its formal orbital normal form will be denoted by $\mathscr{E}_{p,\lambda}$.

The general theory gives not only formal, but also certain geometrical corollaries for the field v. Namely, the theory of invariant manifolds implies that the field v possesses an invariant holomorphic curve passing through the origin and tangent to the eigenvector with nonzero eigenvalue. This curve may be chosen as the w-axis. By a polynomial transformation preserving the w-axis, it is possible to normalize the $(p + 1)$-jet of v at the origin. Then the equation determined by v takes the form

$$\dot{w} = -w + g(z, w), \qquad \dot{z} = zf(z, w),$$
$$j_0^p f = z^p, \qquad j_0^1 g = 0. \tag{3.2}$$

Normalizing the field on the w-axis, one obtains the equality $g(0, w) = 0$, so that $g(z, w) = zQ(z, w)$. This form is used in Paper III. Moreover, the equation (3.2) can be approximated by its normal form in all points of y-axis. Indeed, the transformation

$$(z, w) \mapsto (zh_1(w) + z^2 h_2(w) + \cdots + z^p h_p(w), w)$$

with h_j holomorphic at the origin, takes (3.2) to

$$\dot{z} = z^{p+1} f_1(z, w), \qquad \dot{w} = -w + g_1(z, w),$$

which is orbitally equivalent to the system

$$\dot{z} = z^{p+1}, \qquad \dot{w} = -w + g_2(z, w), \qquad j_0^1 g_2 = 0. \tag{3.3}$$

This method can be termed as the "normalization along the invariant manifold"; it was systematically used by H. Dulac [9]. The existence of such a transformation can be proved by constructing a finite number of successive approximations.

Thus the *preliminary normal form* for a degenerate elementary singularity is constructed, in terms of which the main result of this section can be formulated. Consider the "rotated nice p-covering" of the punctured neighborhood U of the origin on z-axis by the sectors

$$S_j = \{z : |\arg z + \pi/2p - \pi j/p| < \alpha\}, \qquad \alpha \in (\pi/2p, \pi/p).$$

Denote the product of any sector S with vertex at the origin of the z-plane and of the disk $D = \{|y| < \varepsilon\}$ by $\widetilde{S}$; such products will be referred to as *extended sectors*.

SECTORIAL NORMALIZATION THEOREM FOR VECTOR FIELDS [15]. *For each germ of the form* (3.3) *and each sufficiently small* U, ε *there exists a normalizing cochain* $H = \{H_j\}$: *the holomorphic map* H_j *is defined in the extended sector* $\widetilde{S}_j$ *of the rotated p-covering and conjugates the system* (3.3) *with the system*

$$\dot{z} = z^{p+1}, \qquad \dot{w} = -w f(z). \tag{$\dagger$}$$

The function f is holomorphic at $z = 0$, $f(0) = 1$. The transformations H_j preserve the z-coordinate: $H_j(z, w) = (z, w + \widetilde{H}_j(z, w))$. *Moreover, $\widetilde{H}_j = o(z^{p+1})$ as $z \to 0$, $z \in \widetilde{S}_j$. The transformation satisfying the above list of properties is unique.*

Though the modern proof of this theorem is not very complicated, it goes beyond the limits of the present survey.

To obtain the normal form (3.1), one only needs to divide ($\dagger$) by the function f and find the holomorphic transformation $(\mathbb{C}, 0) \to (\mathbb{C}, 0)$, $z \mapsto h(z)$ conjugating the z-component of the new field with the standard field $w_{p,\lambda} \partial/\partial z$. For simplicity denote the substitution $(z, w) \mapsto (h(z), w)$ also by h. In this notation the normalizing mappings take the form $h \circ H_j$ in $\widetilde{S}_j$.

3.2. Orbital analytic classification: the Martinet-Ramis moduli. The normalized field has several important properties. Here we point out only one of them: the field possesses a first integral of the form

$$u(z, w) = w \exp(t(z)), \qquad t(z) = -p^{-1} z^{-p} + \lambda \ln z, \tag{3.4}$$

holomorphic in every extended sector $\widetilde{S}_j$. The phase curves of the normalized equation different from the singular point have the form

$$w(z) = c \exp(-t(z)) = c z^{-\lambda} \exp(p^{-1} z^{-p}).$$

The closed sectors on the z-axis in which

$$\operatorname{Re} z^{-p} \to \infty \qquad (\text{respectively}, \operatorname{Re} z^{-p} \to -\infty)$$

will be called the *sectors of jump* (respectively, *fall*). In the sectors of jump all the solutions $w(z)$ grow exponentially, while in the sectors of fall they exponentially converge to zero as $z \to 0$. Every sector of the rotated nice covering intersects the adjoining sectors by the sectors of jump and fall alternatively. The intersection $S_1 \cap S_{2p}$ is the sector of jump (it contains the germ of the positive semiaxis $(\mathbb{R}^+, 0)$). The first integral behaves in the opposite way: it grows exponentially in the extended sectors of fall and decreases in those of jump, if $z \to 0$, provided that w is bounded away from 0.

After such preparations have been made, consider as usual the transition functions of the normalizing atlas constructed for equation (3.3). The holomorphic maps $h \circ H_j$ defined in $\widetilde{S}_j$ preserve the foliation $w = \text{const}$ and take phase curves of the initial system (3.3) to those of the normalized system (3.1). The transition functions

$$\Phi_j = (h \circ H_{j+1}) \circ (h \circ H_j)^{-1}$$

preserve both the z-coordinate and the foliation into phase curves of the system (3.1). These curves can be "cnumcratcd" by values that the first integral u takes on them. The function u is multivalued; its restriction to $\widetilde{S}_1$ (where $\arg z$ is close to 0) differs from its restriction to S_{2p} (to which it can be continued through the chain of sectors $\widetilde{S}_1, \ldots, \widetilde{S}_{2p-1}$) by the multiplier $\nu = \exp 2\pi i \lambda$. Denote by u_j the result of such a continuation on the sector $\widetilde{S}_j$ (the functions u_j and u_{j+1} serve as the analytic continuation one of the other for $1 \leqslant j \leqslant 2p-1$, and $u_{2p} = \nu u_1$). In the same manner (3.4) defines the functions t_j with $t_{2p} = t_1 + 2\pi i \lambda$.

From the properties of Φ_j described above it follows that for each j there exists a holomorphic function φ_j defined in some neighborhood of the origin and such that

$$u_{j+1} \circ \Phi_j = \varphi_j \circ u_j.$$

Hence

$$\Phi_j(z, w) = (z, \varphi_j(u)e^{-t_{j+1}(z)}). \tag{3.5}$$

From the sectorial normalization theorem it follows that $\Phi_j \to 0$ as $z \to 0$. Let us prove that the corrections $\Phi_j - \text{id}$ exponentially tend to zero and investigate, at the same time, what kind of functions φ_j may appear in this construction.

First assume that $S_j \cap S_{j+1}$ is the sector of fall, j is odd. Then in this sector $u_j = u_{j+1}$, $t_j = t_{j+1}$ (since these equalities are violated only in the intersection $S_1 \cap S_{2p}$, the sector of jump), and for $z \to 0$, $w \not\to 0$ one gets $u \to \infty$. If in the Taylor expansion $\varphi_j(u) = \sum \alpha_{jn} u^n$ all the coefficients at nonlinear terms vanish, then

$$\Phi_j(z, w) = (z, \alpha_{j0} e^{-t_j(z)} + \alpha_{j1} w).$$

Since $\Phi_j - \mathrm{id}$ tends to zero and $\operatorname{Re} t_j$ to $+\infty$ as $z \to 0$, we conclude that $\alpha_{j1} = 1$, and $\Phi_j - \mathrm{id}$ decreases exponentially.

If there is at least one nonvanishing nonlinear term in the expansion of φ_j, then $\Phi_j - \mathrm{id} \to \infty$ as $z \to \infty$, which is a contradiction. Hence in the sectors of fall for j odd we have

$$\varphi_j(u) = u + a_j, \qquad a_j = \alpha_{j0}, \tag{3.6}$$

$$\Phi_j(z, w) = (z, w + a_j e^{-t_j(z)}), \tag{3.7}$$

and $\Phi_j - \mathrm{id}$ is *exponentially flat*, that is, it tends to zero together with all its derivatives, each one admitting an estimate of the form $c\exp(-\delta/|z|)$, $\delta, c > 0$.

Let now $S_j \cap S_{j+1}$ be the sector of jump, j being even. First put $j \ne 2p$. Then $u_{j+1} = u_j$, $t_{j+1} = t_j$ in $\widetilde{S}_j \cap \widetilde{S}_{j+1}$. In the sectors of jump $\exp t_j(z) \to 0$ as $z \to 0$. So the assumption $\Phi_j - \mathrm{id} \to 0$ implies

$$\varphi_j(0) = 0, \qquad \varphi_j'(0) = 1 \tag{3.8}$$

for even j. Then $\varphi_j(u) = u(1 + \widetilde{\varphi}_j(u))$, $\widetilde{\varphi}_j(0) = 0$, $\widetilde{\varphi}_j$ being a holomorphic function. From (3.5) one obtains

$$\Phi_j(z, w) = (z, w(1 + \widetilde{\varphi}_j(we^{-t_j(z)}))). \tag{3.9}$$

The function $1 + \widetilde{\varphi}_j \circ u$ is exponentially flat in the sector of jump; the same is true for $\Phi_j - \mathrm{id}$.

Finally let $j = 2p$. Then in $\widetilde{S}_1 \cap \widetilde{S}_{2p}$ one has $u_{2p} = \nu u_1$, $\nu = \exp 2\pi i\lambda$. Therefore by (3.5)

$$\Phi_{2p}(z, w) = (z, w\varphi_{2p}(u_{2p})/u_1). \tag{3.10}$$

From the condition $\Phi_{2p} - \mathrm{id} \to 0$, it follows that

$$\varphi_{2p}(0) = 0, \qquad \varphi_{2p}'(0) = \nu^{-1}.$$

From the exponential flatness of u in the sector of jump it follows that the correction $\Phi_{2p} - \mathrm{id}$ is also exponentially flat.

Renumber the collection $\{\varphi_j\}$ in order to stress the distinction between the functions defining the transition in sectors of jump and fall. Recall that those sectors alternate. So enumerate them starting from zero as

$$S_0^+, S_0^-, \dots, S_{p-1}^+, S_{p-1}^-, \qquad S_0^+ = S_{2p} \cap S_1;$$

"+" stands for jump, "−" for fall. The collection φ of functions denoted according to this principle by $\varphi_j^\pm$, $j = 0, \dots, p-1$, satisfies the following conditions:

$$\varphi = (\varphi_0^+, \varphi_0^-, \dots, \varphi_{p-1}^+, \varphi_{p-1}^-) = (\varphi_{2p}, \varphi_1, \dots, \varphi_{2p-1}),$$

$$\varphi_j^+(0) = 0, \qquad \varphi_j^{+\prime}(0) = \begin{cases} 1, & j \ne 0, \\ e^{-2\pi i\lambda}, & j = 0, \end{cases} \tag{3.11}$$

$$\varphi_j^-(u) = u + a_j, \qquad a_j \in \mathbb{C}.$$

REMARK. Any automorphism preserving both the z-coordinate and the foliation defined by the system (3.1) at the same time preserves the field (3.1). It is easy to show that such an automorphism must be of the form $(z, w) \mapsto (z, Cw)$, $C \in \mathbb{C}^*$. Hence the normalizing map $h \circ H_j$ and the integral $u_j = u \circ h \circ H_j$ are uniquely defined up to multiplication by a constant. This circumstance motivates the following

DEFINITION. Two collections φ and $\widetilde{\varphi}$ of the form (3.11) are called *equivalent* if they are conjugated by a linear transformation:

$$\varphi \sim \widetilde{\varphi} \iff \exists C \in \mathbb{C}^* : \varphi \circ C = C\widetilde{\varphi}. \tag{3.12}$$

This equivalence relation coincides with (2.4); this is not accidental as will be shown below. The equivalence class (3.11), (3.12) corresponding to the germ v, and constructed as explained above will be denoted by ε_v.

3.3. Classification theorem.
Now we can formulate a theorem describing classes of analytic orbital equivalence in the space of germs of holomorphic vector fields. As before, we restrict ourselves to the class $\mathscr{E}_{p,\lambda}$ of formally equivalent germs.

THEOREM. 1. Invariance. *Every germ v of a holomorphic vector field at a degenerate elementary singular point (a germ of a class $\mathscr{E}_{p,\lambda}$) is associated to an equivalence class ε_v (3.11), (3.12). Analytically equivalent germs are associated to the same class.*

2. Equimodality and equivalence. *Two germs with coinciding invariants are orbitally analytically equivalent.*

3. Realization. *Any collection (3.11) can be realized as the functional invariant for an appropriate germ of class $\mathscr{E}_{p,\lambda}$.*

4. Analytic dependence on parameters. *If a germ $v \in \mathscr{E}_{p,\lambda}$ depends analytically on a finite-dimensional parameter, then so does the invariant associated with the germ.*

PROOF. 1. The *invariance* of the class ε_v is proved as in 2.6 (also see the remark at the end of 3.2). An independent proof is contained in Corollary 1, in 3.4 below.

2. *Equimodality and equivalence.* Let two germs of class $\mathscr{E}_{p,\lambda}$ have the same functional modulus. This means that two collections φ, ψ of the form (3.11) constructed for these germs in the manner described above are equivalent in the sense of (3.12). Denote by $v, \widetilde{v}$ the preliminary normal forms (3.3) for these germs. We may multiply the second component of the normalizing charts for the germ $\widetilde{v}$ by an arbitrary constant; by an appropriate choice of this constant, we can ensure the coincidence of φ and ψ. Then the transition maps (3.5) for the atlases normalizing v and $\widetilde{v}$ also must coincide. Define a biholomorphic mapping conjugating v and $\widetilde{v}$ as follows: the point $p \in \widetilde{S}_j$ is mapped to the point $\widetilde{p} \in \widetilde{S}_j$ if both of them have the same coordinates in the corresponding normalizing charts. This mapping is

correctly defined on the union of the extended nice sectors since the normalizing atlases have the same transition functions between the charts with the same numbers. The map thus constructed, initially defined in a neighborhood of the origin with the axis removed, can be extended to this axis by the bounded extension theorem. Evidently this map solves the conjugation problem.

3. *Realization.* As in the case of one-dimensional holomorphisms, this is one of the key points in the investigation. Its general pattern remains the same for all classification problems including those concerning germs of one-dimensional maps tangent to the identity (see 2.6 above) and resonant vector fields (4.6 below). Namely, using the formulas (3.7), (3.9), the collection (3.11) is transformed into a set of maps Φ_j of transition functions for an atlas which is not yet constructed. These maps have the following properties:

(1) they preserve the z-coordinate and the normalized field (3.1);
(2) they differ from the identity by corrections which are exponentially flat on the w-axis.

The disjoint union of $2p$ copies of the neighborhoods of the origin in $\mathbb{C}^2$ with the field (3.1) on them is then considered. Let two points

$$p \in \widetilde{S}'_j \times \{j\}, \qquad q \in \widetilde{S}'_{j+1} \times \{j+1\}$$

from the "diminished" extended sectors be identified if and only if $q = \Phi_j(p)$. The abstract manifold obtained as the quotient space is denoted by $\widetilde{S}$.

Denote by $\widetilde{\widetilde{S}}_j$ the domains corresponding to the sectors $\widetilde{S}_j \times \{j\}$ with the natural charts on them denoted by $\underline{z}_j$. Thus an atlas on $\widetilde{S}$ appears with charts $\underline{z}_j$ and transition functions Φ_j. On this manifold, define a holomorphic vector field $\widetilde{v}$, which coincides with the normal form (3.1) in any chart $\underline{z}_j$. This field is well defined, because the transition maps preserve the field (3.1).

Now it is necessary to "recognize" a neighborhood of the origin with deleted w-axis in the abstract manifold $\widetilde{S}$ and verify that $\widetilde{v}$ is indeed a representative of a germ belonging to the class $\mathscr{E}_{p,\lambda}$ with the prescribed modulus.

First construct a diffeomorphism

$$H_0 \colon \widetilde{S} \to H_0(\widetilde{S}) = W \subseteq \mathbb{C}^2 \setminus \{w = 0\}, \qquad \overline{W} \ni 0,$$

as in 2.6. Consider a partition of unity $\{\theta_j\}$ on the manifold $\widetilde{S}$ subordinate to the covering $\{\widetilde{\widetilde{S}}_j\}$:

$$\sum_j \theta_j = 1, \quad \theta_j \in C^\infty(\widetilde{S}), \quad \operatorname{supp} \theta_j \subset \widetilde{\widetilde{S}}_j.$$

The functions θ_j can be chosen so that their derivatives grow no faster than polynomials in $|z|^{-1}$ as $z \to 0$. Put
$$H_0 = \sum_j \theta_j \underline{z}_j.$$
This map is a diffeomorphism (as shown below) and provides its target space with an almost complex structure induced from the complex structure on $\widetilde{S}$.

REMARK ON ALMOST COMPLEX STRUCTURES. As explained in 2.4, an almost complex structure on a manifold is defined by a collection of 1-forms $\omega_1, \ldots, \omega_n$, which are assumed to be $\mathbb{C}$-linear (or of type $(1, 0)$) in the sense of this structure. Recall that an almost complex structure is called *integrable* if it is "the trace of a lost complex structure", i.e., if there are functions $z_1, z_2, \ldots, z_n$ whose differentials span the same subbundle of the cotangent bundle as $\omega_1, \ldots, \omega_n$. This means that there exists a matrix-valued function F such that
$$\omega = F\,dz, \quad z = (z_1, \ldots, z_n), \quad \omega = (\omega_1, \ldots, \omega_n). \tag{3.13}$$
Hence $d\omega = dF \wedge dz$, the vector-valued form on the right-hand side being the linear combination of pairwise products of forms of the types $(1, 0)$ and $(0, 1)$ (the latter ones come from the complex conjugate subbundle generated by ω). Note that in this expansion forms of type $(0, 2)$ are absent, since there are no pairwise products of forms of type $(0, 1)$. This is precisely the necessary condition for integrability of almost complex structures: the collection $d\omega$ must admit decomposition into a linear combination of forms of types $(2, 0)$ and $(1, 1)$ only (recall that type is defined in terms of the subbundle ω).

For smooth collections ω, the above condition proves to be sufficient.

NEWLANDER-NIRENBERG THEOREM [28]. *Suppose that a C^∞-smooth almost complex structure satisfies the above necessary condition for integrability. Then it is indeed integrable, and local charts holomorphic in the sense of the corresponding complex structure are infinitely differentiable.*

This concludes the remark on almost complex structures in the multidimensional case.

We continue the proof of the realization part of the main theorem. Let us examine the structure induced by H_0 on the domain $W = H_0(\widetilde{S})$. The function z (the first coordinate of the charts $\underline{z}_j$) is well defined on $\widetilde{S}$, since all transition maps preserve it. Hence the almost complex structure induced on W by H_0 is spanned by the forms dz and ω, the latter being constructed as follows. First, let us construct a form $\widetilde{\omega}$ of type $(1, 0)$ on $\widetilde{S}$ so that for all j the form $\widetilde{\omega}$ is close to the form dw_j in the chart $\underline{z}_j$ on the domain $\widetilde{\Omega}_j$. To do this, fix an arbitrary j and consider the transition function Φ_j from the chart $\underline{z}_j = (z, w_j)$ to the chart $\underline{z}_{j+1} = (z, w_{j+1})$; the

function Φ_j is defined on the intersection $\widetilde{\Omega}_j \cap \widetilde{\Omega}_{j+1} = \widetilde{\Omega}_{j\,j+1}$. The mapping $\Phi_j - \mathrm{id}$ tends to zero exponentially as $z \to 0$ in $\widetilde{\Omega}_{j,j+1}$. Let $\Phi_j = (z, \Phi_{j2})$; then
$$dw_{j+1} = \frac{\partial \Phi_{j2}}{\partial z} dz + \frac{\partial \Phi_{j2}}{\partial w_j} dw_j.$$
Put
$$f_{1j} = \frac{\partial \Phi_{j2}}{\partial z}, \qquad f_{1j} = \frac{\partial \Phi_{j2}}{\partial w_j} - 1.$$
The functions f_1 and f_2 tend to zero exponentially together with all their derivatives in $\widetilde{\Omega}_{j,j+1}$. Denote by F_{1j}, F_{2j} the extensions of the functions f_{1j}, f_{2j} to all of $\widetilde{\Omega}_j$ that have the same asymptotic property and vanish on $\widetilde{\Omega}_{j-1,j}$. Put
$$\widetilde{\omega}_j = F_{j1} dz + (1 + F_{j2}) dw_j.$$
This form is defined everywhere on $\widetilde{\Omega}_j$, is of type $(1,0)$ and coincides with dw_j and dw_{j+1} on $\widetilde{\Omega}_{j-1,j}$ and $\widetilde{\Omega}_{j,j+1}$, respectively. Hence the form $\widetilde{\omega}$ defined by $\widetilde{\omega}|_{\Omega_j} = \widetilde{\omega}_j$ for all j is well defined on $\widetilde{S}$, and its type is $(1,0)$. Put
$$\omega = (H_0^{-1})^* \widetilde{\omega}.$$
ω thus defined is the required form.

Let us prove that the difference $\omega - dw$ in the domain $W = H_0 \widetilde{S}$ decreases exponentially together with all its derivatives as $z \to 0$. Indeed, all the differences $\Phi_j - \mathrm{id}$ in the charts z_j possess the same property. Hence the maps $H_0 - \mathrm{id}$ and $H_0^{-1} - \mathrm{id}$ written in the charts z_j and (z, w), respectively, possess the same property as well. Indeed, the derivatives of the functions θ_j involved in the expression for H_0 increase no faster than a negative power of z; however, exponential decrease overcomes polynomial growth. Hence the difference $\omega - dw$ is exponentially flat at the origin. This argument also proves that H_0 is a diffeomorphism.

Now extend the above almost complex structure to the deleted w-axis: the expression for dz is preserved; the form ω is extended to dw. The extended structure is smooth in the entire neighborhood of the origin in $\mathbb{C}^2$. The integrability conditions provided by the Newlander-Nirenberg theorem hold for $w \neq 0$, because the almost complex structure induced by the diffeomorphism outside the axis is integrable by definition. By reasons of continuity, the integrability conditions also hold on the removed axis, so the structure turns out to be integrable in the entire neighborhood. By diminishing the size of the sectors and the disks defining $\widetilde{S}_j$ (if necessary), one may assume that there is a map G defined on $\overline{W}$ which is biholomorphic in the sense of the extended almost complex structure. Without loss of generality z can be taken as the first component of G. The map G is smooth, and the composition $H = G \circ H_0$ possesses a $\mathbb{C}$-linear differential by construction. Therefore it is biholomorphic, and there exists an embedding $H\colon \tilde{S} \to \mathbb{C}^2$.

4. The analytic dependence of the functional invariant on the parameters follows from the theorem on the analytic dependence on parameters of the charts given by the sectorial normalization theorem, see 3.1. The last result is known to the experts, but, as far as I know, its proof is not published.

The first analytic dependence is used in Paper III.

3.4. Monodromy. Consider the monodromy maps for the initial vector field v of the class $\mathscr{E}_{p,\lambda}$ (3.3) and its normal form (3.1), corresponding to the loop on the separatrix $w = 0$. Denote them by Δ and Δ_0 respectively. The map Δ_0 is embeddable. Indeed, system (3.1) has separated variables and can be explicitly integrated: the integration yields

$$\Delta_0(z) = g_{w_{p,\lambda}}^{-2\pi i} = g_{-2\pi i w_{p,\lambda}}^1, \qquad w_{p,\lambda} = z^{p+1}(1+\lambda z^p)^{-1}. \tag{3.14}$$

Let $t(z)$, as before, be the function straightening the field $w_{p,\lambda}\partial/\partial z$, see (2.4), $u = w \exp t(z)$ being the first integral of the normalized system. Then the restriction

$$\tau_j^* = u|_{S_j \times \{1\}}$$

to the jth sector of the rotated nice p-covering of the disk $\{0 < |z| < \varepsilon\} \times \{1\}$ maps the space of orbits of $\Delta_0 = g_{w_{p,\lambda}}^{2\pi i}$ onto the doubly punctured sphere $\mathbb{C}^* \times \{j\}$ and takes different values on different orbits.

The linear substitution $z \mapsto cz$, $c^p = 2\pi i$, takes the field $w_{p,\lambda}$ to the field $w_{p,\alpha}$ with $\alpha = \lambda/2\pi i$. Hence the germ Δ_0 is of the formal type $\mathscr{A}_{p,\alpha}$.

Now consider the monodromy map Δ for the initial system. By narrowing the sector $S_j \times \{1\} \subset \Gamma$ slightly (and denoting the result by $S_j' \times \{1\} \subset \Gamma$), one may assume that the real curve starting at $(z, 1) \in S_j' \times \{1\}$ and finishing at $(\Delta(z), 1)$, which determines the monodromy, entirely belongs to the extended sector $\tilde{S}_j$. The map $h \circ H_j$ that conjugates the initial vector field $\tilde{v}$ with its formal normal form (3.1) is $\tilde{S}_j$ (see the end of 3.1). The function $\tau_j = u \circ h \circ H_j|_{S_j \times \{1\}}$ maps the orbits of Δ to points of the sphere $\mathbb{C}^* \times \{j\}$. Therefore the transition functions φ_j given by (3.11) are the functional moduli of the monodromy map in the sense described in 2.3. This explains the coincidence of the equivalence relations (2.4) and (3.12). Thus we have proved

THEOREM. 1. *The monodromy map Δ for a germ of a vector field $v \in \mathscr{E}_{p,\lambda}$ belongs to the formal class $\mathscr{A}_{p,\alpha}$ with $\alpha = \lambda/2\pi i$.*

2. *The functional modulus* (3.11), (3.12) *of the germ v coincides with the functional modulus μ_Δ^* of the map Δ.*

COROLLARY 1. *Consider two germs of holomorphic vector fields having degenerate elementary singularities, together with their monodromies corresponding to holomorphic invariant submanifolds. These germs of fields are* *analytically orbitally equivalent if and only if the corresponding monodromies are analytically equivalent.*

Functional moduli of maps tangent to the identity were defined in two different forms, (2.3) and (2.10). So, besides (3.11), there is a functional modulus for germs $v \in \mathscr{E}_{p,\lambda}$ which has the form (2.10). Let us renumber the transition functions $\tilde{\Phi}_j$ in (2.10) in the same manner as the functions φ_j at the end of 3.2:

$$(\tilde{\Phi}_{2p}, \tilde{\Phi}_1, \ldots, \tilde{\Phi}_{2p-1}) = (\Phi_0^+, \Phi_0^-, \ldots, \Phi_{p-1}^+, \Phi_{p-1}^-).$$

The map $\exp \circ 2\pi i$ takes orbits of the unit shift $\mathrm{id}+1$ into points. So the transition functions $\tilde{\Phi}_j^\pm$ of the normalizing atlas for the germ Δ are related to the identifying maps $\varphi_j^\pm$ appearing in the construction of the space of orbits by the formula

$$\exp 2\pi i \circ \Phi_j^\pm = \varphi_j^\pm \circ \exp 2\pi i \tag{3.15}$$

(see the commutative diagram (2.14)).

Taking this into account, the equalities (3.11) for φ_j^- yield

$$\Phi_j^- = \frac{\ln(e^{2\pi i t} + a_j)}{2\pi i} = t + \sum_{-1}^{-\infty} \frac{(-a_j)^l}{l} e^{2\pi i l t}, \qquad \mathrm{Im}\, t \ll -1. \tag{3.16}$$

COROLLARY 2. *The classification theorem from 3.3 remains valid if in its formulation collection (3.8) is replaced by either the collection*

$$(\Phi_0^+, \Phi_0^-, \ldots, \Phi_{p-1}^+, \Phi_{p-1}^-)$$

with Φ_j^- *defined by (3.16) and*

$$\Phi_j^+(t) = t + \sum_1^\infty a_{j,l} \exp 2\pi i l t, \qquad j = 1, \ldots, p-1,$$

$$\Phi_0^+(t) = t - 2\pi i \lambda + \sum_1^\infty a_{0,l} \exp 2\pi i l t$$

subject to the condition (3.16), or by the set of Fourier coefficients

$$\begin{aligned} \{a_{j,l} \mid l \in \mathbb{N},\ j = 0, 1, \ldots, p-1\}, \\ \{a_{j,-l} \mid l \in \mathbb{N},\ j = 0, 1, \ldots, p-1\} \end{aligned} \tag{3.17}$$

defining a convergent series with the condition $a_{j,-l} = (a_{j,-1})^l / l$. *The set* $\{a_{j,-n}\}$ *in (3.17) is well defined by the set* $\{a_{0,-1}, \ldots, a_{p-1,-1}\}$ *and may be replaced by the latter.*

The last form is especially suitable for the computation of functional moduli for germs of class $\mathscr{E}_{p,\lambda}$ (see Paper III). Sometimes we shall write $a_{j,l}^D$ instead of $a_{j,l}$ in order to recall that the classification of degenerate singularities is considered.

3.5. Topological classification. We finish this section by stating topological classification results for germs of vector fields having degenerate elementary singularities in $(\mathbb{C}, 0)$.

DEFINITION. Two holomorphic vector fields having singularities at the origin $0 \in \mathbb{C}^2$ are said to be orbitally *topologically* equivalent in a neighborhood of the origin, if there exists an orientation-preserving homeomorphism (defined in this neighborhood) which preserves the origin and takes (complex) phase curves of the first field into those of the second.

A specific trait of the topological classification described below is that the equivalence class of a field cannot be determined in terms of any finite jet at the origin.

Recall that a vector field having degenerate elementary singularity always possesses a holomorphic invariant manifold; this manifold is tangent to the eigenvector with nonzero eigenvalue. On the other hand, in general there is no holomorphic central manifold tangent to the null space of the linearization at the origin, as it is illustrated by the famous Euler example

$$\dot{z} = z^2, \qquad \dot{w} = w - z. \tag{3.18}$$

The formal Taylor series for the solution tangent to the z-axis at the origin diverges.

Denote by $\mathscr{E}_p$ the set of all germs of holomorphic vector fields on the complex 2-plane, having a degenerate elementary singularity of multiplicity $p+1$ at the origin. As before, $\mathscr{E}_{p,\lambda}$ stands for the subset of germs from $\mathscr{E}_p$, formally orbitally equivalent to (3.1).

THEOREM 1. *An equation of the class $\mathscr{E}_p$ has a holomorphic central manifold if and only if in the collection (3.7), (3.9) defining the Martinet-Ramis modulus all the coefficients a_j vanish.*

This is an easy corollary to the sectorial normalization theorem. In the extended sector of jump there exists a unique solution containing the origin in its closure. Call it the *sectorial separatrix*. The condition $a_j = 0$ is necessary and sufficient for all the sectorial separatrices to be parts of a unique phase curve. Completing this curve by the origin leaves it holomorphic by the removable singularity theorem.

Equations of class $\mathscr{E}_p$ that possess (do not possess) a holomorphic central manifold are denoted by $\mathscr{E}_p^0$ ($\mathscr{E}_p^1$, respectively). The topological classification of equations of the second class is much simpler than that of the first class.

The topological classification of equations of the class $\mathscr{E}_p^1$ is completely determined by indicating the adjacent sectorial separatrices which continue each other, being parts of a single phase curve. More precisely, define a map T which takes a germ v into the string $T(v) = \varepsilon = (\varepsilon_1, \ldots, \varepsilon_p)$ consisting of zeros and ones: $\varepsilon_j = 0$, if in the corresponding collection (3.7) all the a_j

vanish, otherwise $\varepsilon_j = 1$. Two ordered collections of p elements are said to be *equivalent*, if one can be obtained from the other by a cyclic permutation. The set of all equivalence classes is denoted by $\mathscr{L}_p$.

THEOREM 2. *Any two germs of class $\mathscr{E}_p^1$ are topologically equivalent if and only if their T-images are equivalent in the above sense. The class $T^{-1}(\varepsilon)$ is adjacent to the class $T^{-1}(\tilde{\varepsilon})$ if and only if a string equivalent to $\tilde{\varepsilon}$ can be obtained from ε by replacing some zeros by ones.*

For the sake of brevity, everywhere in this section we shall use the term "equivalence" instead of "orbital topological equivalence".

COROLLARY 1. *All equations from the class $\mathscr{E}_1^1$ are equivalent to the Euler equation* (3.18).

Indeed, the absence of a holomorphic central manifold implies that the map T takes the equation (3.18) (with $p = 1$) to the string (1).

Now we investigate the class $\mathscr{E}_p^0$. It can be proved that the topological equivalence class of the monodromy transformation corresponding to the holomorphic central manifold is a modulus of this classification. The rest depends essentially on the arithmetical nature of the number λ. Recall that a real λ is called a *Liouvillian number*, if it is irrational and admits an anomalously good approximation by rationals, which means that the Bryuno condition [6, 39] is violated (the precise form of this condition is of no importance now). Otherwise the irrational λ is called a *Diophantine number*.

THEOREM 3. *Consider two germs of class $\mathscr{E}_p^0$ with the formal invariants which are either nonreal, or real, coinciding, and not Liouvillian. Then these germs are equivalent if and only if their monodromies corresponding to the holomorphic central manifolds are topologically equivalent.*

The monodromy Δ corresponding to the central manifold can be easily expressed in terms of the Martinet-Ramis moduli: it is equal to their product

$$\Delta = \varphi_0^+ \circ \varphi_{p-1}^+ \circ \cdots \circ \varphi_1^+. \tag{3.19}$$

This follows immediately from the construction of the moduli (see 3.2). In the same subsection one can find the relations

$$\varphi_j^{+'}(0) = 1 \quad \forall j \neq 0, \qquad \varphi_0^{+'}(0) = \exp(-2\pi i \lambda),$$

which imply that the multiplier $\mu = \Delta'(0)$ of the monodromy is equal to $\exp(-2\pi i \lambda)$. For $\operatorname{Im} \lambda < 0$, $\operatorname{Im} \lambda > 0$, $\operatorname{Im} \lambda = 0$, the linearization of the monodromy is a contraction, an expansion, or a rotation, respectively. Denote the corresponding subclasses of $\mathscr{E}_p^0$ by $\mathscr{E}_{p,+}^0$, $\mathscr{E}_{p,-}^0$ and $\mathscr{E}_{p,0}^0$.

COROLLARY 2. *The subclasses $\mathscr{E}_{p,+}^0$ and $\mathscr{E}_{p,-}^0$ are two distinct classes of orbital topological equivalence. Their representatives may be chosen in the* *form*

$$\dot{z} = z^{p+1}(1 \pm iz^p)^{-1}, \qquad \dot{w} = w.$$

Indeed, local contractions and local expansions constitute two classes of topological equivalence.

COROLLARY 3. *For germs of the subclass $\mathscr{E}_{p,0}^0$, the number λ is a topological invariant independently of its arithmetic nature.*

By the Naĭshul′ theorem [27], the multiplier of a holomorphic map $(\mathbb{C}, 0) \to (\mathbb{C}, 0)$ is its topological invariant provided that the modulus of the multiplier is equal to 1. This implies that the fractional part of λ is a topological invariant for the subclass $\mathscr{E}_{p,0}^0$. A detailed proof requires some additional considerations.

THEOREM 4. 1. *If λ is a Diophantine irrational, then $\mathscr{E}_{p,\lambda}$ is the only class of topological equivalence within the set $\mathscr{E}_p^0$.*

2. If $\lambda = m/n$ is rational, then $\mathscr{E}_{p,\lambda}$ splits into countably many topological equivalence classes: every equation from $\mathscr{E}_{p,\lambda}$ is equivalent to either (3.1) or the equation

$$\dot{z} = z^{p+1}, \qquad \dot{w} = w(1 + \lambda z^p) + z^l w^{kn+1} \qquad (3.20)$$

for some $k \in \mathbb{N}$ with $l = -r + 2p + 1$; here r stands for the remainder of the division of km by p. Two equations (3.20) with different k belong to different equivalence classes.

Indeed, maps $(\mathbb{C}, 0) \to (\mathbb{C}, 0)$ with multipliers equal to $\mu = e^{-2\pi i}$, λ being a Diophantine number, are topologically (or, which is the same in our case, analytically) equivalent to multiplication by μ. This follows from the Siegel theorem and its generalizations [6].

Germs of maps $(\mathbb{C}, 0) \to (\mathbb{C}, 0)$ tangent to rotations by angles rationally comparable with π constitute countably many topological equivalence classes for any fixed multiplier μ [8, 31]. In the case $\mu = 1$, this classification was described in 2.1.

§4. One-dimensional maps with resonance linear terms and saddle resonant vector fields on the plane

The primary subject of this section is the analytic classification of germs of conformal mappings $(\mathbb{C}, 0) \to (\mathbb{C}, 0)$ whose linear part is the rotation by an angle rationally comparable with π. Such linear parts are called *resonant*. Further, we give an orbital analytic classification of *complex resonance saddles*, i.e., germs of vector fields at the singular point on the complex 2-plane whose eigenvalues have a negative rational ratio (the term comes from the theory in the real case). The latter classification can be reduced to the former one, which in turn follows from the classification developed in §2 above.

The results described below were obtained by J. Martinet and J. P. Ramis [26] (a detailed text), and independently by S. M. Voronin, P. M. Elizarov and the author [18, 14] (these papers contain formulations and only a part of the proof; no sequel to the paper [14] was published, since the paper [26] had already appeared by that time).

4.1. Outline of the results. Let $f: (\mathbb{C}, 0) \to (\mathbb{C}, 0)$ be the germ of a conformal mapping with resonant linearization, i.e., having the multiplier

$$f'(0) = \alpha_{m,n} = \exp(2\pi i m/n). \tag{4.1}$$

Then the superposition power $f^{[n]}$ is a germ with identity linear term. The analytic classification of such germs was given in §2.

THEOREM 1. *Germs of conformal maps $f: (\mathbb{C}, 0) \to (\mathbb{C}, 0)$ with multipliers of the form* (4.1) *are analytically equivalent if and only if their nth superposition powers are.*

Not every map tangent to the identity can be realized as the nth power of a germ of the class (4.1). The subset of all germs which can be represented in this way is of infinite codimension in the entire space of germs; it is described by a certain conditions imposed on the Ecalle-Voronin moduli.

The analytic classification of resonance saddles can be relatively simply reduced to that of one-dimensional maps with resonant linear terms. The general theory implies that a holomorphic vector field with a negative ratio of eigenvalues has exactly two phase curves which remain holomorphic after their completion by the singular point; these curves are tangent to the eigenvectors of the linearization of the field. They are called *complex separatrices*. To every closed loop encircling the singularity on a complex separatrix, associate the monodromy transformation of the one-dimensional transversal onto itself. If the ratio of the eigenvalues is equal to $-m/n$, then the multiplier of the monodromy corresponding to one of the two separatrices is equal to $1/\alpha_{m,n}$. It is well known that

$$\begin{array}{c}\textit{orbital analytic equivalence of germs of vector fields}\\ \textit{implies analytic equivalence of the corresponding monodromies}\end{array} \quad (*)$$

(see [16]). It turns out that for formally equivalent resonance saddles the converse is also true [14]. Finally, any germ of the form (4.1) can be realized as the monodromy transformation for some resonance saddle with the ratio of the eigenvalues equal to $-m/n$. This remark concludes the second classification.

Let us proceed with the classification of one-dimensional maps.

4.2. Formal classification of one-dimensional resonant maps. Any germ having $\alpha_{m,n}$ as multiplier is formally equivalent to the time 1 map for a holomorphic flow. The field generating this flow must have a nonzero linear term if $\alpha_{m,n} \neq 1$. But it is more convenient to decompose vector fields into the sum of a linear field and a nonlinear one with zero linear terms, commuting with the former field (for a germ which is the time 1 map for a field, this corresponds to decomposition into a linear rotation and a nonlinear map tangent to the identity and commuting with the rotation).

THEOREM 2. *Let* $f: (\mathbb{C}, 0) \to (\mathbb{C}, 0)$ *be the germ of a conformal mapping with multiplier* $\alpha_{m,n}$ *which is not formally equivalent to a linear rotation. Then there exist an integer* k *and a complex number* λ *such that* f *is formally equivalent to the composition of* f_0, *which is the time* $1/n$ *map for the field*

$$w_{nk,\lambda} = z^{nk+1}(1+\lambda z^{nk})^{-1} \partial/\partial z, \tag{4.2}$$

and a (linear) *multiplication by* $\alpha_{m,n}$.

REMARK. The field $w_{nk,\lambda}$ is invariant under multiplication by $\alpha_{m,n}$ for all m, k, λ.

PROOF. For simplicity denote $w_{nk,\lambda}$ by w. Then

$$f_0 = \alpha_{m,n} g_w^1(z) = \alpha_{m,n}\left(\operatorname{id} + w + \frac{1}{2}\frac{\partial w}{\partial x}w + \cdots\right)(z)$$

$$= \alpha_{m,n} z\left(1 + \frac{1}{n}z^{nk} + \left(-\frac{\lambda}{n} + \frac{1}{2n^2}(nk+1)\right)z^{2nk} + \cdots\right).$$

On the other hand, by a formal substitution any germ can be transformed into a formal power series commuting with the linear part of the germ [2]. For the multiplier (4.1) this series takes the form

$$z \mapsto \left(\alpha_{m,n} + \sum_1^\infty a_j(z^n)^j\right)z.$$

Not all coefficients of the series vanish, because f is not formally linearizable. Let a_k be the first nonzero term in the series. Then f is formally equivalent to the germ

$$z \mapsto \alpha_{m,n} z(1 + n^{-1}z^{nk} + \cdots).$$

The first nonzero coefficient before a nonlinear term may be given an arbitrary value by using a linear change of variable. Subsequent formal transformations can remove all the terms which can be obtained by commutation with the Taylor polynomial $\alpha_{m,n} z(1 + n^{-1}z^{nk})$. A straightforward calculation shows that in this way we get the polynomial

$$z \mapsto \alpha_{m,n} z(1 + n^{-1}z^{nk} + \alpha z^{2nk})$$

for some α. Thus the formal type of the previous polynomial will not be changed by the addition of higher order terms. The same polynomial normal form corresponds to the germ $\alpha_{m,n} g_w^1$ with an appropriate λ. Therefore the initial germ is formally equivalent to the map $\alpha_{m,n} g_w^1$. □

The germ $\alpha_{m,n} g_w^1$ will be called *R-embeddable* (the letter R reminds us that this is the composition of the embeddable germ and a rotation). The equivalence class of the germ $\alpha_{m,n} g_{w_{nk,\lambda}}^1$ is denoted by $\mathscr{A}_{m,n,k,\lambda}$.

REMARK. A germ formally equivalent to a rotation by a rational angle is analytically equivalent to it. The elementary proof of this well-known fact is omitted to save space.

4.3. Modification of the functional moduli for germs tangent to the identity. It is necessary to modify the construction of the functional moduli given in §2 for the following reasons. Let $f \in \mathscr{A}_{m,n,k,\lambda}$. The germ $f^{\{n\}}$ has an identical linear term and commutes with f. The corresponding orbit space consists of $2p$ doubly punctured spheres, glued as this was described in 2.3. Here we have $p = kn$. The map f swaps orbits of the map $f^{\{n\}}$, hence it defines a map of the quotient space. But in contrast with §2, where all the linear parts of the maps were the identity, the quotient map in our case permutes the spheres. In the definitions from 2.3, the sphere $\mathbb{C}^* \times \{1\}$ is distinguished from the others, and a collection obtained from (2.3), (2.3′) by permutation of the subscripts, is not of type (2.3), (2.3′) at all if the subscript 1 is replaced by $j \neq 1$. This originates in the inconvenient choice of charts on the spheres $\mathbb{C}^* \times \{j\}$ which transform maps φ_j into functions. We shall modify this choice. Namely, let us require that all the identifying maps φ_j have coinciding multipliers at the origin equal to the inverses of the ones at infinity (e.g., the map $\zeta \mapsto 2\zeta$ has multiplier equal to 2 at the origin and $1/2$ at infinity).

Let us choose the charts ω_j on the spheres $\mathbb{C}^* \times \{j\}$ so that they differ from ζ_j chosen before, by factors:

$$\omega_j = \nu^j \zeta_j, \qquad \nu = \exp(2\pi^2 \lambda/p).$$

Then the identifying maps will take the form represented by the following germs:

$$\widetilde{\varphi}_j : (\mathbb{C}, 0) \times \{j\} \to (\mathbb{C}, 0) \times \{j+1\}, \quad \begin{cases} \widetilde{\varphi}'_j(0) = \nu, & j \text{ odd;} \\ \widetilde{\varphi}'_j(\infty) = \nu, & j \text{ even.} \end{cases} \quad (4.3)$$
$$\widetilde{\varphi}_j : (\mathbb{C}, \infty) \times \{j\} \to (\mathbb{C}, \infty) \times \{j+1\},$$

The equivalence relation remains as before: two collections are called *equivalent* if and only if one can be obtained from the other by multiplication by some common (for all j) constant both in the source and in the target spaces. The functions forming the collection (4.3) are related to those from (2.3) by

$$\widetilde{\varphi}_j = \nu^{j+1} \circ \varphi_j \circ \nu^{-j}.$$

The moduli constructed in 2.6 are modified in the same manner. Let $\{S_j\}$ be the nice covering of the punctured disk by sectors, t_j denoting the restriction of the "time" function (2.5) on S_j. Let us choose the chart $\widetilde{\tau}_j = \tau_j + \text{const}$, where $\tau_j = t_j \circ H_j(z) : S_j \to \mathbb{C} \times \{j\}$, in such a way that

$$\tau_j - \widetilde{\tau}_j \to c_0 j \quad \text{as } z \to 0, \quad 2pc_0 = -2\pi i\lambda, \quad (4.3')$$

and τ_j conjugates the initial germ with the shift $\tau_j \mapsto \tau_j + 1$. The last equality in (4.3′) is implied by the relation (2.10′) between the formal and the analytic classifications of germs $(\mathbb{C}, 0) \to (\mathbb{C}, 0)$ tangent to the identity. The transition functions Φ_j (not to be confused with the notation of 2.5!) for the new normalizing atlas $\{\tilde{\tau}_j\}$ are related to the old ones (2.10) by the formulas

$$\Phi_j = \tilde{\Phi}_j \circ (\mathrm{id} + \gamma j) - \gamma(j+1), \qquad \gamma = \pi i \lambda / p. \qquad (4.4)$$

Moreover, one has

$$\Phi_j \circ (\mathrm{id} + 1) = \Phi_j + 1, \qquad j = 1, \ldots, 2p.$$

The functions Φ_j are defined in a sufficiently high upper halfplane for j odd and in a lower one for j even. The equivalence relation remains as before:

$$\{\Phi_j\} \simeq \{\Psi_j\} \iff \exists c : \Phi_j + c = \Psi_j(\mathrm{id} + c). \qquad (4.5)$$

4.4. Analytic classification of conformal mappings $(\mathbb{C}, 0) \to (\mathbb{C}, 0)$ **with resonant linear parts.** Let f be a germ of the class $\mathscr{A}_{m,n,k,\lambda}$. Then $f^{[n]}$ is a germ of the class $\mathscr{A}_{p+1}$, $p = nk$. Let (4.3) and (4.4) be the corresponding functional moduli. Then the map $\pi_* f$ rearranges the spheres $\mathbb{C}^* \times \{j\}$ in the same manner as the rotation by the angle $2\pi m/n$ rearranges the sectors of the nice p-covering:

$$\pi_* f : \mathbb{C}^* \times \{j\} \to \mathbb{C}^* \times \{j + 2km\}.$$

Before analyzing this map, consider the germ f in the charts $\tau_j = t_j \circ H_j(z)$ normalizing its power $f^{[n]}$ (recall that this power in the given charts is the unit shift). Since f commutes with $f^{[n]}$, it transforms the normalizing chart into one retaining this property. Indeed, the map f written in the normalizing chart takes the form $\tilde{f}$ which commutes with the unit shift $\mathrm{id} + 1$. Its difference from the identity is 1-periodic and defined in the "extended" sectors $\tilde{S}_j$. Therefore $\tilde{f} - \mathrm{id}$ is a constant (compare with 2.6). Thus for every j one has

$$\tilde{\tau}_{j+2km} \circ \tilde{f} = \tilde{\tau}_j + C_j.$$

Hence

$$\Phi_{j+2km} = (\tilde{\tau}_{j+1} + C_{j+1}) \circ (\tilde{\tau}_j + C_j)^{-1} = \Phi \circ (\mathrm{id} - C_j) + C_{j+1}.$$

Zero order terms for the differences $\Phi - \mathrm{id}$ are the same for all j (see 4.3). Therefore the last formula implies that $C_{j+1} = C_j = C$. Next,

$$\tilde{\tau}_{j+2p} \circ f^{[n]} = \tilde{\tau}_j + Cn.$$

On the other hand, $\tilde{\tau}_{j+2p} = \tilde{\tau}_j$, since the enumeration is mod $2p$, and

$$\tilde{\tau}_{j+2p} \circ f^{[n]} = \tilde{\tau}_j + 1$$

by the definition of normalizing charts. So

$$\tilde{\tau}_{j+2km} \circ \tilde{f} = \tilde{\tau}_j + C, \qquad C = 1/n. \qquad (4.6)$$

Thus the functional modulus of the germ $f^{[n]}$ is determined by its first $2k$ functions $\Phi_1, \ldots, \Phi_{2k}$. The remaining functions can be found from the relation

$$\Phi_{j+2km} = \Phi \circ (\mathrm{id} - 1/n) + 1/n. \tag{4.7}$$

The map $\pi_* f$ is now analyzed using the commutation relation

$$\exp \circ 2\pi i \tilde{f} = \pi_* f \circ \exp 2\pi i$$

and (4.6). From this relation it immediately follows that $\pi_* f$ is reduced to multiplication by $\exp(2\pi i/n)$.

Now we can describe the moduli space of analytic classification of germs $(\mathbb{C}, 0) \to (\mathbb{C}, 0)$ of the class $\mathscr{A}_{m,n,k,\lambda}$. The modulus is the class of equivalent collections of $2k$ functions

$$\Phi_j = \begin{cases} \tau + C + \sum_1^\infty a_{jl} \exp 2\pi i l \tau & \text{for } j \text{ odd}, \ 1 \leqslant j \leqslant 2k-1, \\ \tau + C + \sum_{-1}^{-\infty} a_{jl} \exp 2\pi i l \tau & \text{for } j \text{ even}, \ 2 \leqslant j \leqslant 2k, \end{cases} \tag{4.8}$$
$$C = -\pi i \lambda/p$$

(the last equality follows from (4.3′)). The functions with odd numbers are defined in a high upper half-plane, while those with even numbers in a sufficiently low one. The equivalence assumed is as follows: one collection turns into the other after conjugating by the same shift

$$\Psi \simeq \Phi \iff \exists C: \Psi_j \circ (\mathrm{id} + C) = \Phi_j + C. \tag{4.9}$$

The collection (4.8) can be replaced by the corresponding set of Fourier coefficients $\{a_{jl} \mid l \in \mathbb{Z} \setminus \{0\}, j = 1, \ldots, k\}$. The equivalence (4.9) means that

$$\{\tilde{a}_{jl}\} \simeq \{a_{jl}\} \iff \exists a: \tilde{a}_{jl} = a^l a_{jl}, \ a \neq 0. \tag{4.10}$$

This form is more convenient for the computation of functional moduli and will be used for this purpose in Paper III.

The classification theorem for germs of class $\mathscr{A}_{m,n,k,\lambda}$ can now be stated exactly as in §2.

THEOREM 3. 1. Invariance. *Any germ of the class $\mathscr{A}_{m,n,k,\lambda}$ is associated with a unique equivalence class* (4.8), (4.9).

2. Equimodality and equivalence. *Equivalent germs are associated with the same equivalence class* (4.8), (4.9), *and conversely*.

3. Realization. *Any collection* (4.8) *can be realized as the functional modulus of a certain germ of the class $\mathscr{A}_{m,n,k,\lambda}$.*

4. Analytic dependence. *The functional modulus depends analytically on the parameters within a fixed class of formal equivalence.*

PROOF. The proof is completely analogous to that in 2.6:

1. Invariance of the equivalence class (4.8), (4.9) follows immediately from the construction of the moduli.

2. Equivalence of two germs f, g associated with equivalent collections is proved in the standard manner: the collection (4.8) is completed to "full" size $(\Phi_1, \ldots, \Phi_{2p})$ using (4.7). So one gets a complete normalizing atlas for the germ $f^{[n]}$. The same procedure works for the germ $g^{[n]}$. By shifting all charts of the first atlas by an appropriate constant, one may assume that the transition functions for both atlases are equal. The map realizing the desired equivalence between the germs is defined as the identification of the points from the corresponding sectors of the nice coverings, the identified points having equal coordinates in the normalizing charts. This map is well defined on the intersections of the charts, due to the assumed coincidence of the transition functions. The map can be holomorphically extended to the fixed point by the removable singularity theorem. Thus the conjugation between $f^{[n]}$ and $g^{[n]}$ is constructed.

The map f written in the charts $\tilde{\tau}_j$ of the normalizing atlas has the form

$$\tilde{f}: \tilde{\tau}_j \mapsto \tilde{\tau}_{j+2km} = \tilde{\tau}_j + 1/n,$$

and the same with g. So the map conjugating the iterative powers conjugates the initial germs $\tilde{f}$, $\tilde{g}$ at the same time. This proves the second assertion of the theorem.

3. Realization is proved as in 2.6 with some minor modifications.

4. Analyticity follows immediately from the analogous assertion for iterative powers. □

Some applications of this theorem, such as the description of embeddable maps, the superposition root extraction, the description of centralizers, are presented in §1 of Paper II. The theorem also yields the orbital analytic classification of resonance saddles (see 4.6).

Let us start with formal classification.

4.5. Resonance saddles and their monodromies: formal classification. Recall that by definition, resonance saddles are germs at the origin of holomorphic fields in $(\mathbb{C}^2, 0)$ having a negative rational ratio of eigenvalues. Such fields possess two complex separatrices, as this was described in 4.1. Any positive loop encircling the singularity on any separatrix generates the monodromy transformation $(\mathbb{C}, 0) \to (\mathbb{C}, 0)$. Here we describe relations between formal classifications of resonance saddles and their monodromies. Without loss of generality, we consider germs with the diagonal linear part.

We shall say that two germs of fields are *equivalent*, if they are analytically orbitally equivalent and the corresponding transformation is tangent to the identity at the origin. First we consider the formal classification of maps with this equivalence relation, then proceed with the analytic classification.

FORMAL CLASSIFICATION THEOREM [5, 7, 19]. *A germ of a resonance saddle with ratio of eigenvalues equal to* $-m/n$ *is formally orbitally equivalent to the linear germ or to the germ*

$$\dot{z} = z, \qquad \dot{w} = w(-m/n + u^k(1+\alpha u^k)^{-1}) \qquad (4.11)$$

for some natural number k and complex α, where $u = z^m w^n$ stands for the resonant monomial (the first integral of the linearized system).

Denote by $\mathscr{B}_{m,n,k,\alpha}$ the class of germs formally equivalent to (4.11). Consider the monodromy corresponding to the separatrix $Oz = \{w = 0\}$ of the normalized system. At points of the z-axis the field (4.11) differs from a linear one by terms of order ≥ 2 in w. So the equation in variations along Oz is the same for both systems, which implies coincidence of the multipliers. Phase trajectories of the linear system have the form $w = cz^{-m/n}$, so the multiplier equals $\exp(-2\pi i m/n)$.

To find higher order terms of the monodromy expansion, consider the function $v = z^{m/n} w$ on its Riemann surface, which is an n-fold branched covering over $(\mathbb{C}^2, 0) \setminus \{z = 0\}$. This is a first integral of the linear field, which on the curve $\Gamma = \{z = 1\}$ lying on the "first" sheet (with $\arg z = 0$) coincides with w. Its values on the "second" sheet, after continuation over the unit circle $z = e^{i\varphi}$, $w = 0$, on the separatrix Oz, differ from the restriction $w|_\Gamma$ by the scalar factor $\tilde{v} = e^{2\pi i m/n} w$. On the other hand, the evolution of v along the trajectories of (4.11) is described by the quotient system

$$\dot{v} = w_{p,\alpha}(v), \qquad p = kn, \qquad w_{p,\alpha} = v^{p+1}/(1+\alpha v^p).$$

So $\tilde{v} \circ \Delta(w) = g_{w_{p,\alpha}}^{2\pi i} w$. But $\tilde{v} \circ \Delta(w) = \exp(2\pi i m/n)\Delta(w)$. Finally,

$$\Delta(w) = \exp(-2\pi i m/n) g_{w_{p,\alpha}}^{2\pi i} w.$$

Transform the shift $\mathrm{id} + 2\pi i$ into $\mathrm{id} + 1/n$. We have $g_{w_{p,\alpha}}^{2\pi i} = g_{2\pi i n w_{p,\alpha}}^{1/n}$. The field $2\pi i n w_{p,\alpha}$ can be transformed into the standard one by the scale change: $w_1 = cw$, $c^p = 2\pi i n$. In the new coordinates, this field takes the form

$$w_{p,\lambda}(w_1) = w_1^{p+1}(1+\lambda w_1^p)^{-1}, \qquad \lambda = \alpha/2\pi i n.$$

Finally we obtain

$$\Delta(w) = \exp(-2\pi i m/n) g_{w_{p,\lambda}}^{1/n}, \qquad p = kn, \quad \lambda = \alpha/2\pi i n. \qquad (4.12)$$

This finishes the calculation of the monodromy for the normalized system.

The monodromy transformation for the germ of class $\mathscr{B}_{m,n,k,\alpha}$ is formally equivalent to the germ (4.12). This is easily proved using the "normalization along the invariant manifold" due to H. Dulac. Namely, an equation of the above class is orbitally analytically equivalent to the equation

$$\dot{z} = z, \qquad \dot{w} = w(-m/n + u^k(1+\alpha u^k)^{-1} + z^N w^N f_N(z,w)) \qquad (4.12')$$

for any given N, where f_N is a function holomorphic at the origin. For N sufficiently large, the last term has no influence on the $(2p+1)$-jet of the monodromy transformation for the corresponding equation, and this jet is the same as that of the germ (4.12). But the formal type of a germ with resonant linear part and nonlinearity of order $p+1$ is determined by its $(2p+1)$-jet.

Note in conclusion that the formal type of the monodromy transformation uniquely determines the formal orbital type of a resonance saddle up to an integer parameter. More precisely, a germ having the monodromy (4.12) is formally orbitally equivalent to a germ of class $\mathscr{B}_{\tilde{m},n,k,\alpha}$; the number n is uniquely determined by the multiplier of the monodromy, while $\tilde{m}$ is defined only up to a multiple of n:

$$\tilde{m} = m + ln, \quad l \in \mathbb{Z}, \quad k = p/n, \quad \alpha = 2\pi i n \lambda. \tag{4.13}$$

Freedom in the choice of l is essentially used in Paper II when "the generalized Thom conjecture" is discussed.

4.6. Resonance saddles: orbital analytic classification. For germs of this kind the classification theorem is analogous to that of 4.4.

THEOREM. *The classification theorem in* 4.4 *remains true if the class of germs* $\mathscr{A}_{m,n,k,\alpha}$ *is replaced by the class* $\mathscr{B}_{m,n,k,\alpha}$. *Namely, the class of equivalent collections* (4.8) *with the equivalence relation* (4.9) *produces a modulus of analytic classification of germs in* $\mathscr{B}_{m,n,k,\alpha}$; *moreover, assertions* 1–4 *of Theorem* 3 *in* 4.4 *are true.*

Let us prove this theorem. The monodromy transform is an invariant of the analytic classification of germs of resonance saddle vector fields, hence the first assertion of the theorem follows. Assertion (∗) in 4.1 implies the second assertion. The fourth assertion follows from the theorem on the analytic dependence of separatrices and the monodromy on the parameter and the fourth assertion of the theorem in 4.4. It remains to prove the third assertion: each collection (4.8) can be realized as the invariant of an appropriate germ within an arbitrary class of formal equivalence $\mathscr{B}_{m,n,k,\alpha}$ with the relation $C = -\alpha/2\pi n^2$ (see (4.8), (4.13)). To do this, we use the method developed in 2.6 and 3.3.

Using the transition functions as identifying maps, compose a "cut" neighborhood of the origin in $(\mathbb{C}^2, 0)$ (that is, a neighborhood with the axis $z = 0$ deleted) from special domains related to the formal type of the equation. These domains are constructed as follows. Consider the nice p-covering of the punctured disk $\Gamma = \{1\} \times \{|w| \leq \varepsilon\}$ by the sectors S_j. Let V be the domain on the Riemann surface of the function $\ln z$ defined by the inequalities

$$0 < |z| \leq 1, \quad \arg z \in [0, 2\pi n).$$

Denote by U the set $\{0 < |z| \leq 1\} \times \{|w| \leq \varepsilon\}$. Consider the saturation $\tilde{S}_j$ of the sector S_j by phase curves of the normalized system (4.11) with initial points in the sector S_j. These phase curves are extended over the domain V until they leave the neighborhood U. Recall that the saturation is the union of all phase curves intersecting the domain.

EXAMPLE. An analogous construction in the simpler case when the system (4.11) is replaced by its linearization can be described in detail. The saturation S_j^0 of the sector S_j by phase curves of the system $\dot{z} = z$, $\dot{w} = (-n/m)w$ is a "thick toric knot of type $(n, -m)$". More precisely, for any point $\omega \in S_j$, the domain U intersects the corresponding phase curve in the annulus which covers the domain

$$|\omega/\varepsilon|^{n/m} \leqslant |z| \leqslant 1$$

n times and consists of toric knots of type $(n, -m)$; one of them is

$$z = e^{2\pi i n \varphi}, \quad w = \omega e^{-2\pi i m \varphi}, \quad \varphi \in [0, 2\pi].$$

Let $p = kn$. Each domain S_j^0 intersects the transversal $\Gamma = \{z = 1\}$ in n sectors of the nice p-covering: $S_j, S_{j+2km}, \ldots$. The union $\bigcup_{j=1}^{2k} S_j^0$ contains the entire domain U.

REMARK. After a phase curve of the system (4.11) is continued n times over a positively oriented circle in the z-plane, it returns to the point on the transversal obtained from the initial one by applying the nth power of the monodromy $f^{[n]}$.

Now consider system (4.11) with the monodromy map $f_0 = \Delta$ given by (4.12). The straightening map (2.5) takes the map $f_0^{[n]}$ in each sector S_j to the unit shift $\mathrm{id} + 1$. The inverse map takes the functions $\tilde{\Phi}_j$ of the given collection (4.8) commuting with the shift to functions Φ_j commuting with $f_0^{[n]}$ and defined in the intersections $S'_j \cap S'_{j+1}$, where S'_j stands for the slightly diminished sector S_j. The described collection

$$\{\Phi_j \mid j = 1, \ldots, 2k\} \tag{4.8'}$$

also constitutes the modulus of analytic classification of germs of the formal type $\mathscr{A}_{m,n,k,\lambda}$. It remains to prove the realization assertion for this last collection.

For any $j = 1, \ldots, 2k$, extend the transition functions Φ_j to the intersection of the domains $\tilde{S}'_j \cap \tilde{S}'_{j+1}$, where $\tilde{S}'_j$ is obtained from $\tilde{S}_j$ by diminishing its base S_j to the size of S'_j. In order to do this, consider two maps: the solution map g for the system (4.11), and the inverse map denoted by h. The germ of g at the point $(1, 0)$ is defined by the assignment

$$(z, \omega) \mapsto (z, \varphi(z, \omega)),$$

where $w = \varphi(z, \omega)$ is the phase curve of the system (4.11) starting at $(1, \omega)$; $h = g^{-1}$. Denote by h_j the continuation of the germ h on the domain $\widetilde{S}_j$, and by g_j the restriction of the solution map g on the domain $h(\widetilde{S}_j)$. Finally let Φ_j^1 denote the map Φ_j extended cylindrically: $\Phi_j^1(z, \omega) = (z, \Phi_j(\omega))$. The transition functions $\Phi_j: S_j' \cap S_{j+1}' \to S_{j+1}$ (whose images are proper subsets of S_{j+1}) now can be extended to the map Ψ_j permuting the phase curves of (4.11):

$$\Psi_j = g_{j+1} \circ \Phi_j^1 \circ h_j. \tag{4.14}$$

LEMMA. *The map Ψ_j converges to the identity together with all its derivatives as $z \to 0$.*

We give an outline of the proof. The map Φ_j^1 exponentially converges to the identity together with all its derivatives (see 2.5). All derivatives of the maps h_j and g_{j+1} have no more than polynomial growth (of order depending on the number of the derivative). If we replace h_j and g_{j+1} by analogous maps for the linearized system, then the last assertion becomes evident. Its proof in the general case is based on integrability of the normalized system and uses the formulas from 4.5. Exponential decrease overcomes polynomial growth, implying the statement of the lemma. □

Denote by $\underline{z}_j$ the chart in the domain $\widetilde{S}_j$ induced by the embedding $\widetilde{S}_j \hookrightarrow \mathbb{C}^2$. Construct a manifold $\widetilde{U}$ by identifying points $p \in \widetilde{S}_j'$ with their Ψ-images $\Psi_j(p) \in \widetilde{S}_{j+1}'$. The identifying maps take the function z into itself and the phase curves of (4.11) one into another. So the vector field $\widetilde{v}$ on $\widetilde{U}$ can be defined, which coincides with (4.11) in each chart $\underline{z}_j$.

Let us prove that there exists a biholomorphic map H of some domain $\widetilde{\widetilde{U}} \subset \widetilde{U}$ onto a domain $W \subset \mathbb{C}^2 \setminus \{z = 0\}$, such that the closure $\overline{W}$ contains 0, and the field $v = H_*\widetilde{v}$ can be holomorphically extended to a field defined in $\overline{W}$ and formally equivalent to (4.11). To do this, first construct a diffeomorphism $H_0: \widetilde{U} \to \mathbb{C}^2 \setminus \{z = 0\}$, and then "correct" it by composing it with a certain G such that the result $H = G \circ H_0$ is biholomorphic. Consider a partition of unity $\{\theta_j\}$ on $\widetilde{U}$ subordinate to the covering by domains $\widetilde{S}_j' \times \{j\}$. As in 2.6 and 3.3, the functions θ_j are smooth, $\sum \theta_j = 1$ and $\operatorname{supp} \theta_j \subset \widetilde{S}_j'$. These functions can be chosen so that they grow no faster than polynomials in $|z|^{-1}$ together with their derivatives as $z \to 0$. Note that the first coordinate function z of all charts $\underline{z}_j$ is well defined on the entire set $\widetilde{U}$, since all the identifying maps preserve it. Let

$$H_0(p) = \sum \theta_j \underline{z}_j(p).$$

The map H_0 in each chart differs from the identity by exponentially flat corrections; an analogous statement was proved in 2.6. So the almost complex structure induced by $H_0(\widetilde{U})$ from the complex structure on $\widetilde{U}$ can be smoothly extended to the removed axis precisely as this was done in 3.3.

By the Newlander-Nirenberg theorem, this almost complex structure is integrable. Let $G: (\mathbb{C}^2, 0) \to (\mathbb{C}^2, 0)$ be a germ of a diffeomorphism holomorphic in the sense of this structure. Without loss of generality one may assume that the first coordinate function of G is z. The germ G is smooth, and the composition map $H = G \circ H_0$ is biholomorphic and defined in some domain $W \subset \widetilde{U}$ with the closure $\overline{H(W)}$ containing the origin in its interior. The desired map H is constructed.

Let $v = H_*\widetilde{v}$, where $\widetilde{v}$ is the field defined above. It remains to check that the formal type and the functional moduli of the germ of v at the origin are the prescribed ones. The procedure is standard. Let us start with linearization. Denote
$$C = \lim_{z \to 0} \partial H / \partial z_j.$$
This limit as a real linear operator exists because of the smoothness of both maps G and H_0. Let $\widetilde{C}$ denote the "realification" of the linear part of the field (4.11) at the origin. Then the realification ${}^\mathbb{R}\partial v/\partial z(0)$ is obtained from $\widetilde{C}$ by conjugation with C. The eigenvalues of $\widetilde{C}$ are $1, 1, -n/m, -n/m$; the operator ${}^\mathbb{R}\partial v/\partial z(0)$ has the same spectrum. Hence the eigenvalues of $\partial v/\partial z(0)$ are $1, -n/m$, as prescribed.

The number p is the topological invariant of the monodromy transformation; it is the same for both v and (4.11). The real number α is uniquely determined by the functional modulus, which we now begin considering.

The chart $\underline{z}_j$ on the domain $\widetilde{S}'_j$ conjugates the equation associated with the field $\widetilde{v}$ with that associated with (4.11). Its restriction to the transversal Γ conjugates the nth power of the monodromy for the field $\widetilde{v}$ with $f_0^{[n]}$. Hence the restrictions $H_j = \underline{z}_j|\Gamma$ form a normalizing atlas in the sense of 2.6, and (4.8') are its transition functions by construction. Therefore the field $\widetilde{v}$ and its monodromy map have the prescribed functional modulus. The formal invariant α is reconstructed from the modulus as shown above in (4.12), (4.8). This finishes the proof of the realization theorem for resonance saddles.

The analytic classification of resonant saddles is now complete. $\square$

4.7. A criterion of reducibility to resonant normal form. It is well known that the germ of a resonance saddle at the origin is formally equivalent to the field
$$\dot{z} = z(\lambda_1 + \widehat{f}(u)), \qquad \dot{w} = w(\lambda_2 + \widehat{g}(u)), \qquad (4.15)$$
$$\lambda_2/\lambda_1 = -m/n, \qquad m, n \in \mathbb{N}, \quad u = z^m w^n$$
(the fraction m/n is assumed to be irreducible). A. D. Bryuno [6] proved that the formal equivalence of such equations implies their analytic equivalence if and only if $\widehat{f}/\widehat{g} = -m/n$. This requirement is equivalent to the following: the initial equation is formally orbitally equivalent to a linear system.

We have considered germs of fields lacking this property. For such germs we may state a necessary and sufficient condition for the convergence of a series normalizing them to the form (4.15). Namely, it is easy to prove that equation (4.15) is formally orbitally equivalent to (4.11). For the field (4.11), the monodromy map is R-embeddable, as shown in 4.5. The functional modulus for R-embeddable maps is trivial: the corresponding collection $\{\widetilde{\Phi}_j\}$ consists only of shifts.

Conversely, any R-embeddable map is realized as the monodromy for a germ of type (4.11): this follows from the realization theorem and the theorem on orbital analytic equivalence of germs with equivalent monodromies. Thus we have proved

THEOREM. *The germ of a resonance saddle on the plane is analytically equivalent to its formal normal form* (4.15) *if and only if its monodromy is R-embeddable, and the corresponding functional modulus is trivial.*

This theorem shows how scarce are convergent series putting resonance saddles into the formal normal form (4.15).

REFERENCES

1. L. H. Ahlfors, *Lectures on quasiconformal mappings*, Van Nostrand, Toronto, New York, and London, 1966.
2. V. I. Arnold, *Supplementary chapters in the theory of ordinary differential equations*, "Nauka", Moscow, 1978; English transl., *Geometrical methods in the theory of ordinary differential equations*, Springer-Verlag, Berlin and New York, 1988.
3. V. I. Arnold and Yu. S. Il'yashenko, *Ordinary differential equations*, Itogi Nauki i Tekhniki: Sovremennye Problemy Mat.: Fundamental'nye Napravleniya, vol. 1, VINITI, Moscow, 1985, pp. 7–150; English transl., Encyclopaedia of Math. Sci., vol. 1, Springer-Verlag, Berlin and New York, 1988.
4. G. D. Birkhoff, *Collected mathematical papers*, vol. 1, Amer. Math. Soc., Providence, R.I., 1950.
5. R. I. Bogdanov, *Local orbital normal forms of vector fields in the plane*, Trudy Sem. Petrovsk. **5** (1979), 51–84. (Russian)
6. A. D. Brjuno [Bryuno], *Analytic form of differential equations*, Trudy Moskov. Mat. Obshch. **25** (1971), 119–262; **26** (1972), 199–239; English transl. in Trans. Moscow Math. Soc. **25** (1971); **26** (1972).
7. _____, *Local method in nonlinear analysis of differential equations*, "Nauka", Moscow, 1979. (Russian)
8. C. Camacho and P. Sad, *Topological classification and bifurcations of holomorphic flows with resonances in* $\mathbb{C}^2$, Invent. Math. **67** (1982), 447–472.
9. H. Dulac, *Recherches sur les points singuliers des équations différentielles*, J. Ecole Polytech. **2** (1904), 1–125.
10. J. Ecalle, *Sur les functions résurgentes*, I, II, Publ. Math. d'Orsay, Université de Paris-Sud, Orsay, 1981.
11. P. M. Elizarov, *Orbital topological classification of analytic differential equations in the neighborhood of a degenerate elementary singular point*, Uspekhi Mat. Nauk **40** (1985), no. 5, 253–254; English transl. in Russian Math. Surveys **40** (1985)
12. _____, *Orbital topological classification of analytic differential equations in the neighborhood of a degenerate elementary singular point on the two-dimensional complex plane*, Trudy Sem. Petrovsk. **13** (1988), 137–165; English transl. J. Soviet Math. **50** (1990).

13. _____, *Orbital analytic nonequivalence of saddle resonant vector fields in* $\mathbb{C}^2$, Mat. Sb. **123** (1984), no. 4, 543–549; English transl. in Math. USSR-Sb. **51** (1985)
14. P. M. Elizarov and Yu. S. Il'yashenko, *Remarks on the orbital analytic classification of germs of vector fields*, Mat. Sb. **121** (1983), 111–126; English transl. in Math. USSR-Sb. **49** (1984)
15. H. Hukuhara, T. Kimura, and T. Matuda, *Equations différentielles ordinaires du premier ordre dans le champ complexe*, Math. Soc. of Japan, Tokyo, 1961.
16. Yu. S. Il'yashenko, *The topology of phase portraits of analytic differential equations on the complex positive plane*, Trudy Sem. Petrovsk. **4** (1978), 83–136. (Russian)
17. _____, *In the theory of normal forms of analytic differential equations when the Brjuno condition fails divergence is the rule, convergence is the exception*, Vestnik Moskov. Univ. Ser. I Mat. Mekh. **1981**, no. 2, 10–16; English transl. in Moscow Univ. Math. Bull. **36** (1981).
18. _____, *Singular points and limit cycles of differential equations on the real and complex plane*, Preprint, Computer Center AN SSSR, Pushchino, Moscow Region, 1982. (Russian)
19. _____, *Dulac's memoir "On limit cycles" and related questions of the theory of local differential equations*, Uspekhi Mat. Nauk **40** (1985), no. 6, 41–78; English transl. in Russian Math. Surveys **40** (1985)
20. _____, *Finiteness theorems for limit cycles*, I, Uspekhi Mat. Nauk **45** (1990), no. 2, 143–200; English transl. in Russian Math. Surveys **45** (1990)
21. _____, *Finiteness theorems for limit cycles*, Amer. Math. Soc., Providence, R.I., 1992.
22. Yu. S. Il'yashenko and A. G. Khovanskiĭ, *Galois group, Stokes operator, and the Ramis theorem*, Funktsional. Anal. i Prilozhen. **24** (1990), no. 4, 31–42; English transl. in Functional Anal. Appl. **24** (1990)
23. I. Kaplansky, *An introduction to differential algebra*, Hermann, Paris, 1957.
24. B. Malgrange, *Travaux d'Ecalle et de Martinet-Ramis sur les systèmes dynamiques*, Séminaire Bourbaki, vol. 1981/1982, Astérisque, vol. 92–93, Soc. Math. France, Paris, 1982, pp. 59–73.
25. J. Martinet and J. P. Ramis, *Problème de modules pour des équations différentielles non linéaires du premier ordre*, Inst. Hautes Études Sci. Publ. Math. (1982 **55**, pp. 63–164).
26. _____, *Classification analytique des équations différentielles non linéaires résonantes du premier ordre*, Ann. Sci. École Norm. Sup. (4) **16** (1983), no. 4, 571–621.
27. V. A. Naĭshul', *Topological invariants of analytic and area-preserving maps and their application to analytic differential equations in* $\mathbb{C}^2$ *and* $\mathbb{C}P^2$, Trudy Moskov. Mat. Obshch. **44** (1982), 235–245; English transl. in Trans. Moscow Math. Soc. **1984** no. 2.
28. A. Newlander and L. Nirenberg, *Complex analytic coordinates in almost complex manifolds*, Ann. of Math. **65** (1957), 391–404.
29. J.-P. Ramis, *Phénomène de Stokes et resommation*, C. R. Acad. Sci. Paris Sér. I Math. **301** (1985), no. 4, 99–102.
30. _____, *Phénomène de Stokes et filtration Gevrey sur le groupe de Picard-Vessiot*, C. R. Acad. Sci. Paris Sér. I Math. **301** (1985), no. 5, 165–167.
31. A. A. Shcherbakov, *Topological classification of germs of conformal mappings with identical linear part*, Vestnik Moskov. Univ. Ser. I Mat. Mekh. **1982**, no. 3, 52–57; English transl. in Moscow Univ. Math. Bull. **37** (1982)
32. _____, *On germs of maps analytically nonequivalent to their formal normal form*, Funktsional. Anal. i Prilozhen. **16** (1982), no. 2, 94–95; English transl. in Functional Anal. Appl. **16** (1982)
33. Y. Sibuya, *Simplification of a system of linear ordinary differential equations about a singular point*, Funkcial. Ekvac. **4** (1962), no. 1, 29–56.
34. _____, *Stokes phenomena*, Bull. Amer. Math. Soc. **83** (1977), 1075–1077.

35. C. L. Siegel, *Vorlesungen über Himmelsmechanik*, Springer-Verlag, Berlin, Gottingen, and Heidelberg, 1957.
36. G. G. Stokes, *On the discontinuity of arbitrary constants, which appear in divergent developments*, Trans. Cambridge Phil. Soc. **10** (1857).
37. S. M. Voronin, *Analytic classification of germs of conformal maps* $(\mathbb{C}, 0) \to (\mathbb{C}, 0)$ *with identical linear part*, Funktsional. Anal. i Prilozhen. **15** (1981), no. 1, 1-17; English transl. in Functional Anal. Appl. **15** (1981)
38. _____, *Analytic classification of pairs of involutions and its applications*, Funktsional. Anal. i Prilozhen. **16** (1982), no. 2, 21-29; English transl. in Functional Anal. Appl. **16** (1982)
39. J.-C. Yoccoz, *Théorème de Siegel, nombres de Bruno et polynômes quadratiques*, Preprint, Centre Math. de l'Ecole Polytechnique, Palaiseau, 1988.

CHAIR OF DIFFERENTIAL EQUATIONS, DEPARTMENT OF MECHANICS AND MATHEMATICS, MOSCOW STATE UNIVERSITY, MOSCOW 119899

Translated by S. YAKOVENKO

Finitely Generated Groups of Germs of One-Dimensional Conformal Mappings, and Invariants for Complex Singular Points of Analytic Foliations of the Complex Plane

P. M. ELIZAROV, YU. S. IL'YASHENKO, A. A. SHCHERBAKOV, AND S. M. VORONIN

Introduction

In this paper we present an almost complete analytic classification for finitely generated groups of germs of conformal mappings $(\mathbb{C}, 0) \to (\mathbb{C}, 0)$ and describe functional invariants for complicated, that is, nonelementary, singular points of differential equations on the complex plane. The second part of this study is far from being complete. The entire contents are distributed almost evenly between original and survey results. We present and develop investigations started by Moussu, Cerveau, and Ramis [CM, M].

The classification for groups of germs of conformal mappings $(\mathbb{C}, 0) \to (\mathbb{C}, 0)$ depends heavily on algebraic properties of the groups involved. We distinguish three classes of groups: abelian, solvable nonabelian, and nonsolvable groups. Recall that a group is called *solvable* if its derived series terminates, that is, ends in the unit after a finite number of steps, the first group of the derived series coinciding with the original one, while each subsequent group is the commutant of the preceding one. For brevity, below and in §§1–3 we often say "a germ" instead of "a germ of a conformal mapping $(\mathbb{C}, 0) \to (\mathbb{C}, 0)$".

Commutative groups of germs of conformal mappings (excluding so-called Liouvillian groups) have a simple classification: they are either analytically equivalent to subgroups of the group of linear germs, or similar to commutative subgroups of the group $\mathscr{A}$ described in 2.7 of Paper I. Here

$$\mathscr{A} = \mathrm{id} \bigcup_{p \geq 1} \mathscr{A}_{p+1},$$

$$\mathscr{A}_{p+1} = \{f: (\mathbb{C}, 0) \to (\mathbb{C}, 0) \mid z \mapsto z + az^{p+1} + \cdots, a \neq 0\}.$$

1991 *Mathematics Subject Classification.* Primary 35C35, 58D27.

©1993 American Mathematical Society
1051-8037/93 $1.00 + $.25 per page

A group of germs is called *Liouvillian* if the multipliers for all its elements violate the Bryuno condition [Br]. By the Yoccoz theorem [Y] for germs with such multipliers, formal linearizing transformations are as a rule divergent. Abelian groups are studied in §1; the end of this section can be regarded as a continuation of Paper I: it contains the solution to the problem of taking the root and describing the centralizer for germs with a resonant multiplier (a root of unity).

Main examples of solvable groups are provided by subgroups of the group

$$G_s(p) = \{\nu g^t_{z^{p+1}} \mid \nu \in \mathbb{C}^*, \, t \in \mathbb{C}\}, \qquad p \in \mathbb{N} \text{ is fixed};$$

here s stands for "solvable". In 2.1 we verify that the commutant of the above group is abelian. Quite unexpectedly, it turns out that all solvable finitely generated groups of germs $(\mathbb{C}, 0) \to (\mathbb{C}, 0)$ are formally equivalent to subgroups of the above group (Theorem 2.2). A solvable group is said to be *exceptional* if it is a skew product of $\mathbb{Z}_2$ by a commutative normal subgroup of rank 1, and *typical* otherwise. For typical solvable groups, one can replace the formal equivalence in Theorem 2.2 by the analytic one (Theorem 2.3). The analytic classification for exceptional groups contains functional moduli; it is given by Theorem 2.4.

For nonsolvable groups of germs, the formal equivalence implies the analytic one. This surprising result is due to Moussu and Cerveau [CM] and depends heavily on a theorem due to Ramis [R]. Section 3 contains a new proof of this fact based on the Phragmen-Lindelöf principle for cochains [I2].

Observe that as the germs that we classify (and the corresponding groups) become more and more complicated, the analytic classification first coincides with the formal one, then sharply differs, and then coincides again. To be more detailed, a hyperbolic germ (such that the absolute value of the multiplier differs from 1) is always analytically equivalent to its linear part. The same is true for germs satisfying the Bryuno condition (see 1.2 below). For germs with "almost resonant" multipliers, the analytic classification differs from the formal one. Quite recently, Yoccoz and Perez-Marco obtained a number of deep results in studying the corresponding generated dynamics [Y, PM]. For germs with resonant multipliers (roots of unity), the analytic classification differs from the formal one and involves functional moduli; the above classification forms the basis of this volume. Further on the situation changes. For abelian noncyclic groups of germs, the analytic classification coincides with the formal one in all cases when the rank of the group is greater than 1, and as a rule does not coincide with it when the rank is 1. For solvable groups, the classifications in question coincide considerably more often; they can differ only for exceptional groups, that is, for those having a commutative normal subgroup of rank 1 and index 2. For nonsolvable groups, both classifications always coincide. However, in this case (in contrast to the previous ones) the formal classification at present is far from complete.

In the second part we study foliations of $(\mathbb{C}^2, 0)$ into analytic curves in a neighborhood of a singular point 0 having a finite number of separatrices.

By a *separatrix* we mean a leaf such that by adding the singular point one obtains a Riemann surface (possibly singular) homeomorphic to a disk.

To each separatrix there corresponds a monodromy transformation: the first return mapping, or the holonomy along a loop going around the singular point; it is the germ of a conformal mapping $(\mathbb{C}, 0) \to (\mathbb{C}, 0)$. Naturally, the analytic equivalence class for the above mapping is an invariant of the analytic classification for the corresponding foliation. Yet the collection of monodromy transformations for the entire finite set of separatrices is not a complete collection of invariants. The thing is that a nonelementary singular point has a "hidden" holonomy (Moussu calls it vanishing). Namely, there exists a "good blowing up" of the complex singular point, that is, a finite collection of σ-processes that split the complex singular point into elementary ones.

The analytic classification for elementary singular points (excluding "Liouvillian saddles", for which it is not obtained yet) is presented in Paper I. In particular, it is mentioned there that the above classification is determined by the analytic classification of the corresponding monodromy transformations. Being shrunk, that is, under the inverse σ-process, some of these transformations turn to monodromy transformations for the separatrices, while some others do not. The latter form the latent holonomy; as in the case of monodromy transformations, their analytic equivalence class is an invariant of the analytic classification for the corresponding foliation.

All the monodromy transformations corresponding to elementary points of a good blowing up form the vanishing holonomy of the foliation.

In §4 we study the analytic classification of germs of vector fields on the plane at the singular point provided the linear term is of "nilpotent Jordan cell" type, while the nonlinearity is in general position. The set of such germs is denoted by J_*. It turns out that the analytic classification of germs of class J_* coincides with the formal one. This fact was discovered in [CM] for germs of class J_* with some additional restrictions; here it is proved in full generality. In the paper [M], Moussu discovered two analytically nonequivalent germs from J_* having trivial (identity) monodromy; he posed the following question: find the number of such germs. It turns out that there exists a countable set of such germs. Their analytic classification based on results from §2 is presented in §4.

In §5 we study the vanishing holonomy for arbitrary complex singular points of analytic vector fields on the complex plane. These studies are far from being complete. The main result of the section is the realization theorem; according to this theorem, for a wide class of foliations, any collection of germs of mappings $(\mathbb{C}, 0) \to (\mathbb{C}, 0)$ with natural normalization conditions is realized as the collection of vanishing holonomy transformations for some foliation of $(\mathbb{C}^2, 0)$ into analytic curves having a finite number of separatrices. In contrast to "Jordan cell" type foliations considered in §4, in general two foliations having equivalent collections of vanishing holonomy transformations can be analytically nonequivalent. Additional discrete invariants are pointed out in §5. S. M. Voronin conjectured that there exist numerical invariants distinct from vanishing holonomy transformations. Thus the problem of finding a complete set of invariants for the analytic classification of foliations of $(\mathbb{C}^2, 0)$ having a finite number of complex separatrices remains to be "widely open". In 1989 J. F. Mattei communicated to us that he knows the realization theorem mentioned above.

In §6 we give the proof of the Yoccoz–Perez-Marco realization theorem: any germ $\Delta \colon (\mathbb{C}, 0) \to (\mathbb{C}, 0)$ with multiplier equal to $e^{i\varphi}$, $\varphi \in \mathbb{R}$, may be realized as a monodromy mapping for a complex saddle.

After this paper was written, the authors learned that some of the results of §2 were obtained independently by Isao Nakai.

§1. Abelian groups of germs of conformal mappings

1.1. Statement of results. The analytic classification of cyclic groups is reduced to that of generating elements. A short survey of the corresponding results is presented in 1.2.

We preface the analytic classification for noncyclic abelian groups by the formal one. For a group G of germs of mappings $(\mathbb{C}, 0) \to (\mathbb{C}, 0)$, denote by Λ_G the group of multipliers for these germs. Now let us define Liouvillian and resonant multipliers.

For each irrational α, denote by $\{p_n/q_n\}$ the sequence generating the corresponding continued fraction. We say that α is *poorly rationally approximated*, or *Diophantine*, if the Bryuno condition [Br]

$$\sum q_n^{-1} \log q_{n+1} < \infty$$

holds, and that it is *finely rationally approximated*, or *Liouvillian*, otherwise.

DEFINITION. A multiplier $\lambda = \exp 2\pi i \alpha$ is said to be *Liouvillian* if α is a Liouvillian irrational number, and *non-Liouvillian* otherwise. It is called *resonant* if α is a rational number.

REMARK. A multiplier whose absolute value differs from 1 is always non-Liouvillian.

A standard example of an abelian group of germs of mappings $(\mathbb{C}, 0) \to (\mathbb{C}, 0)$ is provided by the group

$$G_a(p) = \{\lambda g_v^t \mid v = z^{p+1}(1 + \alpha z^p)\partial/\partial z,\ t \in \mathbb{C},\ \lambda \in \mathbb{C},\ \lambda^p = 1\}, \quad (1.1)$$

where α is fixed and a stands for "abelian".

The formal classification for noncyclic abelian groups depends on properties of the intersection $G \cap \mathscr{A}$, in other words, on the rank of the group $G \cap \mathscr{A}$.

THEOREM 1.1. *Assume that G is a noncyclic abelian group of germs.*

1. *Let there exist a p such that $G \cap \mathscr{A}_{p+1} \neq \mathrm{id}$. Then G is formally equivalent to a subgroup of the group $G_a(p)$.*

2. *Let $G \cap \mathscr{A} = \mathrm{id}$. Then there exists a change of variables that simultaneously reduces all the germs from G to a linear normal form.*

The next two theorems give the analytic classification of abelian groups of germs, excluding Liouvillian ones.

THEOREM 1.2. *Assume that G is a noncyclic abelian group of germs, and Λ_G is the corresponding group of multipliers.*

1a. *Let $\operatorname{rank} G \cap \mathscr{A} > 1$ ($G \cap \mathscr{A}$ is a noncyclic group). Then G is analytically equivalent to a subgroup of the standard group $G_a(p)$ for some p.*

1b. *Let $\operatorname{rank} G \cap \mathscr{A} = 1$ ($G \cap \mathscr{A}$ is a nontrivial cyclic group). Then there exist natural numbers p, q and an integer k such that G is generated by the two germs $h \in G \cap A_{p+1}$ and g: $g'(0) = \lambda = \exp(2\pi i/q)$, while $g^{[q]} = h^{[k]}$ and q divides p.*

2. *Assume that $\operatorname{rank} G \cap \mathscr{A} = 0$ ($G \cap \mathscr{A} = \mathrm{id}$) and Λ_G has non-Liouvillian multipliers. Then G is analytically equivalent to its linearization Λ_G.*

REMARKS. 1. The case $k = 0$ in statement 1b corresponds to a linear germ g.

2. The case when $\operatorname{rank} G \cap \mathscr{A} = 0$ and all $\lambda \in \Lambda_G$ are Liouvillian is not completely studied at present. Recently Perez-Marco obtained the following surprising result [PM]:

There exists a continuum of pairwise commuting Liouvillian germs that are not analytically equivalent to their linear parts. The multipliers for the above germs form a Cantor-type set on the unit circle.

Theorem 1.2 states that in cases 1a and 2 the analytic classification for noncyclic abelian groups coincides with the formal one. In case 1b these classifications are distinct. The following theorem describes the functional moduli for the analytic classification in case 1b.

THEOREM 1.3. *Let g and h be the pair of germs from statement 1b of Theorem 1.2. More precisely, suppose that $p > 1$, $q \geq 1$, k, s are integers, $h \in A_{p+1}$, g commutes with h, $g'(0) = \lambda = \exp(2\pi i/q)$, while $g^{[q]} = h^{[k]}$ and q divides p: $p = qs$. Then the following statements are true.*

1. *The modulus. To each pair g, h as above there corresponds a modulus for the analytic classification of pairs, that is, a class of equivalent collections of $2s$ 1-periodic functions $\psi = (\psi_1, \ldots, \psi_{2s})$, where the functions with odd numbers are defined in a sufficiently high upper half-plane, while those with*

even ones in a sufficiently low lower half-plane. Two collections $\psi = \{\psi_j\}$ *and* $\tilde{\psi} = \{\tilde{\psi}_j\}$ *are equivalent if and only if there exists a constant* $c \in \mathbb{C}$ *such that* $\psi \circ (\mathrm{id} + c) = \tilde{\psi}$.

2. Equimodality and equivalence. *Pairs with equivalent collections are analytically equivalent.*

3. Realization. *To each collection of integers* p, q, k, s *and holomorphic functions* ψ_j *as described above there corresponds a pair of germs* (g, h) *with the properties described above.*

1.2. Analytic classification for germs of conformal mappings $(\mathbb{C}, 0) \to (\mathbb{C}, 0)$.

A. *A hyperbolic germ of a conformal mapping* $(\mathbb{C}, 0) \to (\mathbb{C}, 0)$ *(that is, such that the absolute value of its multiplier differs from* 1) *is analytically equivalent to its linear term.*

B. *If the absolute value of the multiplier equals* 1 *and the multiplier itself is non-Liouvillian and nonresonant (i.e., is not a root of unity) then the germ is analytically equivalent to its linear part* (Bryuno [Br]).

C. *If a multiplier is Liouvillian, then there always exists a germ of a conformal mapping with the same multiplier that is not analytically equivalent to its linear part* (Yoccoz [Y]). *This proves that the Bryuno condition on multipliers is a necessary and sufficient condition for the existence of the analytic linearization of the germ.*

D. *The analytic classification for germs with resonant multiplier is presented in Paper* I.

Germs with Liouvillian multipliers were studied in detail by Ecalle [E], who discovered all the moduli for the analytic classification of the above germs. These moduli have a complicated algebraic nature and are characterized by the author as "nonconstructable".

1.3. Formal normal forms for noncyclic abelian groups of germs of mappings $(\mathbb{C}, 0) \to (\mathbb{C}, 0)$. Here we prove Theorem 1.1 from 1.1.

Let G and Λ_G be the same as in the assertion of the theorem.

CASE 1: $G \cap \mathscr{A} \neq \mathrm{id}$. Let $h \in G \cap \mathscr{A}$. By Theorem 1 from Paper I, 2.1, the germ h is formally equivalent to the germ

$$h_0 = g_v^1, \qquad v = z^{p+1}(1 + az^p)^{-1} \partial/\partial z, \qquad p \in \mathbb{N}. \tag{1.2}$$

Below in this section v stands for the field defined by (1.2).

Denote by $\widehat{\mathscr{A}}$ the set of formal series with identical linear terms. For any germ $f \colon (\mathbb{C}, 0) \to (\mathbb{C}, 0)$, $\widehat{f}$ denotes its formal Taylor series.

PROPOSITION 1.1. *Each formal series* $\widehat{h}$ *in* $\widehat{\mathscr{A}}$ *that commutes with* $\widehat{h}_0$ *in the sense of superposition has the form* $\widehat{g}_v^t$ *for some complex* $t \in \mathbb{C}$.

PROOF. The functional equation for the series $\widehat{h}$ commuting with $\widehat{h}_0$ can be solved by the method of undetermined coefficients. By assumption, the coefficient of the linear term equals 1. While finding coefficients for "lower" terms, whose degree is less than $p+1$, one finds that all of them vanish. The coefficient for z^{p+1} can be chosen arbitrarily; then the coefficients for further Taylor terms are determined uniquely. This proves that the commutativity equation for $\widehat{h}$ has a unique solution up to the choice of the coefficient for z^{p+1}. On the other hand, the series $\widehat{g}_v^t$ commutes with $\widehat{h}_0$ for any $t \in \mathbb{C}$ and its coefficient for z^{p+1} equals t. Therefore these series are exactly the solutions of the commutativity equation. □

COROLLARY 1.1. *Suppose that G is the same as in Theorem 1.1, $G \cap \mathscr{A} \neq$ id, and in some formal chart a germ $h \in G \cap \mathscr{A}$ has the form $\widehat{h}_0$. Then all the germs from the subgroup $G \cap \mathscr{A}$ have the form g_v^t, $t \in T_G$ in this chart, where T_G is a subgroup of the additive group of complex numbers.*

COROLLARY 1.2. *Let p be the same as in (1.2). Then $\Lambda_G^p = 1$.*

PROOF. Let $\lambda \in \Lambda_G$, $\lambda \neq 1$, $\lambda = f'(0)$, $f \in G$, $h(z) = z + z^{p+1} + \cdots$. Then
$$f^{-1} \circ h \circ f \colon z \mapsto z + \lambda^{-p} z^{p+1} + \cdots .$$
However, $f^{-1} \circ h \circ f = h$, hence $\lambda^{-p} = 1$. □

COROLLARY 1.3. *Assume that G is the same as in Theorem 1.1, $G \cap \mathscr{A} \neq$ id, and in some formal chart z a germ $h \in G \cap \mathscr{A}$ has the form (1.2). Then all the germs from the group G have the form (1.1) in this chart.*

PROOF. For germs $f \in G \cap \mathscr{A}$ the assertion is true by Corollary 1.1. Let $f \colon f'(0) = \lambda$ be an arbitrary germ from $G \setminus \mathscr{A}$. By Corollary 1.2, $\lambda^p = 1$, hence $f^{[p]} \in G \cap \mathscr{A}$. By Corollary 1.1, $\widehat{f}^{[p]} = \widehat{g}_v^t$ for some $t \in T$ (all the formal series are considered in the chart z). On the other hand, the formal series $f = \nu g_v^{t/p}$ satisfies the equation
$$\widehat{f}^{[p]} = \widehat{g}_v^t \tag{1.3}$$
for any ν such that $\nu^p = 1$. The proof of the corollary will now be complete if we prove the following

PROPOSITION 1.2. *The "formal root" equation (1.3) with the normalization condition $\widehat{f}'(0) = \mu$ has a unique solution for any μ such that $\mu^p = 1$.*

PROOF. The proposition is proved via a recursive calculation of the coefficients for the series $\widehat{f}$. This calculation implies immediately that all formal solutions for (1.3) have the form (1.1). This proves Corollary 1.3 and thus Theorem 1.1 for case 1. □

CASE 2: $G \cap \mathscr{A} = \mathrm{id}$. In this case the group G is isomorphic to the group Λ_G, since a germ from the former group having identity linear term is itself the identity by the assertion.

CASE 2a: *the group Λ_G possesses a nonresonant element, that is, a number such that none of its powers equals* 1. Denote this number by λ. Let $f \in G$ be a germ whose multiplier equals λ. By the Poincaré-Dulac theorem, the germ f is formally equivalent to a linear germ. Assume that z is the corresponding formal chart, g is an arbitrary germ from G, and $\hat{g}$ is the formal series corresponding to g in the chart z. Elementary calculations show that the formal series $\hat{g}$ that commutes with a linear nonresonant one is linear itself. This proves Theorem 1.1 for case 2a.

CASE 2b: *all the elements from Λ_G are resonant*. This case cannot occur within the assumptions of Theorem 1.1: since the group G is finitely generated and isomorphic to Λ_G, in case 2a it turns out to be cyclic, in contradiction to the assumption of Theorem 1.1.

This ends the proof of Theorem 1.1 and completes the formal classification for abelian groups of germs of mappings $(\mathbb{C}, 0) \to (\mathbb{C}, 0)$.

1.4. Analytic normal forms for noncyclic abelian groups of germs of mappings $(\mathbb{C}, 0) \to (\mathbb{C}, 0)$. In this subsection we prove Theorem 1.2. It turns out to be a simple corollary of results from 2.7 of Paper I.

CASE 1a: $G \cap \mathscr{A} \neq \mathrm{id}$, *and the group T_G (see Corollary 1.1) is noncyclic*. Then the centralizer for the germ $h \in G \cap \mathscr{A}$ is noncyclic. By Theorem 3 of Paper I, 2.7, the germ h is embeddable. Therefore there exists an analytic map that conjugates the germ h with the germ g_v^1. By Corollary 1.3 from 1.3, the same map conjugates each germ $g \in G$ with its normal form (1.1).

CASE 1b: *the group $G \cap \mathscr{A}$ is cyclic of rank* 1. In this case one can find a linear change of the formal normalizing chart that guarantees the condition $T_G = \mathbb{Z}$. Let $h \in G \cap \mathscr{A}$ be the germ whose formal series in the above chart equals g_v^1. Then all the germs from $G \cap \mathscr{A}$ are integer powers of h. Choose a generator λ for the group Λ_G. Recall that by Theorem 1.1 we have $\Lambda_G^p = 1$, hence λ can be written as $\lambda = \exp(2\pi i/q)$, where q divides p. The noncyclicity of G implies that $q > 1$, $p > 1$. Let $g \in G$ be a germ whose multiplier equals λ. Then $g^{[q]} \in G \cap \mathscr{A}$, hence $\hat{g}^{[q]} = \hat{g}_v^k$, $g^q = h^k$ for some integer k. Finally, the commutativity of the group G yields

$$G/G \cap \mathscr{A} \approx \Lambda_G.$$

Hence the germs g and h are generators of the group G; this proves Theorem 1.2 for case 1b.

CASE 2: $G \cap \mathscr{A} = \mathrm{id}$, *and the group Λ_G possesses non-Liouvillian multipliers*. Then, as mentioned at the end of 1.3, it possesses a nonresonant multiplier. By the Bryuno theorem, the corresponding germ is analytically equivalent to a linear one. In the normalizing map, the formal series for all the other germs from the group commute with the linear nonresonant one, hence they are linear themselves. □

This completes the proof of Theorem 1.2, and thus completes the analytic classification for abelian groups of germs of mappings $(\mathbb{C}, 0) \to (\mathbb{C}, 0)$, excluding the case 1b when the group T_G is a cyclic group of rank 1. In the latter case the classification is provided by Theorem 1.3 proved in 2.7.

1.5. Appendix: centralizers and taking roots of germs with resonant multipliers.

THEOREM 1.4 (on centralizers). *Let g be a germ with a resonant multiplier and suppose that g is not formally equivalent to a linear germ. Then its centralizer is a commutative group. If the rank for this group exceeds 1, then the group is analytically equivalent to the standard abelian group $G_a(p)$ for some natural p.*

PROOF. Assume that $g^{[n]} = h \in \mathscr{A}_{p+1}$ and f commutes with g. Then f commutes with h. Let z be the formal change that reduces the germ h to the form $h_0 = \nu g_\nu^t$. Then by Corollary 1.3 all the germs from the group generated by f and h have the same form in the above chart (ν and t may be different). Therefore, there exists a formal map that takes the centralizer of the germ g to a subset of the commutative group $G_a(p)$. The first statement of the theorem is proved.

The second statement follows immediately from the first and Theorem 1.2.

THEOREM 1.5 (on taking roots). *Assume that g is a germ, $\exp(2\pi i m/n)$ is its multiplier, and g admits taking superposition roots, that is, there exists a germ f satisfying the equation $f^{[N]} = g$ with $f'(0) = \exp(2\pi i d/(Nn))$. Let $g^{[n]} = h \in \mathscr{A}_{p+1}$. Then the functional modulus for the germ g is a class of equivalent collections of 1-periodic functions ψ_j, $j = 1, \ldots, 2p$, described in 2.6 of Paper I, while $r = p/(Nn)$ is an integer and the following symmetry conditions are true:*

$$\psi_j = \psi_{j+2dr} \circ (\mathrm{id} + 1/(Nn)).$$

The proof of this theorem is similar to arguments used in 2.7 of Paper I, and especially to those of 2.7 below, so we do not go into details.

§2. Solvable nonabelian groups

In this section we present both the formal and the analytic classifications for solvable nonabelian groups of germs of 1-dimensional conformal mappings. In contrast to the abelian case, the above classifications turn out to be complete. We start with an example, which plays an important role in what follows.

2.1. Main example. Consider the set of germs

$$G_s(p) = \{\lambda g_{z^{p+1}}^t \mid \lambda \in \mathbb{C}^*, t \in \mathbb{C}\},$$

where s stands for "solvable". For brevity, denote $\lambda g_{z^{p+1}}^t$ by (λ, t). Let us check that $G_s(p)$ is a group with respect to multiplication, and construct the multiplication table. For any germ of a conformal mapping $f: (\mathbb{C}, 0) \to$

($\mathbb{C}$, 0) (and in more general situations as well), one has the following elementary relation: $f \circ g_v^t = g_{f_*v}^t \circ f$. It implies $\nu \circ g_{z^{p+1}}^t = g_{\nu_* z^{p+1}}^t \circ \nu$, or

$$g_{z^{p+1}}^t \circ \nu = g_{\nu_* z^{p+1}}^{\nu^{-p} t} \circ \nu = \nu \circ g_{z^{p+1}}^{\nu^{-p} t}.$$

Hence G_s is a group with respect to superposition, and the multiplication table for G_s has the following form:

$$(\lambda, t) \times (\mu, s) = (\lambda\mu, t\mu^{-p} + s). \tag{2.1}$$

Evidently, the commutant for $G_s(p)$ equals $\{g_{z^{p+1}}^t \mid t \in \mathbb{C}\}$ and is a commutative group; the group itself is solvable. Below we show that the above example "almost exhausts" the list of solvable groups of germs of mappings $(\mathbb{C}, 0) \to (\mathbb{C}, 0)$.

2.2. Statements of results. In what follows "solvable" stands for "nonabelian solvable". We preface theorems on normal forms for solvable groups by the following solvability criterion.

THEOREM 2.1. *A nonabelian group of germs of conformal mappings from $(\mathbb{C}, 0)$ to $(\mathbb{C}, 0)$ is solvable if and only if the following condition A holds:*

A. *The intersection $G \cap \mathscr{A}$ differs from* id. *There exists a germ $h \in G \cap \mathscr{A}$ and a set of generators $\{g_i\}$ for the group G such that*

A1. *Each generator g_i is either formally equivalent to a linear germ, or commutes with h; at least one of them does not commute with h.*

A2. *The germs $\mathrm{Ad}(g_i)h$ commute with h; further $\mathrm{Ad}(a)b = a^{-1} \circ b \circ a$.*

In what follows we refer to Theorem 2.1S for the sufficiency part of the above theorem, and Theorem 2.1N for its necessity part.

THEOREM 2.2. *A finitely generated solvable group of germs of mappings $(\mathbb{C}, 0) \to (\mathbb{C}, 0)$ is formally equivalent to a finitely generated subgroup of the group $G_s(p)$ defined in 2.1, for some p.*

The subgroup of the group $G_s(p)$ given by Theorem 2.2 will be called the *formal normal form* of the initial group G, despite the fact that it is not determined uniquely by G. The normal form for the group G is denoted by G_{norm}. Denote by T_G the set of values t such that

$$G_{\mathrm{norm}} \cap \mathscr{A} = \{g_{z^{p+1}}^t \mid t \in T_G\},$$

and by Λ_G the group of multipliers for the group G.

DEFINITION. A solvable group G is called *exceptional* if the following two conditions hold:

1. T_G is a cyclic group;
2. Λ_G is a subgroup of the group of roots of unity of degree $2p$; more precisely, $\Lambda_G^p = \{-1, 1\}$.

A nonexceptional solvable group is called *typical*.

EXAMPLE. Let ν be a root of unity of degree $2p$. Then the group generated by the germs ν and $g^1_{z^{p+1}} = h_0$ is either exceptional solvable or commutative. Relations (2.1) imply that the elements of this group have the following form:
$$G^\nu(p) = \{\nu^m g^n_{z^{p+1}} \mid m, n \in \mathbb{Z}\}.$$
If $\nu^p = 1$, then the group $G^\nu(p)$ is commutative, while if $\nu^p = -1$, it is solvable nonabelian (exceptional).

Below we show that all the solvable exceptional groups are formally equivalent to the group $G^\nu(p)$ with $\nu^p = -1$, where ν and p depend on the group.

THEOREM 2.3. *A typical finitely generated solvable group is analytically equivalent to its formal normal form. An exceptional solvable group is generated by two generators $h \in G \cap \mathscr{A}$ and g such that g "anticommutes" with h: $g \circ h = h^{-1} \circ g$, and $\nu^p = -1$, where $\nu = g'(0)$ is the multiplier of g.*

The analytic classification for exceptional solvable groups involves functional moduli described by the following theorem.

THEOREM 2.4. *Let g and h be a pair of germs as in the second assertion of Theorem 2.3. To each such pair there corresponds a functional modulus of analytic classification, that is, a 1-periodic holomorphic function ψ defined in a sufficiently high upper half-plane. Two pairs having the same functional modulus are equivalent. Each function with properties as described above is realized as the functional module for the pair (g, h), $h \in \mathscr{A}_{p+1}$, $g'(0) = \nu$, $\nu^p = -1$, $\mathrm{Ad}(g)h = h^{-1}$. Each such pair generates an exceptional solvable group.*

REMARKS. 1. Theorem 2.4 provides the analytic classification for exceptional solvable groups having a distinguished pair of generators, as in Theorem 2.3.

2. The functional modulus defined in Theorem 2.4 is constructed as follows: it is the functional modulus of the germ h (see 2.7 of Paper I) having additional symmetries implied by the existence of the germ g that anticommutes with h, while $\nu^p = -1$, where ν is the multiplier of g.

The rest of the section consists of proofs of Theorems 2.1–2.4.

2.3. Solvability criterion. Here we prove Theorem 2.1, which turns out to be useful not only for proving Theorem 2.2, but enables us also to identify solvability of the "vanishing holonomy" group for a vector field in $(\mathbb{C}^2, 0)$ whose linear term is of "nilpotent Jordan cell" type and the monodromy transformation corresponding to the unique separatrix is the identity. Together with realization theorems, of which one is Theorem 2.4 while the others are formulated below, this yields the construction for a number of germs of vector fields possessing the above properties. This, in turn, answers in affirmative the following question posed by Moussu [M]: do there exist equations possessing the above properties that are not orbitally analytically equivalent to the foliations $d(z^2 + w^3) = 0$ and $d(z^2 + w^3) + w(2z\,dw - 3w\,dz) = 0$? (The quotation is not exact, since we use a slightly different terminology.)

The proof of Theorem 2.1 goes as follows. First we prove the necessity for condition A, that is, Theorem 2.1N. Next, in 2.4, we prove the following theorem.

THEOREM 2.2'. *Suppose that a finitely generated group of germs of mappings* $(\mathbb{C}, 0) \to (\mathbb{C}, 0)$ *satisfies condition* A *of Theorem 2.1. Then it is formally equivalent to a subgroup of the group* $G_s(p)$ *for some* p.

This immediately implies Theorem 2.1S, since $G_s(p)$ is a solvable group. Theorems 2.1N and 2.2' imply Theorem 2.2. It remains to prove Theorems 2.1N and 2.2'; the former is proved below, the latter in the next subsection.

PROOF OF THEOREM 2.1N. The following proposition clarifies why the commutant of a solvable group of germs of mappings $(\mathbb{C}, 0) \to (\mathbb{C}, 0)$ is commutative.

PROPOSITION 2.1. *A group* G *of germs of conformal mappings that contains two noncommuting germs with identity linear terms is nonsolvable.*

PROOF. Let $f \in G \cap \mathscr{A}_m$, $g \in G \cap \mathscr{A}_n$, and $m = n$. Then $[f, g] \in \mathscr{A}_k$ for $k > m$.

Now let $m \neq n$. Then $[f, g] \in \mathscr{A}_{m+n-1}$. Hence the sequence $f_1 = g$, $f_2 = [f, g], \ldots, f_{k+1} = [f_k, f_{k-1}], \ldots$ does not contain the germ of the identity mapping, therefore the derived series for the group G, that is, G, $[G, G]$, $[[G, G], [G, G]], \ldots$, does not terminate. Hence the group G is not solvable. □

PROPOSITION 2.2. *Proposition* 2.1 *remains true if germs with identity linear parts are replaced by those with resonant linear parts.*

This proposition is implied immediately by the following (proved at the end of the subsection).

PROPOSITION 2.3 (M. Entov). *Suppose that the multipliers for germs* f: $(\mathbb{C}, 0) \to (\mathbb{C}, 0)$ *and* $g: (\mathbb{C}, 0) \to (\mathbb{C}, 0)$ *are primitive roots of unity of degrees* m *and* n, *respectively. Then these germs commute if and only if the same is true for the germs* $f^{[m]}$ *and* $g^{[n]}$.

Let us turn directly to the proof of Theorem 2.1N. Take an arbitrary germ from $G \cap \mathscr{A}$ for h and prove a sharpened condition A1: an arbitrary germ $g \in G$ is either formally linearizable, or commutes with h. Assume the opposite: let g possess none of these properties. Since it is not formally linearizable, its multiplier is resonant. Since it does not commute with h, the same is true for its power belonging to $\mathscr{A}$ (by Proposition 2.3). Hence by Proposition 2.1 the group G is nonsolvable, a contradiction. □

Among the generators of the group G there always exists a nonlinearizable germ commuting with h. Otherwise the entire group G would belong to the centralizer of the germ h, hence, by Corollary 1.3, it would be commutative, a contradiction. This proves condition A1 modulo Proposition 2.3.

Let us prove a sharpened condition A2: for any germ $g \in G$ the germs h and $\mathrm{Ad}(g)h$ commute. This follows immediately from the relation $\mathrm{Ad}(g)h \in G \cap \mathscr{A}$ and Proposition 2.1.

Thus Theorem 2.1N is proved modulo Proposition 2.3. Let us prove the latter. The implication $f \circ g = g \circ f \implies f^{[m]} \circ g^{[n]} = g^{[n]} \circ f^{[m]}$ is evident. The inverse implication follows easily from Proposition 1.2 of 1.3 (on the uniqueness of taking iterative roots). Namely,

$$f^{[m]} \circ g^{[n]} = g^{[n]} \circ f^{[m]} \implies (\mathrm{Ad}(g^{[n]})f)^{[m]} = f^{[m]} \implies \mathrm{Ad}(g^{[n]})f = f;$$

the first implication is evident while the second follows from Proposition 1.2. Similarly,

$$f \circ g^{[n]} = g^{[n]} \circ f \implies (\mathrm{Ad}(f)g)^{[n]} = g^{[n]} \implies \mathrm{Ad}(f)g = g.$$

This proves Proposition 2.3 and thus Theorem 2.1N as well. □

2.4. Formal normal forms for solvable groups. Here we prove the main result of the section, Theorem 2.2.

LEMMA 2.2. *Suppose that the germs h and $\{g_i\}$ satisfy condition A of Theorem 2.1. Then the germ h is formally equivalent to the germ*

$$h_0 = g_{z^{p+1}}^1. \tag{2.2}$$

Moreover, each formal chart z reducing the germ h to form (2.2) simultaneously reduces each germ $g \in \{g_i\}$ to the form

$$g_0 = \nu g_{z^{p+1}}^t, \tag{2.3}$$

where ν and t depend on the germ g and the chart z.

Lemma 2.2 not only proves Theorem 2.2′, but also provides its sharpening: it describes the entire set of normalizing charts instead of proving the existence of such a set.

PROOF OF LEMMA 2.2. *Step* 1. Let $g \in \{g_i\}$ be a formally linearizable germ that does not commute with h. Assume that w is a formal chart such that $\widehat{g} = \nu w$. In this chart, the germ h can be written as a formal series $\widehat{h}$. In the same time, the germ h is formally embeddable: there exists a formal vector field having the trivial 1-jet such that $\widehat{h}$ is its formal exponent. Let $\widehat{u}$ be the formal series for this field in the chart w; then $\widehat{h} = g_{\widehat{u}}^1$.

By condition A2, the formal series $\widehat{h}$ and $\mathrm{Ad}(\nu)\widehat{h} \stackrel{\text{def}}{=} \widehat{f}$ commute. There always exists a formal vector field with trivial linear part that generates a given formal mapping with the identity linear part; this is called the *iterative logarithm* for the mapping. It is known that if two formal mappings with identity linear terms commute, then the same is true for their iterative logarithms.

Let us apply the above argument to the formal series $\widehat{h}$ and $\widehat{f}$. Evidently, $\widehat{f} = g_{\widehat{u}}^1$, $\widehat{\widetilde{u}} = \nu \cdot \widetilde{u} \circ \nu^{-1}$. Therefore the fields $\widehat{u}$ and $\nu \cdot \widetilde{u} \circ \nu^{-1}$ commute. However, two commuting one-dimensional analytic vector fields are proportional. Thus

$$\nu \cdot \widehat{u} \circ \nu^{-1} = \lambda \widehat{u}, \quad \lambda \in \mathbb{C}. \tag{2.4}$$

Moreover,

$$\nu \cdot \widehat{u} \circ \nu^{-1} \neq \widehat{u}, \tag{2.5}$$

since the mappings $g = \nu w$ and h do not commute, thus $\widehat{f}$ and $\widehat{h}$ do not coincide.

Step 2.

PROPOSITION 2.4. *Suppose that a formal vector field $\widehat{u}$ satisfies conditions (2.4), (2.5). Then there exists a formal transformation $\widetilde{z} = \widehat{H}(w)$ reducing the field to the form $\widetilde{z}^{p+1} \partial/\partial \widetilde{z}$ and commuting with multiplication by ν.*

COROLLARY 2.1. *The germs g and h considered in Step 1 can be written as $\nu \widetilde{z}$ and $g_{\widetilde{z}^{p+1}}^1$, respectively, in the chart $\widetilde{z}$.*

COROLLARY 2.2. *Suppose that G is a solvable nonabelian group, $h \in G \cap \mathscr{A}$, $g \in G$, and the germs g and h do not commute, while the germ g is formally linearizable. Then there exists a formal chart $\widetilde{z}$ that reduces the germs g and h to the forms $\nu \widetilde{z}$ and $g_{\widetilde{z}^{p+1}}^1$.*

PROOF. The germs h and $\mathrm{Ad}(g)h$ commute by Proposition 2.1. Together with the assertion of Proposition 2.2, these are the only conditions used in the argument of Step 1. Now Corollary 2.2 follows from Corollary 2.1. □

PROOF OF PROPOSITION 2.4. Consider the following two cases: the multiplier ν is nonresonant, or it is resonant.

CASE 1: $\nu^j \neq 1$, $j \in \mathbb{N}$. Let $\widehat{u} = w^{p+1} \sum a_j w^j \partial/\partial w$. Then

$$\nu \cdot \widehat{u} \circ \nu^{-1} = \nu^{-p} w^{p+1} \sum a_j \nu^j w^j \partial/\partial w.$$

Hence by (2.4), $a_j \nu^j = a_j$ for any $j \in \mathbb{N}$. Therefore $a_j = 0$ for all $j \in N$, and thus $\widetilde{z} = w$ is the required chart.

CASE 2: ν is a primitive root of unity of degree q. Then by the previous displayed equality for $\nu \cdot \widehat{u} \circ \nu^{-1}$, we have

$$\widehat{u} = w^{p+1} \sum a_{jq} \widetilde{w}^{jq}, \qquad a_0 = 1. \tag{2.6}$$

Besides, relation (2.5) yields $\nu^p \neq 1$. Hence q is not a divisor of p. Let us look for a formal change $\widetilde{z} = \widehat{H}(w)$ in the form $\widehat{H} = w\widehat{F}(w^q)$. Evidently, this formal change commutes with the multiplication by ν.

The change $\tilde{z} = \widehat{H}(w)$ must satisfy the differential equation

$$\frac{d\tilde{z}}{\widehat{u}(\tilde{z})} = \frac{dw}{w^{p+1}}.$$

By (2.6),

$$\frac{1}{\widehat{u}(\tilde{z})} = \sum_{j=0}^{\infty} b_j \frac{\tilde{z}^{jq}}{\tilde{z}^{p+1}}, \qquad jq - (p+1) \neq -1$$

for all j, hence the formal series for $1/\widehat{u}$ has no term of degree -1. Therefore

$$\int \frac{d\tilde{z}}{\widehat{u}(\tilde{z})} = -\frac{\widehat{F}_1(\tilde{z}^q)}{p z^p}.$$

Thus the equation for $\tilde{z}$ admits a solution defined implicitly by the relation $\tilde{z}^p/\widehat{F}_1(\tilde{z}^q) = w^p$. Hence $\tilde{z}^q \widehat{F}_2(\tilde{z}^q) = w^q$, where $\widehat{F}_2^p = 1/\widehat{F}_1^q$, $\widehat{F}_2(0) = 1$. By the implicit function theorem for formal series, $\tilde{z}^q = w^q \widehat{F}_3(w^q)$, $\widehat{F}_3(0) = 1$, or $\tilde{z} = w\widehat{F}(w^q)$, as required.

Step 3. Assume that z is an arbitrary formal map reducing the germ h to the form (2.2), g is an arbitrary germ from the collection $\{g_i\}$. Let us prove that the germ g has the form (2.3) in the chart z. First assume that g does not commute with h. Then by condition A1 it is formally linearizable. Denote by $\tilde{z}$ the chart provided by Corollary 2.2; in this chart the formal series for the germs g and h can be written as $\nu\tilde{z}$ and $g^1_{z^{p+1}}$, respectively. Let $z = \widehat{k}(\tilde{z})$ be the formal "transition function" from $\tilde{z}$ to z. The formal series for the germ h in the charts z and $\tilde{z}$ coincide. Hence the series $\widehat{k}$ commutes with the series $\widehat{g}^1_{z^{p+1}}$. By Corollary 1.3, the first series has the form: $\widehat{k}(\tilde{z}) = g^t_{z^{p+1}}$, $t \in \mathbb{C}$. In the chart $\tilde{z}$ the germ g can be written as $\nu\tilde{z}$. Hence in the chart z it corresponds to the formal series $\widehat{k} \circ \nu\tilde{z} \circ \widehat{k}^{-1} = \nu g^\tau_{z^{p+1}}$ for some $\tau \in \mathbb{C}$ (observe that all the factors in the left-hand side belong to the group $G_s(p+1)$). This yields Lemma 2.2 for germs $g \in \{g_i\}$ that do not commute with h.

Assume now that g commutes with h. Then one can apply to g Corollary 1.3 of 1.3 thus proving Lemma 2.2. □

Theorem 2.2 on formal normal forms for nonabelian solvable groups is proved. Note the difference between the formal classifications for abelian and solvable nonabelian groups. In the former case, a vector field generating germs from these groups having identity linear parts can be formally equivalent to the field $z^{p+1}(1 + az^p)^{-1}\partial/\partial z$ with an arbitrary value of the formal classification modulus $a \in \mathbb{C}$. In the solvable nonabelian case, the formal modulus for a similar vector field necessarily vanishes: $a = 0$. This restriction follows from properties (2.4), (2.5) of the vector field $\widehat{u}$.

2.5. Analytic classification for solvable nonabelian groups. In this subsection we prove Theorem 2.3.

First assume that G is a typical solvable group. Then one of the two properties of exceptional groups is violated.

CASE 1: *the group T_G is noncyclic*. In this case the centralizer for any germ $h \in G \cap \mathscr{A}$ is a noncyclic group. Hence the germ h is embeddable, by Theorem 3 in 2.7, Paper I. Therefore there exists an *analytic* map z reducing the germ h to the form (2.2) (the formal type of the generating vector field is described in Theorem 2.2). By Lemma 2.2, the same map reduces all the germs from the group G to form (2.3).

CASE 2: *the group T_G is cyclic*. As mentioned above, $G \cap \mathscr{A}$ is a normal subgroup of the group G: $\mathrm{Ad}(G)(G \cap \mathscr{A}) = G \cap \mathscr{A}$. This relation together with (2.1) yield $T_G(1 + \Lambda_G^p) \subset T_G$. Hence $\Lambda_G^p \subset \mathbb{Z}$, provided the group T_G is cyclic. If $\Lambda_G^p = 1$, then G is an abelian group; this case is excluded by the assumption. If $\Lambda_G^p = \{-1, 1\}$, then G is an exceptional group; yet by the assumption, G is a typical group. Hence the group Λ_G possesses a nonresonant multiplier ν: $\nu^p \in \mathbb{Z}$, $|\nu| \neq 1$. Assume that $g \in G$ is a germ whose multiplier equals ν, h is an arbitrary germ from $G \cap A$. The germs g and h do not commute, since $\nu^p \neq 1$. Then there exist maps $\tilde{z}$ and w with the following properties: $\tilde{z}$ is a formal map; in this map the germs g and h can be written as $\nu \tilde{z}$ and $g^1_{z^{p+1}}$, respectively; it exists by Corollary 2.2 of 2.4; w is an analytic map; in this map the germ g is linear; the map w exists by a classical result. The transition function $\tilde{z} = \widehat{H}(w)$ is a formal series commuting with the nonresonant linear mapping of multiplication by ν. Hence this series is linear, and the map $\tilde{z}$ is analytic as well. By Lemma 2.2, $\tilde{z}$ reduces all the germs from the group G to normal form (2.3).

Theorem 2.3 for typical solvable groups is proved.

Let now G be an exceptional solvable group. First, let us describe its formal normal form.

PROPOSITION 2.5. *All solvable exceptional groups that are formally equivalent to a subgroup of the group $G_s(p)$ have one and the same formal normal form*:
$$\nu^m g^n_{z^{p+1}}, \qquad m \in \mathbb{Z} \bmod 2p, \ n \in \mathbb{Z}, \ \nu = \exp(r\pi i/p), \qquad (2.7)$$
with r and $2p$ relatively prime.

PROOF. Assume that h is a generator for the group $G \cap \mathscr{A}$ while ν is one for the group Λ_G. By the definition of an exceptional group, $\nu^p = -1$. Let $g \in G$ be a germ whose multiplier equals ν. By Theorem 2.2, there exists a formal chart $\tilde{z}$ such that formal series for h and g in this map can be written as $\nu g^a_{z^{p+1}}$ and $g^1_{z^{p+1}}$, respectively. Consider another formal map z with the transition function $\tilde{z} = g^{-a/2}_{z^{p+1}}$. This chart preserves the formal normal form for the germ h and takes the germ g to νz. The latter follows from relations (2.1):
$$\mathrm{Ad}(g^{-a/2}_{z^{p+1}})(\nu g^a_{z^{p+1}}) = g^{a/2}_{z^{p+1}} \circ \nu \circ g^{a/2}_{z^{p+1}} = \nu.$$

The formal chart z reduces the group G to the form (2.7).

Applying relations (2.1) to normalized germs g and h once more, we obtain $g \circ h = h^{-1} \circ g$. Indeed, for $\nu^p = -1$ we have

$$(\nu, 0) \times (1, 1) = (1, -1) \times (\nu, 0).$$

This proves Theorem 2.3 for solvable exceptional groups. □

REMARK. The above proof provides a motivation for the definition of exceptional groups. Namely, if G is a solvable group and T_G is a cyclic group, then the group of multipliers Λ_G possesses the following property: $\Lambda_G^p \subset \mathbb{Z}$. If $\Lambda_G \not\subset \{-1; 1\}$, then the group G possesses an analytically linearizable germ g whose multiplier is *nonresonant*. If $\Lambda_G = \{-1, 1\}$, then an analytically linearizable germ g still exists in G, yet its multiplier is resonant (equals -1 or a root of -1). In the first case, a formal series commuting with $\hat{g}$ is convergent, while in the second case this is not true in general. This fact causes a significant difference between typical and exceptional groups.

2.6. Analytic classification continued: an exceptional solvable group.

In this subsection we prove Theorem 2.4. Let g and h be the germs as in the assertion. First we describe functional invariants for the pair (g, h). These are invariants of the germ h having additional symmetries caused by existence of the germ g that anticommutes with h. Recall the construction of functional invariants for a germ h belonging to the class $\mathscr{A}_{p+1}$, which was presented in 2.5, 2.6 of Paper I; some of the notation is borrowed from there, and we do not repeat definitions. Let $\{S_j\}$ be a nice p-covering of a punctured disk with sectors, $\tilde{S}_j$ is a sector with the vertex at infinity:

$$\tilde{S}_j = \begin{cases} S^+ \times \{j\} & \text{for odd } j, \\ S^- \times \{j\} & \text{for even } j, \end{cases} \quad j = 1, \ldots, 2p,$$

$$S^+ = \{t \mid |\arg t| < \pi/2 + \varepsilon, |t| > R\},$$

$$S^- = \{t \mid |\arg t - \pi| < \pi/2 + \varepsilon, |t| > R\}.$$

The restriction t_j of the function $t = -p^{-1}z^{-p}$ to S_j defines a conformal mapping $t_j: S_j \to \tilde{S}_j$. Denote by t_j^{-1} the inverse mapping $\tilde{S}_j \to S_j$. By the sectorial normalization theorem, each sector S_j possesses a holomorphic map H_j that conjugates the germ h in a slightly smaller sector $S'_j \subset S_j$ with its formal normal form $g_{z^{p+1}}^1$, provided the radius of S_j is sufficiently small. The function t_j rectifies the vector field z^{p+1}. Therefore the chart $\tau_j = t_j \circ H_j: S'_j \to \tilde{S}_j$ conjugates the germ $h: (\mathbb{C}, 0) \to (\mathbb{C}, 0)$ with the germ of unit shift at infinity: $\tau_j \mapsto \tau_{j+1}$, $j = 1, \ldots, 2p$.

The functional modulus for the germ h is defined by "transition functions" $\tilde{\Phi}_j = \tau_{j+1} \circ t_j^{-1}$ that commute with the shift by 1. The functions $\tilde{\Phi}_j$ are defined in the intersection of $\tilde{S}_j$ with a sufficiently high upper half-plane

$\operatorname{Im} t_j > C > 0$ for j odd, and a sufficiently low lower half-plane $\operatorname{Im} t_j < -C < 0$ for j even. These transition functions are defined by relations (2.10) of Paper I uniquely up to a shift of the argument. They are connected with the modulus λ of the formal classification: a typical germ $h \in \mathscr{A}_{p+1}$ is formally equivalent to the time 1 shift along phase curves of the field $z^{p+1}(1+\lambda z^p)^{-1} \partial/\partial z$; in our case $\lambda = 0$. Relations (2.10) of Paper I now imply that

$$\Phi_j(\tau_j) = \tau_j + \psi_j(\tau_j), \qquad \psi_j(\tau_j + 1) = \psi_j(\tau_j), \qquad (2.7a)$$

$\psi_j \to 0$ for $|\operatorname{Im} \tau_j| \to \infty$ in the corresponding upper or lower half-plane.

Now consider the symmetry relations between transition functions that are caused by existence of the germ g anticommuting with h. Without loss of generality one can assume that the multiplier ν of the germ g has the least positive argument among all elements of the group Λ_G. Since $\nu^p = -1$, one has $\nu = \exp(\pi i r/p)$ with r odd. The mapping g "almost permutes" sectors S_j, by increasing the number j by $r \bmod 2p$.

Denote by $\widetilde{g}$ the mapping g written in the atlas $\{\tau_j\}$. The mapping g anticommutes with the mapping h, which in the above atlas turns into the shift by 1, denoted by $\widetilde{h}$. Let $\widetilde{g}_j$ be the restriction of $\widetilde{g}$ to $\widetilde{S}_j$. Then

$$\widetilde{g}_j(\tau_j + 1) = \widetilde{g}_j(\tau_j) - 1 \qquad (2.8)$$

and $\widetilde{g}_j(\widetilde{S}'_j) \subset \widetilde{S}_{j+r}$, where $\widetilde{S}'_j$ is slightly less than the sector $\widetilde{S}_j$. Relation (2.8) enables us to find the analytic continuation for $\widetilde{g}_j$ along any orbit for the shift by 1 (since each such orbit intersects the sector $\widetilde{S}_j$). Thus the mapping $\widetilde{g}_j$ is defined on the entire plane $\mathbb{C}$, is biholomorphic, and therefore affine. Relation (2.8) implies now that in the charts τ_{j+r}, τ_j one has

$$\tau_{j+r} = \widetilde{g}_j(\tau_j) = a_j - \tau_j.$$

Let us now describe symmetries themselves. The collection of mappings $\widetilde{g}_j$ represents one and the same mapping g in the atlas $\{\tau_j\}$ on the covering $\{\widetilde{S}_j\}$ of a punctured disk (centered at infinity). Therefore the mappings $\widetilde{g}_j$ commute with the gluing mappings $\widetilde{g}_{j+1} \circ \widetilde{\Phi}_j = \widetilde{\Phi}_{j+r} \circ \widetilde{g}_j$, or

$$a_{j+1} - \tau_j - \psi_j(\tau_j) = -\tau_j + a_j + \psi_{j+r}(a_j - \tau_j). \qquad (2.9)$$

Since both functions ψ_j and ψ_{j+r} tend to zero at infinity, one gets $a_j = a_{j+r}$. Therefore in the atlas $\{\tau_j\}$, $\tau_j = \widetilde{H}_j(z)$, one has $\tau_{j+r} = \widetilde{g}_j(\tau_j) = a - \tau_j$.

Next, the normalizing atlas $\{\tau_j\}$ is defined uniquely up to a shift, and a functional modulus for the germ h is the class of equivalent collections $\{\psi_j\}$ with the equivalence relation

$$\{\psi_j\} \sim \{\widetilde{\psi}_j\} \iff \exists c \in \mathbb{C} : \psi_j = \widetilde{\psi}_j(\mathrm{id} + c)$$

(see relation (2.11) of Paper I). The existence of the germ g anticommuting with h enables us to choose a "distinguished" value for c: the atlas $\{\tau_j\}$

should be replaced by $\{\tau_j + c\} = \{\tilde{\tau}_j\}$ in such a way that the mappings $\tilde{g}_j$ become $\tilde{g}_j: \tilde{\tau}_j \mapsto \tilde{\tau}_{j+r}(\tilde{\tau}_j) = -\tilde{\tau}_j$. To do this, it suffices to take $c = -a/2$. Finally, taking into account relations $a_j = 0$, one can rewrite the symmetry relations (2.9) in the form

$$\psi_{j+r}(-\tau_j) = \psi_j(\tau_j). \qquad (2.10)$$

Note that $2p$ and r are relatively prime. Hence the entire collection of functions $\psi_1, \ldots, \psi_{2p}$ is determined by one function (ψ_1).

The description of functional invariants for Theorem 2.4 is completed.

Coincidence of two invariants implies analytic equivalence for pairs. Indeed, in the normalizing atlas the pair (g, h) turns into the pair $(\tilde{g}, \tilde{h})$ defined by the following relations:

$$\tau_j \stackrel{\tilde{h}}{\mapsto} \tau_j + 1, \qquad \tau_j \stackrel{\tilde{g}}{\mapsto} \tau_{j+r}(\tau_j) = -\tau_j. \qquad (2.11)$$

The mappings that conjugate the above two pairs interchange points of the jth sectorial domain having equal values of normalizing coordinates.

The fact that $\tilde{g}$ and $\tilde{h}$ are well defined is guaranteed by the coincidence of the gluing mappings, which are determined uniquely provided the invariant is known. This proves equivalence for equimodal pairs.

Finally, let us prove the realization assertion from Theorem 2.4. By the realization theorem from Paper I, 2.5, any collection (2.7a), in particular, a collection with symmetries (2.10), is realized as a functional modulus for a germ h of class $\mathscr{A}_{p+1}$. As is usually done when proving realization theorems, we can glue the sectors $\tilde{S}_j$ with the help of the mappings $\tilde{\Phi}_j$ to obtain an abstract Riemann surface S. The 1-periodicity for the function ψ_j implies that the mapping $\tilde{h}$ given by (2.11) is well defined on S. The symmetry relations (2.10) imply that the same is true for the mapping $\tilde{g}$ given by (2.11) as well. As was proved in 2.6 of Paper I, there exists a conformal mapping of S to a punctured disk that conjugates $\tilde{h}$ with the germ $h \in \mathscr{A}_{p+1}$. The same mapping takes $\tilde{g}$ to the germ of a conformal mapping g, which is holomorphic at the origin by the removable singularity theorem. The mapping g possesses the following properties:

1) g anticommutes with h; this follows from the similar relation for $\tilde{g}$ and $\tilde{h}$, which is immediately implied by definition (2.11);

2) $g^{[2p]} = \mathrm{id}$ (follows from (2.11)).

Next, g "almost takes" the sector S_j to S_{j+r}. Hence $g'(0) = \exp(r\pi i/p)$. This completes the proof of Theorem 2.4, and thus our investigation of exceptional solvable groups.

2.7. Abelian groups of germs of rank 1: the analytic classification. In this subsection we prove Theorem 1.3. Since the proof is similar to arguments from 2.6, we present it only schematically.

Let g and h be the same as in the assertion of the theorem. Observe that the germ h here is the same as in 2.6, while the germs g are different. Assume that S_j, $\widetilde{S}_j$, H_j, $\tau_j = t_j \circ H_j(z)$ are the same as in 2.6. In the atlas $\{\tau_j\}$, the mapping h turns into the unit shift $\widetilde{h}\colon \tau_j \mapsto \tau_j + 1$, $j = 1, \ldots, 2p$. Denote by $\widetilde{g}$ the mapping g written in the atlas $\{\tau_j\}$. It commutes with $\widetilde{h}$; put $\widetilde{g}_j = \widetilde{g}|_{\widetilde{S}_j}$ to obtain $\widetilde{g}_j(\tau_j + 1) = \widetilde{g}_j(\tau_j) + 1$. For reasons similar to those described in 2.6 (or in [V]), one obtains that $\widetilde{g}_j$ is a shift: $\tau_j \mapsto \tau_j + c_j$. Since
$$\nu = g'(0) = \exp(2\pi i/q) = \exp(2\pi i s/p),$$
the mapping $\widetilde{g}_j$ takes the sector $\widetilde{S}_j$ to the sector $\widetilde{S}_{j+2s}$. Let Φ_j be the maps $\widehat{\Phi}_j$ modified by shifts in order to have equal constant terms, as it was done in 4.3 of Paper I. Let $\Phi_j = \mathrm{id} + \psi_j$, $\psi_j \to 0$ at infinity. Let $\widetilde{\tau}_j$ be the corresponding normalizing charts, see formula (4.3′) of Paper I. The mapping $\widetilde{g}$ respects the gluing mappings $\{\Phi_j\}$; therefore
$$\widetilde{g}_{j+1} \circ \Phi_j = \Phi_{j+2s} \circ \widetilde{g}_j.$$
We now have $g_j\colon \widetilde{\tau}_j \mapsto \widetilde{\tau}_{j+2s}(\widetilde{\tau}_j) = \widetilde{\tau}_j + c_j$, $c_j = c_{j+1}$. Therefore $\widetilde{g}_j(\tau_j) = \tau_j + c$, where c does not depend on j. Next, $\widetilde{g}^{[q]} = h^{[k]}$. Hence $c = k/q$. The symmetry relations for the components Φ_j have the following form:
$$\Phi_j + k/q = \Phi_{j+2s}(\mathrm{id} + k/q), \quad \text{or} \quad \psi_j = \psi_{j+2s} \circ (\mathrm{id} + k/q), \quad \Phi_j = \mathrm{id} + \psi_j.$$
Thus, only $2s$ functions $\psi_1, \ldots, \psi_{2s}$ from the collection $\psi_1, \ldots, \psi_{2r}$ remain independent. This completes the description of the functional modulus for the pair of germs (g, h), and thus the proof of assertion 1 of the theorem.

Equivalence for equimodal pairs is proved exactly as the similar statement from 2.6, and we do not go into details.

The realization statement is proved in a similar way as well, but the mapping $\widetilde{g}$ is now defined by
$$\tau_{j+2s} \circ \widetilde{g} = \tau_j + k/q, \quad j = 1, \ldots, 2p.$$
Theorem 1.3 is proved.

Thus §§1, 2 contain a complete formal description of abelian and solvable groups of mappings $(\mathbb{C}, 0) \to (\mathbb{C}, 0)$. The analytic description for solvable groups is complete as well; the only abelian case that is not considered is the case of a formally linearizable group with Liouvillian multipliers.

Next, let us turn to nonsolvable groups.

§3. Equivalence of the formal and analytic conjugacy for nonsolvable groups of germs of conformal mappings $(\mathbb{C}, 0) \to (\mathbb{C}, 0)$

3.1. Statement and comments. In this section we deal with the following theorem.

CMR THEOREM [CM, R]. *Assume that two nonsolvable groups of germs of conformal mappings* $(\mathbb{C}, 0) \to (\mathbb{C}, 0)$ *are formally equivalent. Then they are analytically equivalent as well, and the conjugating formal series converges.*

For the definition of formal equivalence for groups of germs, see the beginning of 3.2. There we present a proof of the CMR theorem; it is independent of the initial one and is based on the Phragmen-Lindelöf theorem for cochains.

Note that the CMR theorem remains true if one replaces the formal equivalence by the topological one.

SHCHERBAKOV'S THEOREM [S]. *Let two nonsolvable groups of germs of conformal mappings* $(\mathbb{C}, 0) \to (\mathbb{C}, 0)$ *be conformally equivalent, and the conjugating homeomorphism germ preserve orientation. Then this germ is biholomorphic, while the groups are analytically equivalent.*

The proof of the above result is lengthy and delicate; as the proof below, it relies on the sectorial normalization theorem.

The rest of the section deals with the proof of the CMR theorem.

3.2. Germs with fixed points of distinct multiplicities.
By the definition of solvability, for each nonsolvable group of germs of conformal mappings $(\mathbb{C}, 0) \to (\mathbb{C}, 0)$ its commutator is noncommutative, see Proposition 2.1 above. All the germs from this commutator have identity linear parts. Since the commutator is noncommutative, it contains at least two nonidentity germs having different multiplicities of fixed points.

Let G and $\widetilde{G}$ be two formally equivalent nonsolvable groups of germs. By the definition of the formal equivalence, this means that there exist an isomorphism $K: G \to \widetilde{G}$ and a formal series $\hat{h}$ whose constant term is zero, while the linear one is not, such that for any germ $f \in G$ one has

$$\hat{h}^{-1} \circ \hat{f} \circ \hat{h} = \widehat{Kf};$$

the hat over a symbol stands for the corresponding formal series. We emphasize that the series $\hat{h}$ remains the same for all the germs $f \in G$.

Our aim is to prove that the series $\hat{h}$ converges. To do this, take two germs from the group G:

$$f: z \mapsto z + az^{k+1} + \cdots, \qquad g: z \mapsto z + bz^{l+1} + \cdots, \qquad l > k, \ ab \neq 0.$$

The fact that such germs exist in any nonsolvable group G was proved in 2.3. Assume that $n > l + 1 > k + 1$, and let h_n be the nth partial sum of the series $\hat{h}$. Replace the group $\widetilde{G}$ by the analytically equivalent group $\widetilde{\widetilde{G}} = h_n^{-1} \circ \widetilde{G} \circ h_n$. The groups G and $\widetilde{\widetilde{G}}$ are conjugated by the formal series

$$\widehat{H} = \hat{h}_n^{-1} \circ \hat{h}, \qquad \widehat{H} - \mathrm{id} = o(z^n).$$

The following natural isomorphism is defined:

$$\widetilde{K}: G \to \widetilde{\widetilde{G}}, \qquad \widetilde{K}f = h_n^{-1} \circ Kf \circ h_n,$$

and $\widehat{H}^{-1} \circ \widehat{f} \circ \widehat{H} = \widehat{Kf}$. The series $\widehat{h}$ and $\widehat{H}$ are either both convergent, or both divergent; below we prove that the series $\widehat{H}$ is convergent. Let $\widetilde{f} = \widetilde{K}f$, $\widetilde{g} = \widetilde{K}g$. Then

$$\widetilde{f}: z \mapsto z + az^{k+1} + \cdots, \qquad \widetilde{g}: z \mapsto z + bz^{l+1} + \cdots;$$

the difference between f and $\widetilde{f}$ (g and $\widetilde{g}$) lies in the terms whose degrees are higher than $l+1$ ($n > l+1 > k+1$).

3.3. Conjugating cochains and operations with them. Let H_f, H_g, $H_{\widetilde{f}}$, $H_{\widetilde{g}}$ stand for normalizing cochains for the germs f, g, $\widetilde{f}$, $\widetilde{g}$, respectively. Namely, let there exist a linear transformation ν that takes the germs f and $\widetilde{f}$ (g and $\widetilde{g}$, respectively) to germs of the form

$$z \mapsto z + z^{k+1} + \cdots, \qquad z \mapsto z + z^{l+1} + \cdots. \tag{3.1}$$

The normalizing cochains for the latter germs were defined in 2.5, Paper I. The cochains H_f, H_g, etc., are obtained from the above cochains via conjugation by the same linear transformation ν. The normalizing cochains for the germs from (3.1) are collections of holomorphic functions defined in sectors of a nice k- or l-covering. The cochains H_f, H_g, etc., correspond to the same coverings subjected to the linear transform ν. One can choose these cochains in such a way that

$$\begin{aligned} H_f - \mathrm{id} &= o(z^{k+1}), & H_{\widetilde{f}} - \mathrm{id} &= o(z^{k+1}), \\ H_g - \mathrm{id} &= o(z^{l+1}), & H_{\widetilde{g}} - \mathrm{id} &= o(z^{l+1}). \end{aligned} \tag{3.2}$$

The proof of the CMR theorem relies on the fact that the cochain $F_1 = H_{\widetilde{f}}^{-1} \circ H_f$ conjugates the germs f and $\widetilde{f}$, while the cochain $F_2 = H_{\widetilde{g}}^{-1} \circ H_g$ conjugates the germs g and $\widetilde{g}$. To make the above statement precise, one should define operations over functional cochains. Let Θ^1 and Θ^2 be two partitions of a punctured disk into an even number of equal sectors; the numbers of sectors in the partitions may differ. The *product* $\Theta^1 \cdot \Theta^2$ of the above partitions is defined as the partition of the punctured disk into all nonvoid pairwise intersections of sectors forming these partitions. Assume that H^1 and H^2 are two functional cochains, while Θ^1 and Θ^2 are the corresponding partitions of the type described above. The *sum* $H^1 + H^2$ of these cochains is the cochain corresponding to the product $\Theta^1 \cdot \Theta^2$ and defined as follows: to any nonvoid intersection $D_1 \cap D_2$ of two domains belonging to the partitions Θ^1 and Θ^2, respectively, assign the function $f + g$ that equals the sum of functions for the collections H^1 and H^2 corresponding to the domains D_1 and D_2. Similarly one defines the difference, the product, and the superposition for cochains. Sums, differences, and products for cochains defined in one and the same punctured disk are defined in any case; a sufficient condition for the existence of the superposition $H^1 \circ H^2$ in a punctured neighborhood of the origin consists of the following: there exists an angle $\varepsilon > 0$ such that all mappings belonging to the collection H^1 can be extended from sectors of the corresponding partition that have the form

$$\arg z \in [\alpha_j, \beta_j], \qquad \alpha_j < \beta_j,$$

to those of the form

$$\arg z \in [\alpha_j - \varepsilon, \beta_j + \varepsilon];$$

the correction $H^2 - \mathrm{id}$ for the cochain H^2 decreases faster than linearly as $z \to 0$.

Now, the cochains $F_1 = H_{\tilde{f}}^{-1} \circ H_f$, $F_2 = H_{\tilde{g}}^{-1} \circ H_g$ are defined. Observe that the cochains H_f and $H_{\tilde{f}}$ conjugate the germs f and $\tilde{f}$ with one and the same embeddable germ; hence the cochain F_1 conjugates the germs f and $\tilde{f}$ with one another; similarly, F_2 conjugates the germs g and $\tilde{g}$. By construction, $F_1 - \mathrm{id} = o(z^{k+1})$, $F_2 - \mathrm{id} = o(z^{l+1})$. The mappings of the first cochain are holomorphic in sectors of a rotated nice k-covering (see 2.6 in Paper I), while those of the second, in sectors of an l-covering, yet rotated by another angle. Thus $F_1 \circ f = \tilde{f} \circ F_1$, $F_2 \circ g = \tilde{g} \circ F_2$. By the assumption of the theorem, $\widehat{H} \circ \widehat{f} = \widehat{\tilde{f}} \circ \widehat{H}$, $\widehat{H} \circ \widehat{g} = \widehat{\tilde{g}} \circ \widehat{H}$.

3.4. Outline of the proof of the CMR theorem. By the sectorial normalization theorem, the normalizing cochains constructed above (and hence the cochains F_1 and F_2 as well) can be expanded into asymptotic Taylor series as $z \to 0$. Hence

$$\widehat{F}_1 \circ \widehat{f} = \widehat{\tilde{f}} \circ \widehat{F}_1, \qquad \widehat{F}_2 \circ \widehat{g} = \widehat{\tilde{g}} \circ \widehat{F}_2.$$

Proposition 1 on the uniqueness of the conjugating formal series under an additional normalizing condition (to be proved below) yields $\widehat{F}_1 = \widehat{F}_2 = \widehat{H}$. Therefore $F_1 - F_2 = o(z^N)$ as $z \to 0$ for any $N > 0$. For the cochain $F = F_1 - F_2$, one has the following

PHRAGMEN-LINDELÖF'S THEOREM. *Assume that the difference $F = F_1 - F_2$ of the two normalizing cochains defined in 3.3 decreases faster than any power as $z \to 0$. Then F vanishes identically.*

This theorem follows from the Phragmen-Lindelöf theorem for so-called simple cochains [12]. The reduction is outlined at the end of 3.6.

Thus $F_1 \equiv F_2$. Yet the cochains F_1 and F_2 correspond to partitions having different numbers of sectors, while functions for the collections that form cochains can be extended to "wider" sectors. This implies (see Proposition 2 below) that

$$\delta F_1 = 0, \qquad \delta F_2 = 0. \tag{3.3}$$

This means that the coinciding cochains F_1 and F_2 are holomorphic functions on the punctured disk. By the removable singularity theorem, they can be extended holomorphically to the origin. The holomorphic function $F_1 = F_2$ obtained is the sum of the series $\widehat{H}$. This will conclude the proof of the CMR theorem.

Now let us turn to a detailed description.

3.5. Uniqueness for the conjugating series. The series $\widehat{F}_1$ and $\widehat{H}$ satisfy the hypothesis of the following proposition.

PROPOSITION 1. *Assume that two formal series $\widehat{F}$ and $\widehat{H}$ conjugate two formal mappings $\widehat{f}$ and $\widehat{\widetilde{f}}$ having identity linear parts, while the order of smallness for corrections of the conjugating series is higher than that for the conjugated mappings. Then the conjugating series coincide:* $\widehat{F} = \widehat{H}$.

PROOF. It suffices to prove that a formal series $\widehat{\Phi}$ that commutes with $\widehat{f}$ coincides with $z = \mathrm{id}$, provided

$$\widehat{f} = z + az^k + \cdots, \quad a \neq 0, \quad \widehat{\Phi} = z + \alpha z^n + \cdots, \quad n > k.$$

Assume the converse: let $\widehat{f} \circ \widehat{\Phi} = \widehat{\Phi} \circ \widehat{f}$, and let $\alpha \neq 0$. Then the coefficients for terms of degree at most n in the expansions of the left- and right-hand sides coincide identically; the fact that the coefficients for z^{n+1} are equal can be written as $a.n\alpha = \alpha.ka$. Since $n > k$ and $a\alpha \neq 0$, one obtains a contradiction. □

3.6. Triviality of coboundaries.

PROPOSITION 2. *Consider two functional cochains, one of them corresponding to a rotated nice k-covering, while the other to an l-covering, $l > k$. Assume that the angles of pairwise intersections of neighboring sectors for the first covering exceed π/l, and the cochains are equal. Then their coboundaries vanish, and the cochains themselves are holomorphic functions.*

PROOF. The coboundary of the cochain corresponding to a rotated nice l-covering decreases not slower than the function $m(r) = \exp(-c/r^l)$ for some $c > 0$ (see the sectorial normalization theorem in 2.5 of Paper I). By the assumption, for any two neighboring sectors S_j, S_{j+1} of the cover corresponding to the first cochain, there exists a radius belonging to the intersection of sectors corresponding to the second cochain. Since the cochains are equal, the absolute value of the coboundary for the first one can be bounded from above on this radius by the function m. However, the corresponding function ψ for the collection that forms the coboundary of the first cochain is defined in the sector $S_j \cap S_{j+1}$ whose angle exceeds π/l. The absolute value for ψ is bounded from above by the function m. By the classical Phragmen-Lindelöf theorem [Ti], this function vanishes identically. Thus the first cochain has zero coboundary; hence this cochain is holomorphic at the origin, as is the second one, which equals the first. □

Let us apply the above proposition to the proof of the CMR theorem. By the sectorial normalization theorem, the cochains H_f and $H_{\tilde{f}}$ correspond to a rotated covering of a punctured disk by sectors

$$|\arg z - \pi j/k| < \alpha, \qquad \alpha \in (\pi j/2k, \pi j/k),$$

for each α from the above interval and a sufficiently small radius of the disk (depending on α). Hence the cochains F_1 and F_2, which were defined in 3.3 and coincide by the Phragmen-Lindelöf theorem, satisfy the assumption of Proposition 2. This proves relation (3.3), and thus the CMR theorem as well. □

It remains to reduce the Phragmen-Lindelöf theorem from 3.4 to a similar theorem for simple cochains, see [I2]. We shall not reproduce the complete definition of simple cochains, which is rather lengthy. We mention only that any difference F of normalizing cochains may be transformed to a simple one by the following change of coordinates: $z \mapsto \zeta = -\ln z + i\varphi$; the real number φ depends on F. This coordinate change transforms the functions and cochains decreasing faster than any power of z as $z \to 0$ to functions and cochains decreasing faster than any exponent $\exp(-\nu\zeta)$, $\nu > 0$, as $\operatorname{Re}\zeta \to \infty$. The Phragmen-Lindelof theorem for simple cochains with this last rate of decrease implies that they are identically zero in some neighborhood of the positive semiaxis. Returning to the chart z, we see that F is identically zero on some segment with vertex $z = 0$. The classical Phragmen-Lindelöf theorem, used as in the proof of Proposition 2, now implies that $F \equiv 0$.

§4. Orbital analytic classification for germs of "nilpotent Jordan cell" type

In this section we present a complete study of relationships between the formal and the analytic classifications of germs of planar vector fields with linear parts of "nilpotent Jordan cell" type and generic nonlinearities, and provide the analytic classification for the case when it is distinct from the formal one. A significant part of the section is an account, yet schematic, of the papers due to Moussu [M] and Cerveau and Moussu [CM]; original results rely on the material of §2.

4.1. Germs from the class J_* and nice blowing ups.

DEFINITION. A germ of a vector field belongs to the *class* J_* if it can be written as

$$\dot{z} = w + \cdots, \qquad \dot{w} = az^2 + bzw + cw^2 + \cdots, \qquad a \neq 0; \qquad (4.1)$$

the dots in the first equation stand for terms of degrees at least 2, while in the second one for those of degrees at least 3.

EXAMPLE. If a field is analytic and real, a germ of class J_* defines a phase portrait on the real plane that is topologically equivalent to the family

FIGURE 1. Phase portrait of the germ of class J^*: the linear part is a nilpotent Jordan cell, and the nonlinear terms are generic.

$y^2 - x^3 = $ const, see Figure 1. One can find the corresponding calculations in various books and papers, see e.g. [ALGM, Ta]. These calculations can be carried over literally to the complex case.

In general, any germ from the class J_* possesses a complex separatrix with semicubical singularity. This can be constructed as follows.

Perform three steps of the complex σ-process (see, e.g., [AI]) to turn a germ from the class J_* into a complex direction field having three elementary singularities and defined in a neighborhood of three pasted Riemann spheres, see Figure 2d on page 90. Two of the above singular points, O_1 and O_2, are corner points: they belong to intersections of the pasted spheres. The third point, O_3, is not a corner one; it belongs exactly to one of the above spheres, namely, to the one intersecting both others; this sphere will be called *central*.

All the above singular points are resonant saddles, and the ratios for eigenvalues equal $-1/2$, $-1/3$, $-1/6$ for O_1, O_2, O_3, respectively (to find these ratios, almost no calculations are needed, as it is shown in 5.2). Separatrices for the corner singular points belong to the pasted spheres. Being shrunk (subjected to the inverse of blowing up), they disappear into the initial singular point. One of the separatrices for the singular point O_3 belongs to the central pasted sphere, while the other one does not; the latter is taken by shrinking to the unique separatrix of the initial vector field. This separatrix has a semicubical singularity.

All the above is verified by elementary, though rather lengthy, calculations: three steps of the σ-process, linearizing the field at the singular points obtained while blowing up, and studies of the restriction of shrinking to any curve transversal to the central sphere (this is the only significant property of the separatrix for the saddle O_3). It turns out that the image of this restriction, that is, the separatrix for the initial germ, possesses a semicubical singularity. Observe that all the above results rely on very scarce information about the initial field: nothing is known, except equation (4.1). Terms of higher orders, and the values for the coefficients b and c, do not influence the described situation.

4.2. The vanishing holonomy group for germs from the class J_*. The above calculations show, in particular, that for one of the corner singular points, the multiplier ε_2 for the monodromy transformation along a loop belonging to the central sphere equals -1, while for the other it is $\varepsilon_3 = \exp(2\pi i/3)$. A natural question arises: which germs f and g with multipliers ε_2 and ε_3, respectively, can be realized as monodromy transformations for the corner singular points O_1 and O_2? An ingenious purely topological argument, due to Moussu, implies a very simple answer for this question: only those formally equivalent to linear ones. In other words,

$$f^{[2]} = g^{[3]} = \mathrm{id}. \tag{4.2}$$

The thing is that each of noncentral spheres, minus the corresponding corner singular point, is a leaf of the blown up foliation, and hence simply connected. Therefore, for any circuit around the corner singular point on the punctured sphere, the corresponding monodromy transformation is the identity. Hence the vector field itself is formally orbitally equivalent to a linear one. Otherwise its formal normal form would be defined by equation (4.11) of Paper I, and none of the monodromy transformations would be the identity. This proves relation (4.2).

THEOREM 4.1 [M]. 1. *Invariance. The above-described germs f and g satisfying* (4.2), *generating the vanishing holonomy group for a germ of the class J_*, constitute an invariant of the orbital analytic classification for such germs.*

2. *Equimodality and equivalence. Two germs from the class J_* having equivalent pairs of generators* (4.2) *are orbitally analytically equivalent.*

3. *Realization. Each pair of germs* (4.2) *can be realized as a pair of generators for the vanishing holonomy group of a germ of class J_*.*

PROOF. The first assertion of the theorem is a corollary of the general fact that monodromy transformations are invariant. The second one requires certain calculations; these are carried out in paper [M], and we do not present them here. Observe merely that they rely on the specific simplicity of germs of class J_*; in the general case, "vanishing holonomy" transformations do not define the analytic type of a foliation (Voronin's hypothesis). The third assertion is proved in a more general context in [CM]; a related result is discussed in §5.

We do not present the complete proof for Theorem 4.1, restricting ourselves to the previous comments.

A mapping f whose square is the identity is usually called an involution. By analogy, we shall call the germ of a mapping $g: (\mathbb{C}, 0) \to (\mathbb{C}, 0)$ whose cube is the identity a *trivolution*.

4.3. Groups generated by an involution and a trivolution. In this section the term "solvable" is taken in its usual sense: an abelian group is regarded as a particular case of a solvable one.

THEOREM 4.2. *A group G of germs of mappings $(\mathbb{C}, 0) \to (\mathbb{C}, 0)$ generated by an involution f and a trivolution g is solvable iff $(f \circ g)^6 = \mathrm{id}$.*

PROOF. 1. Let $(f \circ g)^6 = \mathrm{id}$, and let us prove that the group G is solvable. Apply the solvability criterion given by Theorem 2.1. Let $h = [f, g] = f^{-1} \circ g^{-1} \circ f \circ g$. If $h = \mathrm{id}$, then G is an abelian group, and the assertion is proved. Now let $h \neq \mathrm{id}$. Evidently, $h \in A_{p+1}$ for some p. Let us check that condition A of Theorem 2.1 holds for the generators $\{f, g\} = \{g_i\}$ and the germ h.

First we verify condition A1. The germs f and g are formally, and even analytically, equivalent to their linear parts, as are all germs whose power is the identity. Therefore condition A1 holds.

Condition A2 must be verified for two pairs: f, h and g, h. One has
$$h_1 = \mathrm{Ad}(f)h = g^{-1} \circ f \circ g \circ f,$$
since $f^{[2]} = \mathrm{id}$. Let us prove the identity $h_1 \circ h = h \circ h_1$; we have
$$(g^{-1} \circ f \circ g \circ f) \circ (f^{-1} \circ g^{-1} \circ f \circ g) = \mathrm{id}, \quad (f^{-1} \circ g^{-1} \circ f \circ g) \circ (g^{-1} \circ f \circ g \circ f) = \mathrm{id}.$$

Next, $h_2 = \mathrm{Ad}(g)h = g^{-1} \circ f^{-1} \circ g^{-1} \circ f \circ g^{[2]}$. Let us prove the identity $h_2 \circ h \circ h_2^{-1} \circ h^{-1} = \mathrm{id}$; we have
$$(g^{-1} \circ f^{-1} \circ g^{-1} \circ f \circ g^{[2]}) \circ (f^{-1} \circ g^{-1} \circ f \circ g)$$
$$\circ (g^{[-2]} \circ f^{-1} \circ g \circ f \circ g) \circ (g^{-1} \circ f^{-1} \circ g \circ f) = (g^{-1} \circ f^{-1})^{[6]} = \mathrm{id}.$$

The first equality above rests on the relations $g^{[2]} = g^{-1}$, $f = f^{-1}$, while the second follows from the assumption. Thus, by Theorem 2.1S, G is a solvable group.

2. Let G be a solvable group. Assume that h, h_1, h_2 are the same as above. By Theorem 2.1N, one has $h_2 \circ h \circ h_2^{-1} \circ h^{-1} = \mathrm{id}$. The above calculation shows that $(f \circ g)^{[6]} = \mathrm{id}$.

Hence the vanishing holonomy group is solvable iff the mapping $f \circ g$ is formally equivalent to a linear one (its sixth power is the identity). For this group, Λ_G is the group of roots of unity of degree 6. The germ h belongs to the group A_{p+1} for some p. If $p \neq 3(2k+1)$ then G is a typical solvable group, since it violates the second requirement of the definition of exceptional groups. In this case G is analytically equivalent to its formal normal form as described in Theorem 2.2. Below we derive from this fact an equivalence theorem for formal and analytic equivalences of germs from the class J_*; for the statement of the criterion, see the next subsection.

4.4. The formal and the analytic classifications for germs from the class J_*. From now on "solvable" means again "nonabelian solvable".

THEOREM 4.3. *Any two germs of class J_* that satisfy condition* B *and are formally equivalent, are analytically equivalent as well.*

This is a generalization of the theorem due to Moussu and Cerveau [CM].

PROOF. First observe that the formal type for a vanishing holonomy group, in particular, for its generators f and g, is defined completely by that of germ of the field $v \in J_*$. Indeed, to define the k-jet for the monodromy transformations of f, g, it suffices to know the field $\tilde{v}$ (obtained from the field v by a nice blowing up) up to a remainder term whose order is greater than the kth power of the distance to the central sphere. In other words, it suffices to know the k-jet for the blown up vector field on the central sphere. On the other hand, for any k there exists an $N(k)$ such that the k-jet of the blown up field on pasted spheres is defined by the $N(k)$-jet of the initial field. This proves that the k-jet for the germs f, g is defined by the $N(k)$-jet for the initial germ v.

REMARK. A likely estimate of the number $N(k)$ in the real case allows us to obtain an upper bound on the order of the topologically sufficient jet for a nice blowing up of an arbitrary nonmonodromic singular point of finite multiplicity; see [K].

Now we shall consider three cases:

(1) $(f \circ g)^6 \neq \mathrm{id}$;
(2) $(f \circ g)^6 = \mathrm{id}$, $[f, g] \in \mathscr{A}_{p+1}$, $p \neq 0 \bmod 3$;
(3) $(f \circ g)^6 = \mathrm{id}$, $[f, g] \in \mathscr{A}_{p+1}$, $p = 0 \bmod 3$.

CASE 1: $(f \circ g)^6 \neq \mathrm{id}$. Then the vanishing holonomy group G is nonsolvable. As was proved above, the formal equivalence for germs v and w from the class J_* implies the formal equivalence for the corresponding vanishing holonomy groups G and $\tilde{G}$. By the CMR theorem stated in 3.1, the formal equivalence for the groups G and $\tilde{G}$ yields their analytic equivalence. By Theorem 4.1, this implies the analytic equivalence of the germs v and w.

CASE 2: $(f \circ g)^6 = \mathrm{id}$, $[f, g] \in \mathscr{A}_{p+1}$, $p \neq 0 \bmod 3$. Then G is a typical solvable group. By Theorem 2.2, the formal equivalence for such groups implies their analytic equivalence. The rest of the proof is the same as in the previous case.

CASE 3: $(f \circ g)^6 = \mathrm{id}$, $[f, g] \in \mathscr{A}_{p+1}$, $p = 0 \bmod 3$. We shall prove that in this case the group G is abelian. Suppose that the converse is true. Then the group G is solvable, and, by Theorem 2.2, there exists a formal chart z such that f and g take the form

$$\widehat{f}(z) = -g^1_{az^{p+1}}, \qquad \widehat{g}(z) = \lambda g^1_{bz^{p+1}}, \qquad \lambda^3 = 1.$$

The transformation $z \mapsto g^1_{cz^{p+1}}$ takes $\widehat{g}$ to a linear germ; therefore we can assume without loss of generality that $b = 0$ and $\widehat{g}(z) = \lambda(z)$. But the germs $\widehat{f}$ and λz commute for $p = 0 \bmod 3$, see (2.1). This contradiction proves the theorem. □

REMARK. The proof of Theorem 4.3 gives a complete list of invariants for the orbital analytic classification of the germs of class J^* with the trivial holonomy group. This is the set of pairs (f, g) that are either linearizable or analytically equivalent to a pair

$$f: z \mapsto -g^1_{z^{p+1}}, \qquad g: z \mapsto \lambda z, \qquad \lambda = e^{2\pi i/3},$$

where $p \not\equiv 1 \mod 2$, $p \not\equiv 0 \mod 3$. Note that the germ $-g^1_{z^{p+1}}$ is an involution iff p is even. Thus the set of pairwise nonequivalent germs of class J_* with trivial holonomy is countable. This answers the question of Moussu mentioned in the Introduction.

The reduction of the analytic classification of germs of class J_* to the formal one is completed.

4.5. Topological classification of germs of vector fields with nilpotent linear part and generic nonlinear terms. Up to now, we dealt only with the formal and the analytic equivalence relations. In this subsection we study the topological classification of germs of vector fields with nilpotent linear part at the singular point.

Two vector fields defined in different neighborhoods of the singular point are called *orbitally topologically equivalent* if there exists an orientation-preserving homeomorphism of these neighborhoods that takes the phase curves of the first equation to those of the second and leaves the singular point fixed.

According to the results presented in Paper I, in general, the formal, the topological, and the analytic classifications of elementary resonant singular points of vector fields on the complex plane are not related to each other. It turns out that for vector fields with nilpotent linear part and generic nonlinear terms all these classifications coincide.

THEOREM 4.3. *Two germs of vector fields from the class J_* are orbitally topologically equivalent if and only if they are analytically equivalent.*

This result is a simple corollary of the topological and analytic classifications of holonomy groups generated by an involution and a trivolution.

THEOREM 4.4. *Topological equivalence of holonomy groups G and $\widetilde{G}$ generated by an involution and a trivolution implies their analytic equivalence.*

PROOF. Assume that $G = \{g, f\}$ and $\widetilde{G} = \{\widetilde{f}, \widetilde{g}\}$ are holonomy groups generated by involutions and trivolutions: $g^3 = \widetilde{g}^3 = f^2 = \widetilde{f}^2 = \text{id}$, $g'(0) = \widetilde{g}'(0) = \varepsilon$, $\varepsilon^3 = 1$. To prove Theorem 4.4, we consider the cases when G is nonsolvable, solvable, and abelian. In the latter two cases, we present representatives for all the orbits of the topological classification explicitly.

1. If G is a nonsolvable group, then the assertion of Theorem 4.4 is a particular case of the main result of the paper [S]. Moreover, the following considerably stronger proposition holds: if G is a nonsolvable group, then any germ of an orientation-preserving homeomorphism H that conjugates G and $\widetilde{G}$ is the germ of a holomorphism.

2. Assume that both G and $\widetilde{G}$ are solvable. Then, by virtue of 4.4, they are typical, and hence analytically equivalent to their formal normal form (see Theorem 2.3). Moreover, there exist analytic changes of coordinates such that one of them takes the germs f and g to $-z$ and $\varepsilon g^1_{z^{p+1}}$, respectively, while the other one takes $\widetilde{f}$ and $\widetilde{g}$ to $-w$ and $\varepsilon g^1_{w^{q+1}}$. The germs g and $\widetilde{g}$ are topologically equivalent. Hence $p = q$, and the normal forms for these germs coincide. Therefore the pairs (f, g) and $(\widetilde{f}, \widetilde{g})$ are analytically equivalent.

3. If both G and $\widetilde{G}$ are abelian, then they are analytically equivalent to the group of roots of unity of degree 6. □

4.6. Formal classification problems. We conclude this section with several problems.

(1) It is known (see [AI, Ta]) that a germ from the class J_* is formally equivalent to the germ $\widehat{F}_1(x)\partial/\partial x + (y + \widehat{F}_2(x))\partial/\partial y$, where $\widehat{F}_1$, $\widehat{F}_2$ are formal series. Are the normalizing changes convergent?

REMARK. If the answer is "yes", then the series $\widehat{F}_1$ and $\widehat{F}_2$ are convergent as well.

(2) If the answer to the previous question is "no", then one should find which formal series $\widehat{F}_1$ and $\widehat{F}_2$ can be realized via normal forms of analytic germs from the class J_*.

(3) It is not difficult to find the normal form for a nonsolvable group with a distinguished germ of class $\mathscr{A}$, provided the exact description of the corresponding moduli space is not required. Any germ with identity linear term defines a formal normalizing chart almost uniquely. The formal series corresponding to generators of the group written in this chart define the formal classification modulus (more precisely, one should speak of a class of equivalent formal series corresponding to different choices of a normalizing chart). The question is: what normalizing series can appear as a result of this construction?

§5. Local invariants for foliations of the complex plane into analytic curves

In this section we investigate foliations of the complex plane in a neighborhood of a singular point having a finite number of complex separatrices; these foliations will be referred to as *nondicritic*. Recall that a leaf is called a *complex separatrix* if it remains analytic after the singular point is added to it. For example, the foliation defined by the equation $z\,dw - \lambda w\,dz = 0$ has infinitely many separatrices for positive rational λ, and exactly two separatrices (they belong to the coordinate axes) otherwise.

We define vanishing holonomy transformations for nondicritic foliations and describe natural requirements on these transformations. The main result of the section is a realization theorem. It states that any collection of germs of holomorphic maps satisfying natural requirements can be realized as a collection of vanishing holonomy transformations.

To give a precise formulation, we must define the notion of nice blowing up.

5.1. Nice blowing ups of nondicritic foliations. Let us list some well-known facts concerning nice blowing ups of singular points for nondicritic foliations. The σ-process performed at such a singular point replaces it by a pasted Riemann sphere. The foliation defined in a punctured neighborhood of the singular point is lifted to a neighborhood of the pasted sphere having a finite number of deleted points; these points belong to the sphere and are singular for the lifted foliation. The pasted punctured sphere is an integral curve for the lifted foliation. The first step of the blowing up process is thus completed.

If the direction field obtained has nonelementary singular points, one must perform new σ-processes at these points. The simultaneous performance of σ-processes at all the nonelementary singular points obtained at step k of the blowing up process is called the $(k+1)$th step.

The Bendikson-Seidenberg-Dumortier theorem states that in a finite number of blowing up steps, any isolated singular point of a nondicritic holomorphic direction field can be split to a finite number of elementary ones; moreover, the following claims are true.

1. All the singular points of the newly obtained direction field have exactly two separatrices.

2. All the pasted spheres with deleted singular points are integral curves of the direction field obtained while blowing up. Intersection points of pasted spheres are singular points of the above foliation (they are called *corner points*; all the other singular points of the foliation obtained are said to be *noncorner points*).

3. At any singular point, the ratio of eigenvalues is not a positive rational number.

The blowing up as described above is said to be *nice*.

5.2. Basic collections of vanishing holonomy transformations. Natural requirements. Consider a nondicritic nice blowing up of an analytic vector field. Let F be the corresponding foliation in a neighborhood of the union of the pasted spheres. Single out a basis point for each of the pasted spheres with deleted singular points. On each sphere, choose numbered loops, beginning at the corresponding basis point, that possess the following properties: exactly one singular point is contained inside each loop, and the product of all the loops in ascending order is contractible.

For each singular point α of the foliation F and each sphere S containing α, define the residue

$$\lambda_\alpha = \operatorname{res}(F, \alpha, S). \tag{5.1}$$

Choose local coordinates (z, w) such that $(z, w)(\alpha) = (0, 0)$, $w|_S = 0$ in a neighborhood of α. Then the foliation F in the above neighborhood is defined by the equation $dw/dz = wf(z, w)$, where f is a meromorphic

function such that $1/f$ does not vanish identically along the line $w = 0$. Put
$$\text{res}(F, \alpha, S) = \text{res}_0(f|_{w=0}).$$

CAMACHO-SAD'S THEOREM [CS]. *The sum of residues taken over all the singular points belonging to the sphere S equals the self-intersection number of the sphere in the two-dimensional surface where the foliation is defined.*

We denote the self-intersection number by i_S:
$$\sum_j \text{res}(F, \alpha_j, S) = i_S. \tag{5.2}$$

The sum is taken over all the singular points of the foliation belonging to S.

EXAMPLES. 1. Let α be a nondegenerate singular point of a nice blowing up F, let λ_1, λ_2 be its eigenvalues, and let λ_1 correspond to the eigenvector of the linearization that is tangent to S at α. Then
$$\text{res}(F, \alpha, S) = \lambda_2/\lambda_1. \tag{5.3}$$

The ratio λ_2/λ_1 is called the *S-characteristic number* of the singular point α.

2. Let α be a degenerate elementary singular point (in short, a complex saddle-node) of the foliation F belonging to the sphere S. Let the corresponding equations at α have the following formal orbital normal form:
$$\dot{z} = z^{k+1}(1 + az^k)^{-1}, \qquad \dot{w} = w.$$

There are two possibilities: either the sphere S is tangent to the kernel of the linearization, or it is not. In the first case
$$\text{res}(F, \alpha, S) = 0, \tag{5.4}$$

while in the second case
$$\text{res}(F, \alpha, S) = a. \tag{5.5}$$

The multiplier ν of the monodromy transformation f corresponding to a simple loop on S going around α in the counterclockwise direction has the following form:
$$\nu = \exp(2\pi i \lambda), \tag{5.6}$$
$$\lambda = \text{res}(F, \alpha, S), \qquad \nu = f'(0). \tag{5.7}$$

3. As an application of the Camacho-Sad theorem, let us show how to find geometrically the characteristic numbers for saddles O_1, O_2, O_3 obtained by a nice blowing up of a germ of class J_*; see §4. The blowing up process has three steps. The circles in Figure 2b, c, d (next page) stand for spheres obtained on the 1st, 2nd, and 3rd steps, respectively; the small circles in Figure 2a, b, c surround the points to be blown up on the current step. Near each sphere, we have written its self-intersection number in a surface obtained by blowing up a neighborhood of the origin. By the Camacho-Sad theorem, the point O_j, $j = 1, 2$, has the following S_j-characteristic number

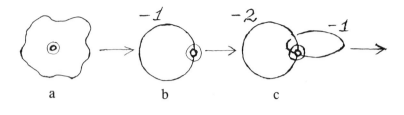

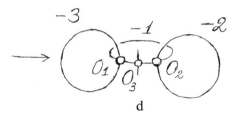

FIGURE 2. Nice blowing up of the germ of class J^*. The self-intersection number is indicated near each sphere.

λ_j: $\lambda_1 = -3$, $\lambda_2 = -2$. Hence the S_3-characteristic numbers for the points O_1, O_2, O_3 equal $-1/3$, $-1/2$, $-1/6$, as was stated before.

DEFINITION. To each singular point α of the nice blowing up F and to the sphere S containing α assign the pair $(f_\alpha, \lambda_\alpha)$, where f_α is the monodromy transformation for the foliation F corresponding to the loop on S chosen above that goes around α, while λ_α is the residue defined by (5.1):

$$\lambda_\alpha = \operatorname{res}(F, \alpha, S).$$

Such a pair is said to be *distinguished*. The collection of all distinguished pairs is called the *basic collection of vanishing holonomy transformations*.

REMARK. Pairs in the basic collection are not independent, as it can be seen from requirement 4 below and the remark at the end of 5.3.

The collection defined above satisfies the following requirements.

1. *The product of all the numbered monodromy transformations corresponding to all the loops (chosen as above) lying on the same sphere is the identity.*

2. *Relations* (5.2)–(5.7) *are satisfied*.

3. *If α is a degenerate singular point and the sphere S is not tangent to the kernel of the linearization at α, then the corresponding distinguished pair is of the form $(f, 0)$, $f'(0) = 1$.*

The modulus μ_f of analytic classification of the germ f is at the same time the modulus of orbital analytic classification of the germ of the foliation

F at α (see 3.4 in Paper I). If α is a corner point, then the second pasted sphere contains its second separatrix. Hence α has a holomorphic central manifold. Therefore, by Theorem 1 of Paper I, 3.5, the collection (3.11) of Paper I takes the form

$$\mu_f = (\varphi_0^+, \mathrm{id}+a_0, \ldots, \varphi_{p-1}^+, \mathrm{id}+a_{p-1}), \qquad a_j = 0. \qquad (5.8)$$

Thus we have two more natural requirements on the elements of a basic collection.

4. *If α is a corner singular point, then the two corresponding distinguished pairs are related to each other, being defined by the two distinct separatrices of the same germ.*

5. *If the residue of a nondegenerate singular point (that is, the ratio of its eigenvalues) is positive and distinct from n and $1/n$ for any natural n (this assumption is included in the definition of a nice blowing up), then f is analytically equivalent to a linear mapping, even when f is a resonant or a Liouvillian germ.*

This fact follows from the Poincaré theorem that states that the corresponding germ of the vector field is analytically equivalent to a linear one.

In what follows, we refer to properties 1–5 as to *natural requirements*.

5.3. Main result: realization theorem for distinguished collections. Before stating the main result, let us describe an abstract vanishing collection. Assume the following data are given:

1. A sequence of σ-processes that blow up a point O on the plane $\mathbb{C}^2$; each blowing up step is performed at a finite number of points; all of these points lie on the curve pasted on the previous step. The pasted curve obtained on the last step is denoted by S_*, while its neighborhood obtained from a punctured neighborhood of O is denoted by M.

2. Each of the pasted spheres S_j possesses a self-intersection number (in the surface M) that will be denoted by ν_j.

3. A finite number of points is deleted from each of the pasted spheres, two corner points among them. One of the labels "degenerate" or "nondegenerate" is assigned to each of these points. One of the labels "tangent to the kernel" or "not tangent to the kernel" is assigned to the germ of the pasted sphere at each deleted point labeled as "degenerate".

4. Loops are chosen on each pasted sphere, as in the beginning of 5.2.

5. To each loop there corresponds a pair (f, λ), where f is the germ of a complex mapping $(\mathbb{C}, 0) \to (\mathbb{C}, 0)$, λ is a complex number. These pairs satisfy the natural requirements of 5.2.

The set of all such pairs is said to be an *abstract distinguished collection*.

THEOREM. *Each abstract distinguished collection can be realized as a basic collection of vanishing holonomy transformations.*

The proof of the theorem is contained in 5.4–5.7 below.

5.4. Realization of germs.
In this subsection we construct the germs of vector fields at singular points having prescribed monodromy transformations.

PROPOSITION 1. *Let λ be a nonpositive number. Then for any distinguished pair (f, λ) satisfying conditions (5.6), (5.7) there exists a germ of a nondegenerate vector field such that λ and f are the residue and the monodromy transformation corresponding to the same separatrix.*

PROOF. The proof consists in references to theorems chosen according to the value of λ.

CASE 1: $\operatorname{Im}\lambda = 0$. Then, by (5.6), (5.7), one has $|f'(0)| \neq 1$. Hence the germ f is holomorphically equivalent to a linear one. Therefore it can be realized as a transformation of the germ that is linear in some suitable chart and has the eigenvalue ratio $\lambda_2/\lambda_1 = \lambda$.

CASE 2: $\operatorname{Im}\lambda = 0$, $\lambda < 0$, λ is rational. In this case Proposition 1 coincides with the Martinet-Ramis realization theorem, see §4 of Paper I.

CASE 3: $\operatorname{Im}\lambda = 0$, $\lambda < 0$, λ is irrational. In this case Proposition 1 coincides with the Yoccoz–Perez-Marco realization theorem, see §6 below. □

REMARK. The latter theorem is nontrivial only for Liouvillian λ. For Diophantine λ it is proved in the same way as in Case 1.

PROPOSITION 2. A. *For any germ $f\colon (\mathbb{C}, 0) \to (\mathbb{C}, 0)$ such that its linear part is the identity, its fixed point is of multiplicity $p + 1$ and its functional modulus equals*

$$\mu_f = [(\varphi_1^+, \operatorname{id}+a_1, \varphi_2^+, \operatorname{id}+a_2, \ldots, \varphi_p^+, \operatorname{id}+a_p)], \quad (5.8')$$
$$\varphi_j\colon (\mathbb{C}, 0) \to (\mathbb{C}, 0), \quad \varphi_1'(0) = \cdots = \varphi_{p-1}'(0) = 1, \quad a_j \in \mathbb{C},$$

there exists a complex saddle-node such that its monodromy transformation corresponding to the hyperbolic invariant manifold (a holomorphic invariant curve that is not tangent to the kernel of the linearization) equals f. If

$$a_0 = a_1 = \cdots = a_{p-1} = 0, \quad (5.9)$$

then the above saddle-node possesses a holomorphic central manifold (a holomorphic invariant curve that is tangent to the kernel of the linearization at the singular point).

B. *For any germ $f\colon (\mathbb{C}, 0) \to (\mathbb{C}, 0)$ there exists a complex saddle-node possessing a holomorphic central manifold with residue λ and monodromy transformation f, provided restrictions (5.6), (5.7) are satisfied.*

PROOF. A. By the Martinet-Ramis realization theorem, see §3 of Paper I, the collection (5.8) is realized as the functional modulus of the germ of a complex saddle-node. For this germ, the monodromy transformation of the hyperbolic invariant manifold possesses the modulus (5.8), hence it is analytically equivalent to f (ibid.). Equalities (5.9) are necessary and sufficient for the existence of a holomorphic central manifold for a complex saddle-node.

B. For $f = \mathrm{id}$ the proposition is trivial. Suppose that $f \neq \mathrm{id}$ and that the multiplicity of the fixed point 0 of the germ f equals $p + 1$. Consider a complex saddle-node such that its functional modulus (5.8) satisfies the following restrictions:

$$\varphi_1 = \cdots = \varphi_{p-1} = \mathrm{id}, \qquad a_0 = a_1 = \cdots = a_{p-1} = 0, \qquad (5.10)$$

while the normal form is given by

$$\dot{z} = z^{p+1}(1 + \lambda z^p)^{-1}, \qquad \dot{w} = -w. \qquad (5.11)$$

It possesses a holomorphic central manifold with residue λ and monodromy transformation

$$\Delta = \varphi_0, \qquad \varphi_0'(0) = \nu = \exp(-2\pi i \lambda).$$

This fact follows from relation (3.19) of Paper I. Put $\Delta = f$ to obtain

$$\varphi_0 = f. \qquad (5.12)$$

By the realization theorem (see §3 of Paper I), there exists a germ whose formal normal form and functional modulus are given by (5.11) and (5.8'), (5.10), (5.12), respectively. □

REMARK. As it was mentioned in 5.2, distinguished pairs of a basic collection are not independent. Moreover, one can always choose one pair out of the two pairs corresponding to distinct separatrices of the same elementary singular point so that the other pair can be reconstructed uniquely (up to replacement of the monodromy transformation by an analytically equivalent one).

First, let us prove this fact for a nondegenerate singular point. Each of the distinguished pairs defines the formal type of the singular point uniquely. In the resonant case this fact follows from relations (4.12), (4.13) of Paper I, while in the nonresonant case it is evident. If the residue of the singular point is nonnegative, then the corresponding germ can be linearized, and the two monodromy transformations are both analytically equivalent to linear ones; the multipliers of the latter transformations are defined by the residues. If the residue of the singular point is negative, the formal equivalence of the germs of the vector fields and the analytic equivalence of the monodromy transformations corresponding to the same separatrix imply the orbital analytic equivalence of the above germs [EI]. Hence the monodromy transformations for the germs that correspond to the second separatrix are analytically equivalent as well.

Now, consider a degenerate elementary singular point. In this case, the functional modulus of the monodromy transformation corresponding to a hyperbolic invariant manifold is also the functional modulus of the corresponding germ of the vector field in a complex saddle-node. Hence the monodromy transformation that corresponds to a holomorphic central manifold (if any) is defined uniquely, up to analytic equivalence, by the other monodromy transformation.

The remaining part of the proof of the theorem can be outlined as follows. One finds punctured spheres that contain distinguished points and a holomorphic foliation in a neighborhood of these spheres so that the foliation has no singularities and has a prescribed monodromy. The holes in the spheres are pasted together from certain domains; a foliation having a singular point with the prescribed monodromy is defined on each of these domains (such foliations are constructed in this subsection). Next, the neighborhoods of the spheres are glued together into one manifold to obtain a neighborhood of the pasted curve, see 5.3. Finally, the obtained neighborhood must be shrunk.

Let us describe the process in details.

5.5. Foliation in a neighborhood of a punctured sphere. Suppose that S is a Riemann sphere, α_i are the distinguished points on S ($i = 1, \ldots, m$), γ_i are the loops that surround these points and have the same endpoint (these loops are described in 5.2), f_i are the germs of conformal mappings $(\mathbb{C}, 0) \to (\mathbb{C}, 0)$ corresponding to the loops γ_i, $f_m \circ \cdots \circ f_1 = \mathrm{id}$. Let D_i be disjoint disks centered at α_i that do not intersect the loops γ_i. Put $\Omega = S \setminus \bigcup D_i$. Let $\widetilde{\Omega}$ be the universal covering of Ω and a be the basis point. Denote by T_γ the covering map $\widetilde{\Omega} \to \widetilde{\Omega}$ corresponding to a loop γ beginning at a. Let $\gamma' \subset \Omega$ be another loop beginning at a. Then $T_{\gamma\gamma'} = T_\gamma \circ T_{\gamma'}$.

Consider the manifold $\widetilde{M} = \widetilde{\Omega} \times \mathbb{C}$ and the group of germs of biholomorphic transformations

$$F_\gamma : (\widetilde{M}, \varphi) \to (\widetilde{M}, \varphi), \qquad (\zeta, \omega) \mapsto (T_\gamma \zeta, f_\gamma^{-1} \omega)$$

on the curve $\varphi = \widetilde{\Omega} \times \{0\}$. This is indeed a group, since

$$F_{\gamma\gamma'} = (T_{\gamma\gamma'}, f_{\gamma\gamma'}^{-1}) = (T_\gamma \circ T_{\gamma'}, f_\gamma^{-1} \circ f_{\gamma'}^{-1}) = F_\gamma \circ F_{\gamma'}.$$

The map $\gamma \mapsto F_\gamma$ is a homomorphism of the fundamental group $\pi_1(\Omega, a)$ into the group of germs $\{F_\gamma\}$. The required manifold is obtained by the "factorization" of a neighborhood of the curve φ in the manifold $\widetilde{M}$ by the action of the group $\{F_\gamma\}$. The word "factorization" is taken in quotation marks since, in general, there is no neighborhood of the curve φ such that the representatives of all the germs F_γ are biholomorphic in this neighborhood.

The exact construction follows below. The closure of the fundamental domain for the group of covering transformations of $\widetilde{\Omega}$ is compact; let it be a curvilinear polygon Φ. A sufficiently small neighborhood U of the polygon Φ in $\widetilde{M}$ turns into a manifold M when the points (ζ, ω) and $F_\gamma(\zeta, \omega)$ that belong to U simultaneously (such points exist only for a finite collection of loops γ) are glued together. This M is exactly the required manifold.

The natural projection $\pi\colon U \to M$ can be extended to a locally biholomorphic mapping $\pi\colon \widetilde{U} \to M$ of the neighborhood $\widetilde{U}$ of the curve φ in $\widetilde{M}$ on M.

A foliation $\mathscr{F}_0$ is defined on the manifold $\widetilde{U}$; its leaves are the "horizontal" curves $(\widetilde{\Omega} \times \{\omega\}) \cap \widetilde{U}$. The curve φ is among the leaves as well. The projection π takes the foliation $\mathscr{F}_0$ to the foliation $\mathscr{F} = \pi_* \mathscr{F}_0$ of the manifold U whose leaves are analytic curves.

PROPOSITION 3. *The foliation constructed above has the prescribed monodromy.*

PROOF. This fact follows easily from the construction. The proof goes as follows.

Let us start with the construction that determines the monodromy. Let $\widetilde{\pi}$ be the natural projection $\widetilde{M} \to \widetilde{\Omega}$ along the second factor. The group $\{F_\gamma\}$ commutes with the above projection. Therefore, a projection $\pi_0\colon U \to \Omega$ such that $\pi_0 \circ \pi = \pi \circ \widetilde{\pi}$ is defined.

Put $\widetilde{\Gamma} = \widetilde{\pi}^{-1} a$, $\Gamma = \pi_0^{-1} a$ (the basis point a on Ω is identified with the basis point on $\widetilde{\Omega}$). Let ω be the natural coordinate on $\widetilde{\Gamma}$. The projection $\pi\colon (\widetilde{\Gamma}, a) \to (\Gamma, a)$ induces a map on Γ, defined by ω as well. The extension of the leaves of the foliation $\mathscr{F}$ to the curves $\gamma \subset \Omega$ can be achieved with the help of the projection π_0.

Suppose that $\gamma \subset \Omega$ is a loop beginning at a, Γ is the transversal to the leaves of the foliation $\mathscr{F}$ defined above. The monodromy transformation $f_\gamma\colon (\Gamma, a) \to (\Gamma, a)$ takes the point (a, ω) to the point $(a, \Delta_\gamma(\omega))$ obtained by the extension of the leaf of $\mathscr{F}$ beginning at (a, ω) to the curve γ (hence the notation $\Delta_\gamma(\omega)$). Denote by γ_ω the lifting of the curve γ beginning at (a, ω) to the leaf.

The lifting of the curve γ_ω to $\widetilde{M}$ is a curve $\widetilde{\gamma}$ beginning at (a, ω) and ending at (T_γ, ω). This means that $\pi(T_\gamma, \omega) = (a, \Delta_\gamma(\omega))$. On the other hand, the projection π glues together images and preimages of the mapping F_γ. By definition, $(T_\gamma, \omega) = F_\gamma(a, f_\gamma \omega)$. Hence $\pi(a, f_\gamma(\omega)) = (a, \Delta_\gamma(\omega))$. Therefore $\Delta_\gamma(\omega) = f_\gamma(\omega)$, as required. □

5.6. Gluing the manifold together. Let us glue together the pieces constructed in 5.4, 5.5 to form a neighborhood of the curve Σ possessing a foliation that has a prescribed collection of basis monodromy transformations. To do this, glue the domains obtained in 5.4 into the holes of the domains obtained in 5.5. The former domains will be called *local*, while the latter ones *nonlocal*. By the definition of an abstract distinguished collection, each separatrix D in a local domain "remembers" to which hole, and in which sphere Ω of a nonlocal domain, it is glued.

Suppose that D and Ω are such a disk and a sphere with holes, B and M are the corresponding local and nonlocal domains with foliations $\mathscr{F}_B$ and $\mathscr{F}$, respectively. Suppose that $\gamma = \partial D$ is a positively oriented circle, $\gamma' \subset \partial \Omega$ is the circle corresponding to γ. Let $\pi_M \colon M \to \Omega$ be the projection generated by $\pi_0 \colon \pi \circ \pi_0 = \pi_M \circ \pi$; π_0 is defined in 5.5. Consider a holomorphic retraction $\pi_D \colon B \to D$; it is of rank 1 everywhere over γ. The domain B can be assumed to be small enough, so as to guarantee that the leaves $\pi_D^{-1} z$, $z \in \gamma$, are disks. Let a and a' be the corresponding points on γ and γ', $\Gamma = \pi_D^{-1} a$, $\Gamma' = \pi_M^{-1} a'$. Let f_γ and $f_{\gamma'}$ be the monodromy transformations for the foliations $\mathscr{F}_B$ and $\mathscr{F}$ corresponding to γ and γ'. These transformations are analytically equivalent. This fact follows from the construction, requirement 4 of 5.2, and the remark at the end of 5.4. Introduce charts w and w' on Γ and Γ' in such a way that the functions that define the transformations f_γ and $f_{\gamma'}$ in these charts coincide. Glue together the solid tori $\pi_D^{-1} \gamma$ and $\pi_M^{-1} \gamma'$ with the help of the mapping Δ having the following properties. The mapping Δ:

takes Γ to Γ' preserving the coordinate: $w' \circ \Delta = w$;

respects the projections π_D and $\pi_M \colon \pi_M \circ \Delta = \Delta \circ \pi_D$;

takes the restriction of the foliation $\mathscr{F}_B$ on $\pi_D^{-1} \gamma$ to the restriction of the foliation $\mathscr{F}$ on $\pi_M^{-1} \gamma'$.

The above properties define the mapping Δ uniquely; it is well defined, since the functions defining f_γ and $f_{\gamma'}$ in the charts w and w' coincide.

The analytic structure on the obtained manifold is defined by the following requirement: *the foliation $\mathscr{F}_B$ and the projection π_B are analytic extensions of the foliation $\mathscr{F}$ and the projection π_M, respectively.*

Denote by $\widetilde{M}$ the manifold obtained by all the above gluings, and by Σ the curve which is the union of the glued spheres.

5.7. The shrinking. Let us prove that there exists a "shrinking"

$$\pi_\Sigma \colon (\widetilde{M}, \Sigma) \to (\mathbb{C}^2, 0) \tag{5.13}$$

that takes Σ to 0. Let us apply the well-known Grauert theorem [G]:

Assume that S is the Riemann sphere and W is its two-dimensional neighborhood such that the self-intersection number of S in W equals -1. Then there exist a manifold W_0 and a holomorphic mapping $W \to W_0$ taking S to a point and biholomorphic outside S. The inverse mapping $W_0 \setminus \{0\} \to W \setminus S$ is a σ-process.

Let us find the required "shrinking". Since the curve S_* in 5.3 is obtained via a composition of σ-processes, it contains several spheres pasted at the final step whose self-intersection numbers in M equal -1. The self-intersection numbers for the corresponding spheres of the curve Σ (to be called exceptional) in $\widetilde{M}$ equal -1 as well. Therefore there exist a manifold $\widetilde{M}_1$ and a mapping $\pi_1 \colon \widetilde{M} \to \widetilde{M}_1$ such that π_1 is biholomorphic outside the exceptional spheres and takes these spheres to points, called distinguished. The inverse mapping of $\widetilde{M}_1$ minus distinguished points to $\widetilde{M}$ minus exceptional spheres is the result of simultaneous σ-processes at the distinguished points. Hence the image $\pi_1 \Sigma$ of the curve Σ under the mapping π_1 is naturally isomorphic to the union of spheres pasted during the first $m - 1$ blowing up steps (here m is the number of blowing up steps required to obtain the manifold M). Self-intersection numbers for the spheres pasted at the $(m - 1)$th step equal -1. To these spheres there correspond exceptional spheres of the curve $\pi_1 \Sigma$ in the surface $\widetilde{M}_1$. Proceeding by induction, one proves the existence of the shrinking (5.13).

Let us find a germ v having a prescribed basic collection of vanishing holonomy transformations. The shrinking π_Σ takes the direction field on the manifold $\widetilde{M}$ constructed in 5.5 to a direction field that is holomorphic outside the origin on $(\mathbb{C}^2, 0)$. By [I1], in a neighborhood of the origin this direction field is defined by a vector field that is holomorphic at the origin. The proof of the theorem stated in 5.3 is completed. □

5.8. Characteristic numbers as additional invariants. Consider a germ of an analytic vector field on the plane at a nondegenerate singular point having two separatrices; assume that f and g are monodromy transformations for these separatrices, while ν and μ are the corresponding multipliers. Is it true that these multipliers define the characteristic numbers λ and λ^{-1} uniquely? This question is answered in the affirmative for all the pairs of eigenvalues except the following four pairs:

$$\lambda_{1,2} = \pm 1 + i, \quad \lambda_{1,2}^{-1} = \pm \frac{1}{2} - \frac{i}{2}; \quad \widetilde{\lambda}_{1,2} = \pm \frac{1}{2} + \frac{i\sqrt{3}}{2}, \quad \widetilde{\lambda}_{1,2}^{-1} = \pm \frac{1}{2} - \frac{i\sqrt{3}}{2}.$$

Each of the two pairs: $\lambda_1^{\pm 1}$ and $\lambda_2^{\pm 1}$ (or $\widetilde{\lambda}_1^{\pm 1}$ and $\widetilde{\lambda}_2^{\pm 1}$) may replace the other in the abstract distinguished collection, thus giving a new collection having the same monodromy transformations and different sets of characteristic numbers. The corresponding vector fields have the same collections of vanishing holonomy transformations, yet fail to be analytically equivalent.

5.9. Problems.

1. *Nonlinear Riemann-Hilbert problem.* Given a Riemann sphere S with a finite number of points deleted and a germ of a mapping $\Delta_i : (\mathbb{C}^n, 0) \to (\mathbb{C}^n, 0)$ at each deleted point such that the product of all the mappings is the identity, find a foliation of a neighborhood $U \subset S \times \mathbb{C}P^n$ of the sphere S into analytic curves such that its singular points coincide with the given deleted points, and its monodromy groups are generated by the given Δ_i.

REMARK. For $n = 1$ this problem is solved in 5.6. The linearization of the problem (given initial conditions, find a variation equation such that its monodromy is defined by differentials of germs Δ_i at the origin) is the Riemann-Hilbert problem. Recently and quite unexpectedly, a negative solution was found for Fuchsian systems by Bolibrukh (see [Bo]). However, the Riemann-Hilbert problem for Fuchsian systems is solved in the affirmative for a wide class of monodromy groups (e.g. for $n = 2$), and for such a class the above nonlinear problem is of interest, even if all the singular points are nondegenerate.

2. *Thom's problem.* Describe the complete set of invariants for the orbital analytic classification of germs of vector fields on the plane having a finite number of separatrices.

§6. Appendix. The realization theorem for the Liouvillian case (after Yoccoz and Perez-Marco)

Any holomorphic germ of a vector field with a real negative ratio of eigenvalues near a singular point in the complex plane has two separatrices (holomorphic invariant curves). Each of them determines a germ of a monodromy transformation of the foliation defined by the field. The multiplier of this germ is equal to $\nu = \exp(2\pi i \lambda)$, where λ is the ratio of the eigenvalues, the one corresponding to the separatrix investigated being in the denominator. The realization problem deals with the inverse situation: given a holomorphic germ $\Delta: (\mathbb{C}, 0) \to (\mathbb{C}, 0)$ such that $|\nu| = 1$, $\nu = \Delta'(0)$, find a germ of a vector field with a real ratio of eigenvalues and monodromy Δ. This problem has a trivial positive answer, if the germ f is analytically equivalent to a linear one. In the resonant case, when $\Delta'(0)$ is a root of unity, the positive answer is given by the Martinet-Ramis theorem, §4 of Paper I. In the Liouvillian case, the positive answer was established recently by Yoccoz and Perez-Marco. A part of the proof below follows the construction presented by Yoccoz at the Conference on complex analytic methods in dynamical systems, IMPA, January, 1992; the rest of the proof makes use of a quasicomplex structure in the way presented in Paper I.

6.1. Statement of the result.

THEOREM. *Let $\alpha \in (0, 1)$ be an irrational number. Then for any germ of a conformal mapping $(\mathbb{C}, 0) \to (\mathbb{C}, 0)$ with $\Delta'(0) = \exp(2\pi i \alpha)$ and any natural k there exists a germ of a holomorphic vector field with ratio of eigenvalues $\lambda_2/\lambda_1 = \alpha - k$ such that the monodromy transformation for the separatrix corresponding to λ_1 equals Δ.*

The proof is given in 6.2–6.6 below. It follows the same ideas as the Martinet-Ramis proof, namely:

We glue together a two-dimensional complex manifold M diffeomorphic to the representative M_0 of the germ $(\mathbb{C}^*, 0) \times (\mathbb{C}, 0)$, with an analytic foliation on it having the prescribed monodromy;

Using a quasicomplex structure and the Nirenberg-Newlander theorem, we prove that M is, in fact, holomorphically equivalent to M_0, and also

The foliation on M_0 can be extended to a neighborhood of the origin in $\mathbb{C}^2$ to obtain the required one.

The proof does not distinguish between the Diophantine and the Liouvillian cases.

6.2. Normalization of a jet of a monodromy transformation. For irrational α, the transformation Δ is formally equivalent to a rotation. Thus, for any natural N, there exists a polynomial change of coordinates that takes Δ to a transformation of the form

$$\Delta: w \mapsto (\text{id} + \varphi) \circ (\nu w), \quad \varphi = o(w^N), \quad \nu = \Delta'(0). \tag{6.1}$$

Choose N large enough (the actual value of N will be indicated later) and suppose that Δ already has the form (6.1).

6.3. The glued manifold with foliation. Take the "covering manifold" $\widehat{M} \subset \mathbb{C}^2$,

$$\widehat{M} = \Pi_\varepsilon \times D_\varepsilon,$$
$$\Pi_\varepsilon = \{\zeta = \xi + i\eta \mid \xi \in (-\varepsilon, 1+\varepsilon), \eta \geq -\varepsilon\}, \quad D_\varepsilon = \{\omega \mid |\omega| < \varepsilon\}.$$

Take the foliation $\mathscr{F}_0$ on $\widehat{M}$ with the leaves $\omega = Ce(\zeta)$, $e(\zeta) = e^{2\pi i(\alpha - k)\zeta}$.

REMARK. $e(\zeta) \to \infty$ as $\operatorname{Im} \zeta \to +\infty$, since $\alpha - k < 0$.

In the domain $L_\varepsilon \times D_\varepsilon$, $L_\varepsilon = \{|\xi| < \varepsilon, \eta \geq -\varepsilon\}$, take a gluing mapping $\widetilde{\Phi}$ with the following properties:

(i) $\zeta \circ \widetilde{\Phi} = \zeta + 1$;
(ii) $\widetilde{\Phi}$ preserves the foliation $\mathscr{F}_0$;
(iii) $\widetilde{\Phi}(0, \omega) = (1, (\text{id} + \varphi)^{-1}(\omega))$.

The germ of $\widetilde{\Phi}$ on $\Pi_\varepsilon \times \{0\}$ is determined uniquely by these properties; for a small ε, it can be extended analytically onto $\Pi_\varepsilon \times D_\varepsilon$. The explicit formula for $\widetilde{\Phi}$ is as follows. Let $(\text{id} + \varphi)^{-1} = \text{id} + \widetilde{\varphi}$. Then

$$\widetilde{\Phi}(\zeta, \omega) = (\zeta + 1, \Phi(\zeta, \omega)), \quad \text{where} \quad \Phi(\zeta, \omega) = e(\zeta)\widetilde{\varphi}(e^{-1}(\zeta)\omega) + \omega.$$

REMARKS. 1. Once more, $\widetilde{\varphi} = o(\omega^N)$ as $\omega \to 0$. Thus,

$$\Phi(\zeta, \omega) - \omega = o(\exp(\alpha - k)(N-1)\eta), \quad \eta = \operatorname{Im} \zeta.$$

Consequently, for any n one can choose the N above large enough so that

$$\Phi(\zeta, \omega) - \omega = o(\exp(-n \operatorname{Im} \zeta)) \quad \text{as} \quad \operatorname{Im} \zeta \to +\infty, \quad \zeta \in \Pi_\varepsilon.$$

The actual value of n will be chosen later.

2. The function

$$\Psi(z, \omega) = \Phi((2\pi i)^{-1} \ln z, \omega) - \omega,$$

taken for the branch $\ln 1 = 0$, tends to zero together with all its derivatives of order $\leq n$ as $z \to 0$ in the sector $\arg z \in 2\pi i(-\varepsilon, 1+\varepsilon)$, provided N is sufficiently large.

Now, glue together the points p and $\widetilde{\Phi}(p)$, $p \in L_\varepsilon \times D_\varepsilon$; the last set will be called the target space of the gluing mapping. If the neighborhood U

of $\Pi_\varepsilon \times \{0\}$, $U \subset M$, is properly chosen, this gluing transforms U into a two-dimensional manifold M. There is a natural projection $\pi: U \to M$. Since the gluing mapping $\widetilde{\Phi}$ preserves the foliation $\mathscr{F}_0$, the projection π defines a foliation on M, namely $\mathscr{F} = \pi_*\mathscr{F}_0$. Denote $\gamma = \pi([0, 1] \times \{0\})$, $z = \exp(2\pi i \zeta)$. The function z is well defined on M.

PROPOSITION 1. *The monodromy transformation of the foliation $\mathscr{F}$ corresponding to the loop γ is equal to Δ.*

PROOF. Let $p \in \Gamma = M \cap \{z = 1\}$. Let ω be the map on Γ induced by the projection $\pi|_{\{\zeta=0\}}$. Let $\widehat{\gamma}$ be the lifting of γ along the lines $z = \text{const}$ onto the leaf of the foliation $\mathscr{F}$ beginning at p. Let $\omega(p) = \omega_0$, and let $\widetilde{\gamma}$ be the curve on the leaf of $\mathscr{F}_0$ defined as the covering of $\widehat{\gamma}$ under the projection π with starting point $(0, \omega_0)$. Then the endpoint q of $\widetilde{\gamma}$ is $(1, \nu\omega_0)$, $\nu = \exp(2\pi i \alpha)$. The point q is the target point of the gluing mapping $\widetilde{\Phi}$ with the source $(0, (\text{id}+\varphi) \circ \nu\omega_0)$ (property (iii) of the gluing mapping). Thus, by (6.1), the projection πq of this point has the coordinate $\Delta(\omega_0)$ in the ω-chart on Γ. □

The gluing process is over. It repeats, in principal details, the construction of Yoccoz and Perez-Marco. Now we shall prove that M may be replaced by M_1, where $M \supset M_1 \supset \pi\Pi_\varepsilon \times \{0\}$, in such a way that M_1 will be biholomorphically equivalent to

$$M_0 = (\mathbb{C}^2, 0) \setminus \{z = 0\}.$$

This is done the same way as in Paper I.

6.4. Embedding of the "diminished M" in $\mathbb{C}^2$.

First, we find a diffeomorphism $G: M \to M_0$. This can be done by means of an explicit formula. We shall define a mapping $\widetilde{G}: U \to \mathbb{C}^2$ with the property

$$\widetilde{G} \circ \widetilde{\Phi} = \widetilde{G}. \tag{6.2}$$

Thus the mapping $G: p \mapsto \widetilde{G} \circ \pi^{-1}p$ will be well defined, although for some p the set $\pi^{-1}p$ consists of two points.

Denote by χ the C^∞-function $\chi: [-\varepsilon, 1+\varepsilon] \to [0, 1]$ that vanishes identically on $[-\varepsilon, \varepsilon]$ and equals 1 identically on $[1-\varepsilon, 1+\varepsilon]$. Denote by Φ_ζ the map $\Phi_\zeta: \omega \mapsto \Phi(\zeta, \omega)$ depending on ζ as a parameter and extended to all of U by the same formula as in the definition. Let

$$\widetilde{G}(\zeta, \omega) = (\exp(2\pi i \zeta), \omega + (\Phi_\zeta^{-1}(\omega) - \omega)\chi(\zeta)).$$

Now, let us prove (6.2). Note that in the source and the target domains of $\widetilde{\Phi}$ the function χ is equal to 0 and to 1, respectively. We have

$$\widetilde{\Phi}(\zeta, \omega) = (\zeta + 1, \Phi_\zeta(\omega)).$$

Thus in the source domain of $\widetilde{\Phi}$, $\widetilde{G}(\zeta, \omega) = (\exp(2\pi i \zeta), \omega)$, and in the same domain,

$$\widetilde{G} \circ \widetilde{\Phi}(\zeta, \omega) = (\exp(2\pi i \zeta), \Phi_\zeta^{-1} \circ \Phi_\zeta(\omega)) = \widetilde{G}.$$

The desired relation (6.2) is proved, and the diffeomorphism G is constructed.

Second, we study the quasicomplex structure induced from the complex structure of M by the mapping G.

PROPOSITION 2. *The number N in (6.1) may be chosen so large that the above structure will be given by 1-forms that admit a C^5-smooth extension to the deleted axis $\{z = 0\}$ by the forms dz, dw.*

This proposition is purely technical and is proved below in 6.6.

The number 5 is motivated by the Newlander-Nirenberg theorem, which allows us to construct a germ $H: (\mathbb{C}^2, 0) \to (\mathbb{C}^2, 0)$ of a C^1-diffeomorphism that is holomorphic in the sense of the above quasicomplex structure.

Let $M_1 \ni 0$ be the domain where H is biholomorphic, $M_2 = G^{-1}M_1$. Thus the mapping of the properly diminished M,

$$\mathcal{H} = H \circ G : M_2 \to \mathbb{C}^2,$$

is a biholomorphic mapping onto its image, which still will be denoted by M_0. Thus M_2 may be completed to become a neighborhood of the origin in $\mathbb{C}^2$.

6.5. Recognition of the foliation. The mapping $\mathcal{H}$ takes the foliation $\mathcal{F}$ on M_1 to the foliation $\widetilde{\mathcal{F}}$ in $(\mathbb{C}^2, 0)$, with the leaf $\{z = 0\}$. This foliation is holomorphic at least outside this leaf and has the prescribed monodromy. We must prove that $\widetilde{\mathcal{F}}$ is holomorphic everywhere and has the prescribed linear part.

PROPOSITION 3. *The foliation $\mathcal{F}_* = (H \circ G)_* \mathcal{F}$ on M_0 can be given by the holomorphic 1-form*

$$\beta = (z + \cdots) dw + (k - \alpha)(w + \cdots) dz,$$

where dots stand for higher order terms.

PROOF. 1. We begin with the construction of the form β. The foliation $\mathcal{F}_* = (H \circ G)_* \mathcal{F}$ is holomorphic outside the line $z = 0$, which is a leaf of this foliation. The foliation $\mathcal{F}_*$ can be extended holomorphically to this line by the Riemann theorem on bounded extensions.

Therefore, it is an analytic one-dimensional foliation in $(\mathbb{C}^2, 0)$ with a unique singular point at the origin. By Proposition 1 of [I1], it can be given by an analytic vector field $v = (v_1, v_2)$ with an isolated singular point at the origin, and thus by the form $\beta = -v_2 dz + v_1 dw$. We want to prove that the form β, probably after multiplication by some constant, has the desired linear part: there exists a c such that

$$c\beta = z\, dw + (k - \alpha)w\, dz + \cdots ; \qquad (6.3)$$

from now on dots stand for higher order terms.

2. The main problem is to prove that the linear part of β is nonzero, and, moreover, nondegenerate. To do this, it is sufficient to prove that the vector field v has index 1 at the origin. Indeed, by the Palamodov theorem [AVGZ], the index of this field is equal to the multiplicity of the mapping given by the functions v_1, v_2. This multiplicity, in turn, is equal to the codimension of the ideal (v_1, v_2) in the ring $\mathcal{O}_0$ of germs of holomorphic functions at the origin. The only ideal of codimension 1 is the maximal one, so the linear part of the field v is nondegenerate.

3. Now we must prove that

$$\operatorname{ind}_0 v = 1. \tag{6.4}$$

We begin with the following remark. Let v and u be two vector fields in $(\mathbb{C}^2, 0)$ with isolated singular point at the origin, both tangent to the same one-dimensional analytic foliation. Then $\operatorname{ind}_0 v = \operatorname{ind}_0 u$. (The fields themselves may be nonholomorphic.)

The remark may be proved in the following way. There is a nonzero function μ on $(\mathbb{C}^2, 0) \setminus \{0\}$ such that $u = \mu v$. Since $(\mathbb{C}^2, 0)$ is simply connected, the function μ admits a logarithm, $\mu = \exp \lambda$. The homotopy

$$u_s = (\exp s\lambda) v, \qquad s \in [0, 1],$$

implies that the indices of all the vector fields u_s are equal and, in particular, of $u_0 = v$ and $u_1 = u$.

4. Now we shall find a vector field u tangent to the leaves of the foliation $\mathscr{F}_*$ and having index 1 at the origin. This is done by straightforward calculations.

First, we shall find a 1-form β_1 on U with the following properties:

(i) β_1 can be descended to M: there exists a form $\widetilde{\beta}$ on M such that

$$\beta_1 = \pi^* \widetilde{\beta}; \tag{6.5}$$

we write $\widetilde{\beta} = \pi_* \beta_1$;

(ii) β_1 is proportional to β_0, and thus gives the same foliation $\mathscr{F}_0$.

Next, we shall calculate the linear part of the form $\beta_2 = zG_* \widetilde{\beta}$ at the origin. The form β_2 is only $\mathbb{R}$-linear, but its linear part appears to be holomorphic and is the required one, as in (6.3). Finally, the holomorphic map H in the Nirenberg-Newlander theorem can be taken with identity linear part at the origin. Thus the form

$$\beta_3 = H_* \beta_2 = (H \circ G)_* \widetilde{\beta} = (H \circ \widetilde{G})_* \beta_1$$

will also be of the form (6.3):

$$j^{-1} \beta_3 = z\, dw + (k - \alpha) w\, dz. \tag{6.6}$$

Let $\beta_3 = A\, dz + B\, dw$. The vector field u tangent to the foliation $\beta_3 = 0$, $u = (-B, A)$, has index 1 at the origin; this follows from (6.6). Note that the forms β and β_3 generate the same foliation $\mathscr{F}_*$. Thus the holomorphic vector field v tangent to the foliation given by β has the same index 1 at the origin as u does. Thus the linear part of β is nondegenerate. Moreover, the forms β and β_3 are proportional, as giving the same foliation. Since the linear parts of both forms are nondegenerate, they coincide after multiplication by an appropriate constant. Together with (6.6), this proves (6.3).

5. It remains to find β_1 with the above properties (i), (ii) and to calculate the linear part of β_2. The foliation $\mathscr{F}_0$ is given by the form

$$\beta_0 = d\omega + 2\pi i(k - \alpha)\omega\, d\zeta.$$

The required form β_1 must be proportional to β_0 and respect the gluing mapping:

$$\widetilde{\Phi}_*\beta_1 = \beta_1. \tag{6.7}$$

So we want to find a function F on U such that $\widetilde{\Phi}_* F\beta_0 = F\beta_0$. Let S and T be the source and the target spaces of $\widetilde{\Phi}$ with the chart (ζ, ω), as in 6.3. Then

$$\widetilde{\Phi}_*\beta_0 = d\Phi + 2\pi i(k - \alpha)\Phi\, d\zeta$$

on S. This form is to be proportional to β_0; thus,

$$\widetilde{\Phi}_*\beta_0 = f\beta_0, \qquad f = \partial\Phi/\partial\omega$$

on S. The function f tends to 1 faster than $\exp(-n\,\mathrm{Im}\,\zeta)$ as $\mathrm{Im}\,\zeta \to +\infty$; it is important now that n can be taken larger than 1. Extend f onto all of U to obtain a function F that has the same asymptotic property and equals 1 identically on T. The form $\beta_1 = F\beta_0$ is the required one, that is to say, (6.7) is fulfilled.

6. Now we prove that

$$j^1\beta_2 = z\, dw + (k - \alpha)w\, dz. \tag{6.8}$$

This will finish the proof of Proposition 3. By the definition of β_1 and β_2 we have $\beta_2 = z.(F \circ \widetilde{G}^{-1}).\widetilde{G}_*\beta_0$. Note that both the function and the form on the right-hand side are nonunivalent on M_0, but their product is. We have

$$\widetilde{G}_*\beta_0 = dw + (k - \alpha)wz^{-1}dz + \cdots, \qquad F \circ \widetilde{G}^{-1} = 1 + \cdots,$$

where the dots stand for $o(z, w)$ in the first formula, and for $o(1)$ in the second. This proves (6.8), and thus Proposition 3. $\square$

6.6. Extension of a quasicomplex structure: proof of Proposition 2. The quasicomplex structure on $M_0 = (\mathbb{C}^2, 0) \setminus \{z = 0\}$ induced by the complex structure on M via the mapping G is given by the forms

$$\omega_1 = (\widetilde{G}^{-1})_* d\zeta, \qquad \omega_2 = (\widetilde{G}^{-1})_* d\omega.$$

The mapping $\widetilde{G}^{-1}$ is not univalent; it is two-fold over the set $\pi S = \pi T \stackrel{\text{def}}{=} P$. Here S and T are the source and the target spaces of the gluing mapping Φ as before. Nevertheless, these forms define the structure quite well, because the space spanned by them is the same on the two leaves.

In order to obtain the univalence, define two forms α_1, α_2 of type $(1, 0)$ on U that can be descended onto M via the projection π. This means that there exist 1-forms $\widetilde{\omega}_1$, $\widetilde{\omega}_n$ on M_0 such that

$$\alpha_1 = \widetilde{G}^* \widetilde{\omega}_1, \qquad \alpha_2 = \widetilde{G}^* \widetilde{\omega}_2. \tag{6.9}$$

To do this, take $\alpha_1 = d\zeta$, $\alpha_2 = F_1 d\zeta + F_2 d\omega$, where the functions F_1 and F_2 will be found later. The first equation in (6.9) holds for $\widetilde{\omega}_1 = dz/2\pi i z$. The form ω_2 in the second equation in (6.9) exists if

$$\widetilde{\Phi}^*(\alpha_2|_T) = \alpha_2|_S. \tag{6.10}$$

Let $\alpha_2|_T = d\omega$. Then (6.10) implies

$$\alpha_2|_S = d\Phi = \frac{\partial \Phi}{\partial \omega} d\omega + \frac{\partial \Phi}{\partial \zeta} d\zeta.$$

Define $F_1 = F$, F is the same as in 6.5. Let F_2 be the extension of $\partial \Phi/\partial \zeta|_S$ to all of U with the following properties:

$$F_2|_T \equiv 0, \qquad F_2 = o(\exp(-n \operatorname{Im} \zeta)) \quad \text{as} \quad \operatorname{Im} \zeta \to +\infty \tag{6.11}$$

together with all its derivatives of order m, where $m = n/2 \in \mathbb{N}$. The function $\partial \Phi/\partial \zeta$ has the property (6.11) on S; thus, the desired extension exists.

The forms α_1, α_2 can be descended from U to M via the projection π, i.e. (6.9) holds. The forms $\widetilde{\omega}_1$, $\widetilde{\omega}_2$ on M are the following:

$$\widetilde{\omega}_1 = \frac{dz}{2\pi i z}, \qquad \widetilde{\omega}_2 = (F_2 \circ \widetilde{G}^{-1}) \frac{dz}{2\pi i z} + (F_1 \circ \widetilde{G}^{-1}) d\widetilde{G}_2^{-1}.$$

Here $\widetilde{G}_2^{-1}$ is the ω-component of $\widetilde{G}^{-1}$. Note that neither the functions F_1, F_2, nor the mapping $\widetilde{G}^{-1}$ are univalent on M_0, but the form $\widetilde{\omega}_2$ is.

Using the asymptotic properties of F_1, F_2 (see 6.5 and (6.11)) we can see easily that the forms $2\pi i z \widetilde{\omega}_1$ and $\widetilde{\omega}_2$ can be extended by dz, dw to the line $z = 0$ up to C^5-smooth functions. This proves Proposition 2.

This is the last point in the proof of the Yoccoz–Perez-Marco theorem.

References

[ALGM] A. A. Andronov, E. A. Leontovich, I. I. Gordon, and A. G. Maĭer, *Qualitative theory of second-order dynamical systems*, "Nauka", Moscow, 1966; English transl., Israel Program for Scientific Translation, Jerusalem; Wiley, New York, 1973.

[AI] V. I. Arnold and Yu. S. Il'yashenko, *Ordinary differential equations*, Contemporary Problems of Mathematics: Fundamental Directions. Dynamical Systems vol. 1, VINITI, Moscow, 1985, pp. 7–150; English transl. in *Encyclopaedia of Modern Mathematics*, vol. 1, Springer-Verlag, Berlin and New York, 1988.

[AVGZ] V. I. Arnold, A. N. Varchenko, and S. M. Gusein-Zade, *Singularities of differential maps*. I, "Nauka", Moscow, 1982; English transl., Birkhäuser, Basel, 1985.

[Bo] A. A. Bolibrukh, *The Riemann-Hilbert problem*, Uspekhi Mat. Nauk **45** (1990), no. 2, 1–49; English transl. in Russian Math. Surveys **45** (1990).

[Br] A. D. Brjuno [Bryuno], *Analytic form of differential equations*, Trudy Moskov. Mat. Obshch. **25** (1971), 119–262; **26** (1972), 199–239; English transl. in Trans. Moscow Math. Soc. **25** (1971); **26** (1972).

[CS] C. Camacho and P. Sad, *Invariant varieties through singularities of holomorphic vector fields*, Ann. of Math. (2) **115** (1982), 579–595.

[CM] D. Cerveau and R. Moussu, *Groups d'automorphismes de* $(\mathbb{C}, 0)$ *et équations différentielles* $ydy + \cdots = 0$, Bull. Soc. Math. France **116** (1988), 459–488.

[E] J. Ecalle, *Sur les fonctions résurgentes*, I, II, Publ. Math. d'Orsay, Université de Paris-Sud, Orsay, 1981.

[EI] P. M. Elizarov and Yu. S. Il'yashenko, *Remarks on the orbital analytic classification of germs of vector fields*, Mat. Sb. **121** (1983), 111–126; English transl. in Math. USSR-Sb. **49** (1984).

[G] H. Grauert, *Über Modifikationen und exzeptionelle analytische Mengen*, Math. Ann. **146** (1962), 331–368.

[I1] Yu. S. Il'yashenko, *Foliations into analytic curves*, Mat. Sb. **88** (1972), no. 4, 558–577; English transl. in Math. USSR-Sb. **17** (1972).

[I2] _____, *Finiteness theorems for limit cycles*, Uspekhi Mat. Nauk **45** (1990), no. 2, 143–200; English transl. in Russian Math. Surveys **45** (1990).

[K] O. A. Kleban, *The order of the topologically sufficient jet for a nonmonodromic singular point on the plane* (to appear).

[M] R. Moussu, *Holonomie évanescente des équations différentielles dégénérées transverses*, Singularities and Dynamical Systems (S. N. Pneumaticos, ed.), North-Holland, Amsterdam, 1985, pp. 161–173.

[PM] R. Perez-Marco, *Centralizers I. Uncountable centralizers for nonlinearizable holomorphic germs of* $(\mathbb{C}, 0)$, Preprint, Inst. Hautes Études Sci., 1991.

[R] J.-P. Ramis, *Confluence et resurgence*, J. Fac. Sci. Univ. Tokyo Sect. IA Math. **36** (1989), 706–716.

[S] A. A. Shcherbakov, *Topological and analytical conjugacy for noncommutative groups of germs of conformal mappings and differential equations on the complex plane*, Trudy Sem. Petrovsk. **10** (1984), 270–296. (Russian)

[Ta] F. Takens, *Singularities of vector fields*, Inst. Hautes Études Sci. Publ. Math. **43** (1974), 47–100.

[Ti] E. Titchmarsh, *Theory of functions*, Oxford Univ. Press, Oxford, 1939.

[V] S. M. Voronin, *Analytic classification of germs of conformal maps* $(\mathbb{C}, 0) \to (\mathbb{C}, 0)$ *with identical linear part*, Funktsional. Anal. i Prilozhen. **15** (1981), no. 1, 1–17; English transl. in Functional Anal. Appl. **15** (1981).

[Y] J.-C. Yoccoz, *Théorème de Siegel, nombres de Bruno et polynômes quadratiques*, Preprint, Centre Math. de l'Ecole Polytechnique, Palaiseau, 1988.

DEPARTMENT OF MATHEMATICS, TVER STATE UNIVERSITY, TVER, RUSSIA

CHAIR OF DIFFERENTIAL EQUATIONS, DEPARTMENT OF MECHANICS AND MATHEMATICS, MOSCOW STATE UNIVERSITY, MOSCOW 119899

Translated by A. VAĬNSHTEĬN

Tangents to Moduli Maps

P. M. ELIZAROV

Introduction

The paper is concerned with the approximate computation of moduli of analytic equivalence for vector fields at elementary singular points in the presence of resonances. Thus the following types of singularities are under consideration: complex saddle-nodes (one zero eigenvalue) and resonance saddles (two eigenvalues with rational negative ratio). The orbital analytic classification of such points is given in Paper I. Formulas for the calculation of functional moduli, written in a different form and established in a different way, were obtained by J. Ecalle [15].

The formal equivalence class for a complex saddle-node or a resonance saddle is determined by one positive integer and one complex parameter. On any class, the moduli map is defined: it takes any germ from this class to its functional modulus of orbital analytic classification. This modulus can be described as a certain tuple of 1-periodic holomorphic functions, or, which is the same, as a collection of converging Fourier series. The problem is to determine the derivative of the moduli map at the point corresponding to a germ coinciding with the formal normal form. The latter is the simplest representative of the formal equivalence class.

The derivative is understood in the sense of Gateaux. Let μ be the moduli map. For the family $v_\varepsilon = v + \varepsilon W$, where v is the formal normal form and W a tangent vector (in the space of germs), the derivative

$$(d\mu(v_\varepsilon)/d\varepsilon)|_{\varepsilon=0}$$

is computed. The Fourier coefficients for the series constituting the functional module, depend on ε and are denoted by $c_j(\varepsilon)$. Their derivatives

$$d_j - (dc_j(\varepsilon)/d\varepsilon)|_{\varepsilon=0}$$

are computed in the present paper.

It turns out that the support of the series W (that is, the set of exponents of monomials entering the Taylor expansion of W with nonzero coefficients)

1991 *Mathematics Subject Classification.* Primary 58D27.

must be split in a countable number of subsets lying on parallel lines in the integer lattice $\mathbb{Z}^2$ on the plane. For the complex saddle-node case these lines are horizontal (it is assumed that the first eigenvalue is zero); for the case of a resonance saddle with ratio of eigenvalues $\lambda_2/\lambda_1 = -m/n$, these lines are parallel to the vector (m, n). The sum of monomials corresponding to any line is expressed in terms of a holomorphic function of one variable. For a natural enumeration of the lines of the splitting, the derivative d_j defined above is a generalized Borel transform of the jth function (see details in §1). If the complex parameter of the formal normal form equals zero, the generalized Borel transform becomes the classical one.

If the support of the germ W lies on one side of the line $y = 1$ (for the saddle-node case) or the line $my - nx = 0$ (for the saddle resonance case), the fields analytically equivalent to $v + W$ accumulate to v. In this case the computation of the derivative of the moduli map yields nonlocal results. For example, the two formally equivalent saddle-nodes

$$v + \alpha z^a w^b \frac{\partial}{\partial w} \quad \text{and} \quad v + \beta z^c w^d \frac{\partial}{\partial w}, \qquad a > p,\, c > p, \qquad (0.1)$$

with

$$v = \frac{z^{p+1}}{1 + \lambda z^p} \frac{\partial}{\partial z} - w \frac{\partial}{\partial w}, \qquad \operatorname{Re} \lambda > 0,$$

are analytically orbitally nonequivalent if $b \neq d$ for all $\alpha, \beta \in \mathbb{C}$, $a, c \in \mathbb{N} + p$. An analogous result holds for resonant saddles of the form (0.1), if we assume

$$v = z \frac{\partial}{\partial z} + w \left(-\frac{m}{n} + \frac{u^p}{1 + \beta u^p} \right) \frac{\partial}{\partial w},$$

$$u = z^m w^n, \quad a, b \in pm + \mathbb{N}, \quad c, d \in pn + \mathbb{N}.$$

Nonequivalence occurs if $an - bm \neq cn - dm$ for all $\alpha, \beta \in \mathbb{C}$.

The above description of our results is on the heuristic level of accuracy. To formulate them in precise form, we need some additional details, so we pass to the precise exposition.

§1. Definitions and main results

1.1. Analytic orbital equivalence of germs of vector fields. Consider a holomorphic vector field

$$v(z, w) = v_1 \frac{\partial}{\partial z} + v_2 \frac{\partial}{\partial w}, \qquad (z, w) \in (\mathbb{C}^2, 0), \qquad (1.1)$$

defined in a neighborhood of the origin of the plane, $v(0) = 0$. Any such field defines a foliation of this neighborhood by its complex one-dimensional phase curves.

DEFINITION. The singular point 0 of the vector field v is called *elementary*, if at least one of the eigenvalues of the linearization of v at the origin is nonzero.

Denote by V the space of germs of holomorphic vector fields having the origin as an elementary singular point.

DEFINITION. Two differential equations defined by holomorphic vector fields with elementary singularities at the origin are called *analytically orbitally equivalent* near this point, if there exists a germ of an origin-preserving holomorphism taking the germ of the foliation defined by the first equation to that defined by the second one.

In other words, the first germ of the field is orbitally equivalent to the second one if they became analytically conjugate after multiplication of one of them by an appropriate invertible germ of a holomorphic function.

DEFINITION. Two germs of the form (1.1) are called *formally orbitally equivalent* if there exist a formal (power series) transformation taking the first series into the second multiplied by a series with nonzero free term.

1.2. Degenerate elementary singular points.

DEFINITION. The singular point 0 of the field (1.1) is called *elementary degenerate*, if it has exactly one nonzero eigenvalue.

The subspace of germs of vector fields of the class V having the origin as their isolated degenerate elementary singularity will be denoted by D.

According to Theorem 1 from Paper I, 3.1, any germ of a holomorphic vector field $v \in D$ is formally orbitally equivalent to

$$v^D_{p,\lambda} = \frac{z^{p+1}}{1+\lambda z^p} \frac{\partial}{\partial z} + w \frac{\partial}{\partial w} \qquad (1.2)$$

for some $p \in \mathbb{N}$, $\lambda \in \mathbb{C}$.

REMARK [1]. Any two germs of the form (1.2) with different invariants p, λ are not formally orbitally equivalent.

Denote by $D_{p,\lambda}$ the equivalence class of germs from D formally orbitally equivalent to (1.2). The latter germ will be called the *formal normal form* for the class $D_{p,\lambda}$.

The analytic orbital classification for germs of the class $D_{p,\lambda}$ was obtained in [1] and described in 3.2–3.4, Paper I. For the sake of completeness, let us recall the description of the moduli of this classification, given in subsection 3.4 of the aforementioned paper.

Corollary 2 of 3.7, Paper I, associates the equivalence class of tuples

$$\{a^D_{j,l} : l \in \mathbb{N} \cup \{-1\}, \, j = 0, \ldots, p-1\} \qquad (1.3)$$

with each germ of the class $D_{p,\lambda}$. These tuples are subject to a single restriction: the series

$$\Phi^+_j = t + \sum_{l=1}^{\infty} a^D_{j,l} e^{2\pi i l t} \qquad (1.4)$$

must be convergent in some lower half-plane; within this limitation they can be arbitrary. The equivalence between tuples is defined as follows: any two of them are said to be equivalent, if the corresponding maps (1.4) are conjugated by the shift $t \mapsto t + C$ with some $C \in \mathbb{C}$. In other words, the two tuples $a_{j,l}^D$ and $\tilde{a}_{j,l}^D$ are equivalent if and only if there exists a $c \in \mathbb{C} \setminus \{0\}$ such that

$$\forall j, l, \quad a_{j,l}^D = c^l \tilde{a}_{j,l}^D. \tag{1.5}$$

The equivalence class of tuples (1.3), (1.5) will be called the *Martinet-Ramis modulus*. The modulus corresponding to the germ $v \in D_{p,\lambda}$ will be denoted by μ_v.

The problem of computing the Martinet-Ramis moduli is a complicated one. Up to the present time these moduli are known only for a few types of vector fields, for example, for fields defined by the Riccati equation [1]. In particular, all the components of the modulus corresponding to the formal normal form $v_{p,\lambda}^D$ are zeros.

In the present paper we compute linear approximations for the Martinet-Ramis moduli for fields of the class $D_{p,\lambda}$ close to the normal form $v_{p,\lambda}^D$. More precisely, certain representatives of the equivalence classes of tuples are computed.

Consider a family of germs $\{v_\varepsilon\} \subset D_{p,\lambda}$ analytically depending on the parameter $\varepsilon \in (\mathbb{C}, 0)$ and such that $v_\varepsilon|_{\varepsilon=0} \equiv v_{p,\lambda}^D$. Denote by $\mu_\varepsilon = \{a_{j,l}^D(\varepsilon)\}$ the family of the representatives of the corresponding Martinet-Ramis moduli, assuming their analytic dependence on the parameter. This assumption does not restrict generality because the moduli themselves analytically depend on the parameter (assertion 4 of the classification theorem, see 3.3 in Paper I). From the triviality of the modulus for the formal normal form $v_{p,\lambda}^D$ it follows that $a_{j,l}^D(0) = 0$ for all j, l. Our goal is to compute the derivatives

$$(d\mu_\varepsilon/d\varepsilon)|_{\varepsilon=0},$$

that is, the variations

$$d_{j,l}^D = d/d\varepsilon|_{\varepsilon=0} a_{j,l}^D(\varepsilon).$$

In order to formulate the results we need some definitions and notation.

DEFINITION. Let $f(\eta) = \sum_{k=1}^{\infty} f_k \eta^k$ be a formal power series. Its *Borel transform* $\mathscr{B}f$ is the series

$$(\mathscr{B}f)(\zeta) = \sum_{k=1}^{\infty} \frac{f_k \zeta^k}{(k-1)!}.$$

By analogy, we will call Borel-type transformation any construction defined as follows. Let $m = \{m_k\}_{k=1}^{\infty}$ be any sequence of complex numbers. We define the *Borel-type transform* as

$$(\mathscr{B}^m f)(\zeta) = \sum_{k=1}^{\infty} f_k m_k \zeta^k. \tag{1.6}$$

Fix any values of the formal invariants p, λ and let $l \in \mathbb{N} \cup \{-1\}$. Denote by $m_k(l)$ the constant of the form

$$m_k(l) = \frac{k}{\Gamma(1 + (k + \lambda l)/p)}, \qquad (1.7)$$

where the expression in the denominator is the Γ-function. By the symbol $\mathscr{B}_D^l$ we denote the Borel-type transformation associated with the sequence $m(l) = \{m_k(l)\}$; the subscript D reminds us that the transformation (1.6), (1.7) appears in the investigation of the fields of the class D. Note that if $(p, \lambda) = (1, 0)$, then $\mathscr{B}_D^l$ coincides with the canonical Borel transformation for all l. We shall speak of the generalized Borel transform if $\lambda = 0$ and p is arbitrary.

Any germ of the class $D_{p,\lambda}$ is analytically orbitally equivalent to a germ $v_{p,\lambda}^D + W$, where $W = P(z, w)\partial/\partial w$; it is convenient to represent the germ of the function $P(z, w)$ holomorphic at the origin in the form

$$P(z, w) = w \sum_{l=-1}^{\infty} \sum_{k=p+1}^{\infty} b_{k,l} z^k w^l.$$

The possibility of such a representation easily follows from (3.3), Paper I; for a proof see [1]. Let

$$v_\varepsilon = v_{p,\lambda}^D + \varepsilon W, \qquad W = P\partial/\partial w, \qquad (1.8)$$

and write the germ P in the form

$$P(z, w) = w \sum_{l=-1}^{\infty} w^l f_l(z), \qquad (1.9)$$

where f_l are holomorphic in a neighborhood of the origin.

Finally let ε_p be a primitive root of minus unity, $\varepsilon_p = \exp(-\pi i/p)$; consider the family of constants

$$e_{j,l} = \operatorname{sgn}(l) p^{-1} l^{-1+\lambda l p^{-1}} e^{-2\pi j \lambda l p^{-1} i}, \qquad l \in \mathbb{N} \cup \{-1\}, \; j = 0, \ldots, p-1.$$

Now all the preparations are done, and we proceed with the formulation of the main results.

We call the map $\mu : v \mapsto \mu_v$ taking each field v to its Martinet-Ramis modulus μ_v the *moduli map*. The parameters $d_{j,l}$ introduced above may be considered as the coordinates on the tangent space to the moduli space at the point $\mu_{v_{p,\lambda}}$. Thus the equality defining $d_{j,l}^D$ determines the (j, l)th coordinate of the image by the action of the linearization of the moduli map μ.

By virtue of the representation (1.8), we may identify the tangent space to the fields of class D at the point $v_{p,\lambda}^D \in D$ with the space of all germs W of the form (1.8), (1.9), the germ $W = 0$ corresponding to the zero of the tangent space.

THEOREM 1. *The derivative (tangent map) $d\mu$ at the point $v_{p,\lambda}^D$ takes the tangent vector W of the form (1.8), (1.9) to the vector with coordinates*

$$d_{j,l}^D \circ (d\mu) \cdot W = e_{j,l} \cdot (\mathscr{B}_D^l f_l)(\varepsilon_p^{2j+1}(l/p)^{1/p}).$$

This theorem, which is proved in §2, allows us to formulate necessary conditions for analytic orbital equivalence of germs of the class D, imposed on their Taylor series, and construct different examples of divergent normalizing series in all classes of formal equivalence (see 1.4 below).

1.3. Saddle resonant singular points.

DEFINITION. The singular point 0 of the field (1.1) is called *resonance saddle* if the ratio of its eigenvalues is negative rational.

In this section we give the explicit formulas for the derivative of the moduli map for the resonance saddle case, analogous to the degenerate case.

Let us start with the formal classification. Denote the subspace of resonance saddle germs by $S \subset V$. According to the theorem in 4.5, Paper I, any germ $v \in S$ having the ratio of eigenvalues $\lambda_2/\lambda_1 = -m/n$, $\gcd(m,n) = 1$, is formally orbitally equivalent to either its linear part, or to a germ of the form

$$v_{p,\lambda}^S = z\left(-\frac{n}{m} + \frac{u^p}{1+\lambda u^p}\right)\frac{\partial}{\partial z} + w\frac{\partial}{\partial w} \qquad (1.10)$$

for some natural p and complex λ, where $u = z^m w^n$ is the resonant monomial. Note that the field (1.10) is orbitally analytically equivalent to (4.11) of Paper I.

REMARK [2]. If two collections (m, n, p, λ) are different, then the corresponding germs are neither analytically orbitally equivalent to each other nor to the linear germ.

In what follows we shall always assume that the ratio of the eigenvalues is fixed once and for all. Denote by $S_{p,\lambda}$ the formal equivalence class of the germ (1.10). The field (1.10) itself will be referred to as the formal normal form for the corresponding class.

By the theorem from 4.6, Paper I, the analytic orbital equivalence of resonance saddles is determined by the moduli, which are the equivalence classes of tuples

$$\{a_{j,l}^S : l \in \mathbb{Z} \setminus \{0\}, \ j = 0, \ldots, p-1\}. \qquad (1.11)$$

The two tuples $\{a_{j,l}^S\}$ and $\{\tilde{a}_{j,l}^S\}$ are said to be *equivalent* if there exists a $c \in \mathbb{C}^*$ such that for all j, l

$$a_{j,l}^S = c^l \tilde{a}_{j,l}^S.$$

For the field analytically orbitally equivalent to its formal normal form all the components of the tuple (1.11) are zeros.

When passing to analytic orbital equivalence, one may assert that any germ is equivalent to a germ of the form

$$z\left(-\frac{n}{m} + \frac{u^p}{1+\lambda u^p}(1+\tilde{P}(z,w))\right)\frac{\partial}{\partial z} + w\frac{\partial}{\partial w}, \quad (1.12)$$

where $\tilde{P}$ is a holomorphic germ of a function divisible by a power of zw as large as we wish (see [2] and the related formula (4.12′) from 4.5, Paper I).

Consider again a one-parameter deformation $\{v_\varepsilon\}$, $\varepsilon \in (\mathbb{C},0)$, of the germ $v_{p,\lambda}^S$:

$$v_\varepsilon = v_{p,\lambda}^S + \varepsilon W, \qquad W = -z\frac{u^p}{1+\lambda u^p}P(z,w)\frac{\partial}{\partial z}, \quad (1.13)$$

where the germ of the function

$$P(z,w) = \sum_{k=pm+1}^{\infty}\sum_{s=pn+1}^{\infty} b_{ks} z^k w^s \quad (1.14)$$

is holomorphic at the origin.

The moduli map is the map taking any resonance saddle germ v to the corresponding classification modulus. For an analytic deformation, we always can choose the tuple $a_{j,l}^S(\varepsilon)$ representing this modulus in such a way that it analytically depends on the parameter. Define the variations

$$d_{j,l}^S = d/d\varepsilon|_{\varepsilon=0} a_{j,l}^S(\varepsilon)$$

in the same way as above. The tangent vector to the deformation (1.13) will be identified with the germ P. In order to express the coordinates of the derivative $d_{j,l}^S \circ (d\mu)\cdot P$, we need some more notation.

Let

$$\tilde{m}_k(l) = k/\Gamma(1+(km-\lambda l)p^{-1}m^{-2}), \qquad l \in \mathbb{Z}, \, l \neq 0,$$

and let $\mathscr{B}_S^l$ be the Borel-type transformation associated with the sequence $\{\tilde{m}_k(l)\}_{k=1}^\infty$. Put

$$e_{j,l}^S = \frac{\operatorname{sgn}}{pm^2}\left(\frac{l}{m}\right)^{\lambda l p^{-1}m^{-2}-1} e^{-2\pi j\lambda l p^{-1}m^{-2}i}, \qquad l\in\mathbb{Z}\setminus\{0\}, \, j=0,\dots,p-1,$$

and let, as before, ε_{pm} stand for the primitive roots of minus unity:

$$\varepsilon_{pm} = \exp(-\pi p^{-1}m^{-1}i).$$

Next, split the support of the function P by means of the lines $ml - nk = s$:

$$J_l = \{(k,s): ms - nk = l, \, k \geq pm+1, \, s \geq pn+1\}.$$

To each function P associate the countable set of functions of one variable:

$$f_l(z) = \sum_{(k,s)\in J_l} b_{k,s} z^k \quad (1.15)$$

so that
$$P(zw^{-n}, w^m) = \sum_{-\infty}^{\infty} f_l(z) w^l. \tag{1.16}$$

THEOREM 2. *The derivative of the moduli map $d\mu$ at the point $v_{p,\lambda}^S$ takes the tangent vector W of the form* (1.13), (1.14) *to the vector with coordinates*
$$d_{j,l}^S \circ (d\mu) \cdot W = e_{j,l} \cdot (\mathscr{B}_S^l f_l)(\varepsilon_{pm}^{2j+1}(lp^{-1}m^{-2})^{1/pm}).$$

For the case $(m, n, p, \lambda) = (1, 1, 1, 0)$ this theorem was proved in [3]. Note that in the latter case $\mathscr{B}_S^L$ coincides with the ordinary Borel transformation. A brief exposition of the proof of Theorem 2 for the general case is given below in §3.

1.4. Analytic orbital nonequivalence of resonance saddles and divergence of the normalizing series. What follows in this subsection is a general observation and refers both to the degenerate elementary and to the resonance saddles classification problems.

Each vector field near its isolated resonant singular point can be put into the Poincaré-Dulac normal form by formal transformations. This normal form contains only resonant monomials and can be simplified further (see 3.1 and 4.5, Paper I). This is why this form is called *preliminary* in what follows. The formal power series transformation putting the field in its normal form is called the *normalizing substitution*, or the *normalizing series*. One of the most important problems investigated in normal form theory is that of finding whether the given field is analytically equivalent to its preliminary normal form or not. Although the divergence of the normalizing series is a rule in this theory and its convergence is an exception [4–6], the known explicit examples of divergence are of very specific nature [7, 8]. The proof of divergence for these cases is reduced to a lower estimate of the Taylor coefficients of the normalizing series and requires considerable computations for each concrete field.

The above question is a particular case of the following one: given any two fields, find whether they are analytically equivalent or not. The first examples of formally equivalent fields which are analytically nonequivalent and have divergent normalizing series were constructed using the functional moduli of analytic classification only in the resonance saddle case, see [9]. For the degenerate elementary case, a similar example was known long ago [10]; the methods used in the latter work are based on certain specific features of degenerate elementary fields linear in the w variable.

The results of the computation of the derivative of the moduli maps as they were stated above, allow us to give examples of vector fields with divergent normalizing series in all the classes of formal equivalence and even to formulate sufficient conditions for analytic nonequivalence of fields in terms of their Taylor coefficients.

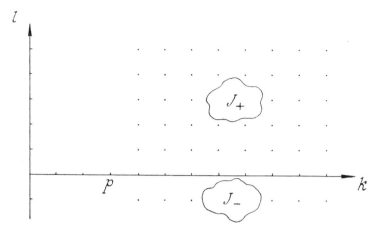

FIGURE 1

Let us proceed now with the precise definitions. First consider the formal equivalence class $D_{p,\lambda}$ and assume p, λ fixed. Denote by J_+ and J_- the following two subsets of the integer two-dimensional lattice (see Figure 1):

$$J_+ = \{(k,l) : (k,l) \in \mathbb{Z}^2, \ k > p, \ l > 0\},$$
$$J_- = \{(k,l) : (k,l) \in \mathbb{Z}^2, \ k > p, \ l = -1\}.$$

In this section the symbol J_D will stand for one of the sets $J_\pm$.

Consider the family of germs $\{v_\varepsilon\} \subset D_{p,\lambda}$, $\varepsilon \in (\mathbb{C}, 0)$, such that

$$v_\varepsilon = v_{p,\lambda}^D + \varepsilon W, \qquad W = wP(z,w)\partial/\partial w; \qquad (1.17)$$

here $wP(z,w) \equiv \tilde{P}$ is a series converging at the origin, and $\mathrm{supp}(P) \subseteq J_D$.

Denote by r the number with the least absolute value such that the line

$$\{(k,l) : (k,l) \in \mathbb{Z}^2, \ l = r\}$$

is the support line for the convex hull of the set $\mathrm{supp}(P) \subseteq \mathbb{Z}^2$. Evidently $r > 0$ if $\mathrm{supp}(P) \subseteq J_+$, otherwise $r = -1$ (see Figure 1).

DEFINITION. The family v_ε will be called *r-nondegenerate* if at least one of the coordinates $d_{j,r}^D \circ (d\mu) \cdot W$ of the image vector is nonzero for $j = 0, \ldots, p-1$.

The germ of the vector field

$$v = v_{p,\lambda}^D + W, \qquad W = \tilde{P}(z,w)\partial/\partial w \qquad (1.18)$$

is called *r-nondegenerate* if the deformation $\{v_\varepsilon = v_{p,\lambda}^D + \varepsilon W\}$ is r-nondegenerate.

REMARK. The occurence of the sets $J_\pm$ in this construction is not accidental: it turns out that the formal normal form $v_{p,\lambda}^D$ lies on the boundary of the set of germs linearly equivalent to the germ $v_{p,\lambda}^D + \varepsilon W$ if $\mathrm{supp}(P) \subseteq J_D$.

In more detail, let $\operatorname{supp} P \subseteq J_+$, and $L_\alpha(z, w) = (z, \alpha^{-1}w)$. Then the normal form $v_{p,l}^D$ is preserved by L_α and lies on the boundary of the set $\{L_{\alpha*}v : \alpha \in \mathbb{C}^*\}$:

$$L_{\alpha*}v \to v_{p,l}^D \quad \text{as } \alpha \to 0.$$

In the case $\operatorname{supp} P \subseteq J_-$ the same is true for $L_\alpha(z, w) = (z, \alpha w)$. Thus one can use Theorem 1 to analyze the analytic classification modulus of the germ v.

THEOREM 3. *The components of any representative of the Martinet-Ramis modulus for an r-nondegenerate germ v of the form* (1.18) *satisfy the following conditions*:

(1) $a_{j,l}^D(v) = 0$ *for all* $j = 0, \ldots, p-1$, *provided that* $(l-r)\operatorname{sgn}(r) < 0$;
(2) $a_{j,l}^D(v) \neq 0$ *for* $l = r$, *provided that* $d_{j,r}^D \circ (d\mu) \cdot W \neq 0$.

This theorem is proved in §4 below.

COROLLARIES. 1. *The normalizing substitution for r-nondegenerate germs of the class D always diverges.*

2. *The condition $r_1 \neq r_2$ is sufficient for two r_i-nondegenerate germs, $i = 1, 2$, to be nonequivalent to each other.*

EXAMPLE. Consider the germ

$$v^{D,k,q} = v_{p,\lambda}^D + z^k w^{q+1} \partial/\partial w, \quad k > p, \; q \in \mathbb{N} \cup \{-1\}, \quad (1.19)$$

and assume that the value $(k + q\lambda)/p$ is not a negative integer. One can easily see that this germ is q-nondegenerate. Indeed, by virtue of the last condition, the coefficient $\tilde{m}_k(q)$ appearing in the definition of Borel-type transformations and given by (1.7) is nonzero; therefore Theorem 1 implies

$$d_{j,q}^D \circ (d\mu)(z^k w^{q+1}) \neq 0 \quad \text{for all } j = 0, \ldots, p-1.$$

Since the germs (1.19) are nondegenerate, both corollaries of Theorem 3 are valid for them, that is:

— the normalizing series always diverges;
— the germ v^{D,k_1,q_1} is not equivalent to the germ v^{D,k_2,q_2} if $q_1 \neq q_2$.

Now we consider the resonance saddle case. Fix the ratio of eigenvalues $\lambda_2/\lambda_1 = -m/n$ and the formal invariants p, λ. Consider the corresponding class $S_{p,\lambda}$ and define the two sets $J^\pm$ as follows (see Figure 2):

$$J^\pm = \{(k,s) : (k,s) \in \mathbb{Z}^2, \; \pm(nk - ms) > 0\};$$

in what follows J^S will stand for any of these sets.

Consider the family of germs $\{v_\varepsilon\} \subseteq S_{p,\lambda}$, $\varepsilon \in (\mathbb{C}, 0)$, of the form (1.13) with $\operatorname{supp}(P) \subseteq J^S$. Denote by r the integer with the least absolute value such that the line

$$\{(k,s) : (k,s) \in \mathbb{Z}^2, \; nk - ms = r\}$$

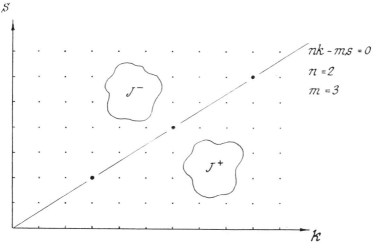

FIGURE 2

supports the convex hull of the set supp(P). Clearly $r > 0$ if $J^S = J^+$ and $r < 0$ if $J^S = J^-$.

DEFINITION. A family of germs (1.13) is called *r-nondegenerate* if at least one of the components $d^S_{j,r} \circ (d\mu) \cdot W$, $j = 0, \ldots, p-1$, is nonzero.

A germ is called *r-nondegenerate* if it occurs in an r-nondegenerate family v_ε (1.13) for $\varepsilon = 1$.

THEOREM 4. *The components* (1.11) *of any representative of the modulus of an r-nondegenerate germ v satisfy the following conditions*:

(1) $a^S_{j,l}(v) = 0$ *for all* $j = 0, \ldots, p - 1$ *provided that* $(l - r)\operatorname{sgn} r < 0$;
(2) $a^S_{j,l}(v) \neq 0$ *for* $l = r$ *provided that* $d^S_{j,r} \circ (d\mu) \cdot W \neq 0$.

The proof of this theorem is analogous to that of Theorem 3 and is omitted. A particular case was proved elsewhere (see [9]).

COROLLARIES. 1. *The normalizing substitution for an r-nondegenerate resonance saddle always diverges.*

2. *The condition $r_1 \neq r_2$ is sufficient for two r_i-nondegenerate germs to be nonequivalent.*

EXAMPLE. Consider a germ (1.13) with $\varepsilon = 1$,

$$P(z, w) = z^k w^q, \qquad k, q > 0, \quad ms \neq nq,$$

and assume that

$$[km - \lambda(mq - nk)]/(pm^2)$$

is not negative integer. Denote this germ by $v^{S,k,q}$. It is r-nondegenerate for $r = mq - nk$. Indeed, by the definition of Borel transformations, the constant $\tilde{m}_k(r)$ is nonzero, therefore by Theorem 2 one has $d^S_{j,r} \circ (d\mu) \cdot W \neq 0$ for all $j = 0, \ldots, p - 1$.

This implies the validity of both corollaries:
— divergence of the normalizing substitution, and
— pairwise analytic orbital nonequivalence of germs v^{S,k_1,q_1}, v^{S,k_2,q_2} with different values of $r_i = mq_i - nk_i$, $i = 1, 2$.

1.5. Solution of the homological equation and the derivative of the moduli map. The results stated below concern the degenerate elementary case. Let $v_{p,\lambda}^D$ be the formal normal form (1.2) for fields of the class $D_{p,\lambda}$, the formal invariants p, λ being fixed.

Consider a vector field
$$v = v_{p,\lambda}^D + W, \qquad W = P(z, w)\partial/\partial w,$$
where P is a germ of the form (1.9) holomorphic at the origin. The problem of putting the germ v into its formal normal form is a particular case of the so-called conjugation problem. The latter consists in determining the change of variables transforming one germ into the other. The substitution $H = \mathrm{id} + h$ taking v into $v_{p,\lambda}^D$ must satisfy the functional equation
$$H_* v_{p,\lambda}^D = (v_{p,\lambda}^D + W) \circ H. \qquad (1.20)$$

With this equation we can associate the so-called *homological equation*
$$[v_{p,\lambda}^D, h] = W, \qquad (1.21)$$
where
$$[v_{p,\lambda}^D, h] = h_* \cdot v_{p,\lambda}^D - (v_{p,\lambda}^D)_* \cdot h$$
is the Poisson bracket of the formal normal form $v_{p,\lambda}^D$ with the vector field h. In normal form theory the solution of many conjugation problems is based on the following heuristic principle: the Newton-Kolmogorov solution method applied to the functional equation converges if and only if solution of the corresponding homological equation is convergent. This principle, being not a theorem in any sense, nevertheless was used in different sources (for example, see [5–8]) and permitted to prove a series of results on the convergence of the above substitutions or, on the contrary, to construct examples of divergence. In the paper [11] certain classes of germs of vector fields on $(\mathbb{C}^n, 0)$ are described, including the class D, for which this principle works in the following way: divergence of solutions of the homological equation implies divergence of the normalizing series.

The question of the convergence of solutions of the homological equation is in general also rather complicated and requires substantial computations for each class of vector fields. The following theorem reduces this question to the investigation of the derivative of the moduli map.

THEOREM 5. *The solution h of the homological equation (1.21) with analytic right-hand side converges if and only if all the components of the derivative of the moduli map for the germ $v_\varepsilon = v_{p,\lambda}^D + \varepsilon W$ are equal to zero.*

REMARK. A similar statement also holds for the resonance saddle case; it is omitted here for the sake of brevity.

This theorem will be proved in §5. The proof is constructive and allows us to find the solution of the homological equation in explicit form.

§2. Computation of the derivative of the moduli map for degenerate elementary singularities

In this section we prove Theorem 1 stated in 1.2. In order to simplify the notation, we shall omit from now on the superscript D indicating the class of the singularity.

According to assertion 2 of Theorem 1, 3.4, Paper I, the functional moduli of analytic orbital classification for germs of the class $D_{p,\lambda}$ coincide with the moduli of analytic equivalence for their monodromies. Thus computation of the derivative in the former case is reduced to that of the moduli map for one-dimensional holomorphisms tangent to the identity. The latter moduli map is described in §2 of Paper I.

The proof of Theorem 1 follows the pattern outlined in Paper I and consists of several steps. First we compute the variation in the parameter ε of the monodromy map for the field $v_\varepsilon = v_{p,\lambda} + \varepsilon W$: we mean the monodromy corresponding to the positively oriented circular loop on the separatrix $z = 0$. Next, we obtain an integral representation of the variation of the functional modulus for the monodromy. Finally, we compute the coordinates of the derivative of the moduli map for the initial germ of the field. In the second and the third steps we use results respectively from 2.5 and 2.8 of Paper I.

2.1. Variation of the monodromy map. Computation of the variation of the monodromy is preceded by some simplifications of the family v_ε. All the substitutions appearing below will be of cylindrical type $(z, w) \mapsto (f(z), w)$; this corresponds to the substitution $z \mapsto f(z)$ for the monodromy.

In order to carry out the computation for the field v_ε, we consider the corresponding autonomous system in $(\mathbb{C}^2, 0)$:

$$\begin{cases} \dot{z} = z^{p+1}(1 + \lambda z^p)^{-1}, \\ \dot{w} = w + \varepsilon w P(z, w), \end{cases} \tag{2.0}$$

where

$$P(z, w) = \sum_{l=-1}^{\infty} \sum_{k=p+1}^{\infty} b_{k,l} z^k w^l.$$

Denote by U_δ the set $\{z \in \mathbb{C} : |z| < \delta\}$ and by V_δ the set $\{w \in \mathbb{C} : |w| < \delta\}$ on the z- and w-planes respectively. In order to simplify the computations we assume that the initial system is well defined on the bidisk $U_\delta \times V_{1+\alpha}$ for some positive δ, α: this can be always achieved by linear rescaling.

The substitution

$$(z, w) \mapsto (R(z), w), \quad \text{where } R: z \mapsto \varkappa = -p^{-1} z^{-p}, \tag{2.1}$$

takes the initial system to the form

$$\begin{cases} \dot{\varkappa} = 1/(1 - \lambda(p\varkappa)^{-1}), \\ \dot{w} = w + \varepsilon w P(R^{-1}(\varkappa), w). \end{cases} \quad (2.2)$$

The latter system is defined in a neighborhood of the singular point in the space $\mathscr{R} \times V_{1+\alpha}$, where $\mathscr{R}$ is the p-fold covering over $(\mathbb{C}, \infty)$ with branch point at infinity.

Let us introduce an atlas on $\mathscr{R} \setminus \{\infty\}$. To do this consider the covering of the punctured neighborhood U_δ of the origin on the z-plane by the sectors

$$S_j = \{z \in U_\delta : |\arg z + \pi/2p - \pi j/p| < \pi/p\}, \quad j = 0, \ldots, 2p-1$$

(in Paper I a likely covering by slightly narrower sectors is called a *rotated nice p-covering* of the punctured neighborhood of the origin).

REMARK. In what follows the parameter δ will be subjected to certain requirements of smallness. Without explicitly pointing this out every time, we shall assume that δ is as small as we need.

The sets $\widetilde{S}_j = R(S_j)$ form the covering of $\mathscr{R} \setminus \{\infty\}$; denote by $\hat{\pi}$ the canonical projection of $\mathscr{R}$ onto $\mathbb{C}$. We introduce the chart $\varkappa_j$ on $\widetilde{S}_j$ by means of the diagram

$$\begin{array}{ccc} S_j & \xrightarrow{R} & \widetilde{S}_j \\ {\scriptstyle \varkappa_j}\downarrow & & \downarrow{\scriptstyle \hat{\pi}} \\ \mathbb{C} & \xrightarrow{\mathrm{id}} & \mathbb{C} \end{array}$$

Thus we have obtained the atlas on $\mathscr{R} \setminus \{\infty\}$ consisting of $2p$ charts.

For the computations it is convenient to represent the local parameter $\varkappa_j$ in polar coordinates: let $z = re^{i\varphi} \in S_j$, $r < \delta$, and

$$\frac{\pi j}{p} - \frac{3\pi}{2p} < \varphi < \frac{\pi j}{p} + \frac{\pi}{2p}. \quad (2.3)$$

Then $\varkappa_j = \hat{\pi} \circ R|_{S_j}$ is given by the formulas

$$\varkappa_j : re^{i\varphi} \mapsto \rho e^{i\psi}, \quad \rho = r^{-p}/p, \quad \psi = -p\varphi + \pi.$$

Denote by C_+ and C_- the complex planes with the deleted positive (respectively, negative) imaginary semiaxis. It is easy to see that by the definition of the local parameter we have

$$\varkappa_j(\widetilde{S}_j) = (\mathbb{C}, \infty) \cap C_\pm,$$

the sign being $+$ if j is even and $-$ otherwise. Denote these two sets by $S_\pm$. Figures 3 and 4 illustrate the case $p = 2$.

The next transformation $(\varkappa, w) \mapsto (Q(\varkappa), w)$ of the system (2.2) is defined in each domain $\widetilde{S}_j \cap V_{1+\alpha}$ separately. In the chart $\varkappa_j$, it has the form

$$Q_j : \varkappa_j \mapsto \zeta = \varkappa_j - (\lambda/p) \ln \varkappa_j. \quad (2.4)$$

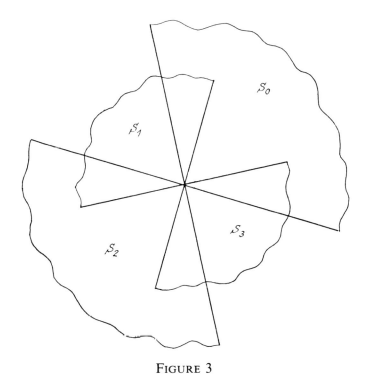

FIGURE 3

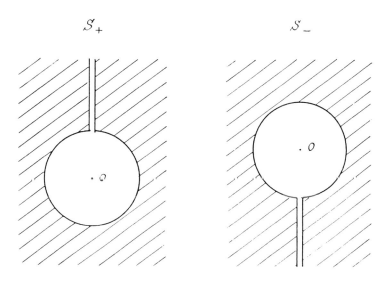

FIGURE 4

We will suppose that the main branch of the logarithm is chosen for Q_0, Q_1 is the analytic continuation of Q_0 and coincides with it in the intersection $\widetilde{S}_0 \cap \widetilde{S}_1$, Q_2 extends Q_1 and so on. Finally the map Q_{2p-1} will differ from Q_0 on the intersection $\widetilde{S}_{2p-1} \cap \widetilde{S}_0$ by the constant $2\pi i \lambda$. Note that a similar construction is part of the proof of the classification theorem for one-dimensional maps, see 2.4, Paper I; the map $Q_j \circ R$ is the rectifying map (2.5) of Paper I.

The transformation described above takes the system (2.2) to the form

$$\begin{cases} \dot{\zeta} = 1, \\ \dot{w} = w + \varepsilon w \widetilde{P}_j(\zeta, w), \end{cases} \qquad \widetilde{P}_j(\zeta, w) = P(R^{-1} \circ Q_j^{-1}(\zeta), w). \qquad (2.5)$$

Now we turn to the computation of the monodromy map. Denote by $\Pi_\Xi^\pm$ the infinite semistrips on the complex plane:

$$\Pi_\Xi^\pm = \{\zeta : |\operatorname{Re}\zeta| < \Xi, \pm \operatorname{Im}\zeta + \Xi > 0\},$$

and let $S_\Xi^\pm$ stand for $S^\pm \setminus \Pi_\Xi^\pm$. Further let $\widetilde{S}_j^\Xi$ be the inverse images on $\widetilde{S}_j \subset \mathscr{R}$ of the set S_Ξ^+ for j even or S_Ξ^- for j odd. Let γ_0 be the unit circle on the w-plane and choose a point $\eta \in \widetilde{S}_j$. By definition, the monodromy map takes the set $\widetilde{S}_j \times \{1\}$ near to itself. The point $(\eta, 1)$ is mapped by the monodromy to $(\Delta_j(\eta), 1)$, the latter point being obtained as the result of the continuation of the solution $(\zeta(t; \varepsilon, \eta), w(t; \varepsilon, \eta))$ with the initial condition

$$(\zeta(0; \varepsilon, \eta), w(0; \varepsilon, \eta)) = (\eta, 1)$$

over the curve γ_0.

The monodromy is well defined, since for Ξ sufficiently large, the curve lying on the solution of the system (2.5) with the initial condition $(\eta, 1) \in \widetilde{S}_j^\Xi \times \{1\}$, which covers the circle γ_0, entirely belongs to the domain $\widetilde{S}_j \times V_{1+\alpha}$ (see Paper I, 3.4).

LEMMA 2.1. *The monodromy map Δ_j for the system (2.5) in the chart $(\widetilde{S}_j, \varkappa_j)$, has the form*

$$\eta \mapsto \eta + 2\pi i - \varepsilon \int_0^{2\pi i} \widetilde{P}_j(t+\eta, e^t)\,dt + \cdots, \qquad (2.6)$$

where $\widetilde{P}_j(\zeta, w)$ is the same as in (2.5); the dots stand for terms of order higher than one in ε.

PROOF. Let us suppose that $(\zeta_1(t; \eta), w_1(t; \eta))$ is the variation of the solution $\chi(t; \varepsilon, \eta) = (\zeta(t; \varepsilon, \eta), w(t; \varepsilon, \eta))$ with respect to the parameter ε for $\varepsilon = 0$ satisfying the initial conditions $\chi(0; \varepsilon, \eta) = (\eta, 1)$. Then

$$\zeta_1(0) = \zeta_2(0) = 0, \quad \zeta_1(t; \varepsilon, \eta) = t + \eta, \quad w(t; 0, \eta) = e^t.$$

The equation in variations has the form
$$\dot{\zeta}_1 = 0, \quad \dot{\zeta}_2 = \zeta_2 + e^t \widetilde{P}_j(t+\eta, e^t), \quad \zeta_1(0) = \zeta_2(0) = 0.$$

Its solution is
$$\zeta_1 \equiv 0, \quad \zeta_2(t, \eta) = \int_0^t \widetilde{P}_j(\tau + \eta, e^\tau) d\tau.$$

Therefore
$$\chi(t; \varepsilon, \eta) = (t+\eta, e^t + \varepsilon \zeta_2(t; \eta) + O(|\varepsilon|^2)), \quad \zeta(t; \varepsilon, \eta) = t + \eta.$$

The first return time T for the solution χ continued over the circle $\zeta = 0$, $w = \exp(i\beta)$, $\beta \in [0, 2\pi]$ to the transversal $w = 1$ is given by a holomorphic function $T(\eta, \varepsilon)$, $T(\eta, 0) = 2\pi i$; this function satisfies the condition
$$e^T + \varepsilon \zeta_2(T, \eta) + O(|\varepsilon|^2) = 1.$$

Hence
$$T(\eta, \varepsilon) = 2\pi i - \varepsilon \zeta_2(2\pi i, \eta) + O(|\varepsilon|^2).$$

The assertion of Lemma now follows from the definition of the monodromy, since $\Delta_j(\eta) = \zeta(T(\eta, \varepsilon); \varepsilon, \eta)$. □

2.2. The variation of Martinet-Ramis moduli. First we recall how the Martinet-Ramis moduli are constructed. The monodromy Δ is put into the normal form $\eta \mapsto \eta + 1$ by holomorphic changes of variables H_j defined in each domain $\widetilde{S}_j^\Xi$ for some sufficiently large Ξ. From now on, we shall always assume that the same Ξ is chosen for all j and is as large as we need.

The maps $\Phi_j^\pm$, $j = 0, \ldots, 2p-1$, forming the $2p$-tuple of the Martinet-Ramis modulus are defined as the transition functions between the normalizing charts:
$$\Phi_j^- = H_{2j+1} \circ H_{2j}^{-1}, \quad \Phi_j^+ = H_{2j+2} \circ H_{2j+1}^{-1}, \quad j = 0, \ldots, p-1;$$

the maps Φ_j^+ and Φ_j^- are defined in the domains $\widetilde{S}_{2j+1}^\Xi \cap \widetilde{S}_{2j}^\Xi$ and $\widetilde{S}_{2j+2}^\Xi \cap \widetilde{S}_{2j+1}^\Xi$ respectively.

The formal normal form for the map (2.6) is the shift $\eta \mapsto \eta + 1$. We look for the normalizing map in the form $H_j = \text{id} + h_j$; in order to determine the correction h_j, we carry out the reparameterization $\varkappa_j \mapsto \varkappa_j / 2\pi i$ and represent the map (2.6) in the form
$$\widetilde{\Delta}_j : \eta \mapsto \eta + 1 - \frac{\varepsilon}{2\pi i} \int_0^{2\pi i} \widetilde{P}_j(\mu + 2\pi i \eta, e^\mu) d\mu + \cdots ; \qquad (2.7)$$

for simplicity we shall denote the integral term by $\varepsilon F_j(\eta)$.

We shall denote by $S_\pm^\Xi$ the images of the sets $S_\Xi^\pm$ under the transformation $\varkappa_j \mapsto (2\pi i)^{-1} \varkappa_j$.

According to the results from 2.5, Paper I (see also [14]), one has

$$h_{2j} = \sum_{k=0}^{\infty} F_{2j} \circ \tilde{\Delta}_{2j}^{[k]}, \quad h_{2j+1} = -\sum_{k=1}^{\infty} F_{2j+1} \circ \tilde{\Delta}_{2j+1}^{[k]}, \quad j = 0, \ldots, p-1;$$

here $\tilde{\Delta}_i^{[k]}$ stands for the kth iteration of the map $\tilde{\Delta}_i$.

From this representation it is easy to obtain an explicit representation for the variation

$$\omega_j^{\pm} = \partial \Phi_j^{\pm}/\partial \varepsilon|_{\varepsilon=0}$$

given by Lemma 2.2 below. Denote by $W_{\mp}$ the two half-planes $\{\zeta \in \mathbb{C} : \pm \operatorname{Im} \zeta > L\}$, $L > 0$, contained in the intersection $S_+^{\Xi} \cap S_-^{\Xi}$.

LEMMA 2.2. *The first order variations of the Martinet-Ramis moduli are given by the expressions*

$$\omega_j^- = \sum_{l=-\infty}^{\infty} F_{2j} \circ (\mathrm{id} + l), \quad \omega_j^+ = \sum_{l=-\infty}^{\infty} F_{2j+1} \circ (\mathrm{id} + l);$$

by choosing an appropriate L, *we may assume without loss of generality that these equalities hold in* W_- *and* W_+ *respectively.*

The proof is an elementary consequence of the above formulas for h_{2j}, h_{2j+1} and $\tilde{\Delta}_j$.

2.3. Coefficients of the Borel-type transform. Let us fix some j, where $0 \leqslant j \leqslant p-1$, and find the Fourier coefficients for the functions $\omega_j^{\pm}$. We have

$$\omega_j^{\pm}(\zeta) = \sum_{r=1}^{\infty} c_{j,r}^{\pm} e^{\mp 2\pi i r \zeta};$$

the coefficients c are given by the formula

$$c_{j,r}^{\pm} = \int_{\mp i\beta}^{\mp i\beta+1} \omega_j^{\pm} e^{\pm 2\pi i r \mu} d\mu$$

for some $\beta > 0$ sufficiently large.

REMARK. The formula (3.16), Paper I, implies that $c_{j,r}^- = 0$ for $r \geqslant 2$ and all j. Disregarding this *a priori* information, we will determine the coefficients for all $r \geqslant 1$.

Let us represent the function $P(z, w)$ from (1.8) and (2.0) as the series

$$P(z, w) = \sum_{s=-1}^{\infty} w^s f_s(z), \qquad (2.8)$$

where f_s are functions in one variable holomorphic near the origin.

LEMMA 2.3.

$$c_{j,r}^+ = -\frac{1}{2\pi i}\int_{\gamma_0^+} e^{r\theta}\tilde{f}_{r,2j+1}(\theta)\,d\theta, \qquad (2.9)$$

$$c_{j,r}^- = -\frac{1}{2\pi i}\int_{\gamma_0^-} e^{-r\theta}\tilde{f}_{-r,2j}(\theta)\,d\theta, \qquad (2.10)$$

where

$$\tilde{f}_{r,2j+1}(\theta) = f_r \circ R^{-1} \circ Q_{2j+1}^{-1}, \quad \tilde{f}_{-r,2j}(\theta) = f_{-r} \circ R^{-1} \circ Q_{2j}^{-1},$$

and $\gamma_0^\pm$ are the straight lines $(-i\infty \mp 2\pi\beta, i\infty \mp 2\pi\beta)$.

The proof of this and the next lemmas is purely technical; it will be postponed till §6.

REMARK. The above formulas imply that the coefficient $c_{j,r}^-$ depends only on f_{-r}, $r \geq 1$. Since $f_{-r} \equiv 0$ for $r > 1$, the value of the corresponding coefficients could be nonzero only for $r = 1$ (see Figure 1), which agrees with the previous remark.

Find $c_{j,r}^\pm$. Let $f_s = \sum_{k=p+1}^\infty b_{k,s} z^k$ and denote by $c_{j,r,k}^\pm$ the values of the integrals (2.9) and (2.10), in which the function f_r is substituted by the monomial z^k.

LEMMA 2.4.

$$c_{j,r,k}^+ = e_{j,r}\varepsilon_p^{k(2j+1)} \frac{k(r/p)^{k/p}}{\Gamma(1+(\lambda r + k)/p)},$$

$$c_{j,r,k}^- = e_{j,-r}\varepsilon_p^{k(2j+1)} \frac{k(-r/p)^{k/p}}{\Gamma(1+(-\lambda r + k)/p)}, \qquad (2.11)$$

where $e_{j,l}$ and ε_p are the constants defined in 1.2.

Substituting (2.11) in the equality

$$c_{j,r}^\pm = \sum_{k=p+1}^\infty b_{k,\pm r} c_{j,r,k}^\pm,$$

and taking into account the definition

$$d_{j,l} \circ (d\mu) \cdot W = \begin{cases} c_{j,l}^+ & \text{if } l \geq 1, \\ c_{j,-l}^- & \text{if } l = -1, \end{cases}$$

we finally obtain the assertion of Theorem 1. The theorem is thus proved modulo Lemmas 2.3 and 2.4.

§3. Computation of the moduli map derivative in the resonance saddle case

In this section Theorem 2 is proved. The pattern of the proof coincides with that of Theorem 1, differing by technical details only, so the exposition is rather brief.

To be definite we shall assume that the separatrix of the field v_ε (1.13) under consideration is a punctured neighborhood of the origin in the w-plane. The monodromy corresponding to the positive circular loop of radius 1 on the separatrix is the germ of a map $(\mathbb{C}, 0) \to (\mathbb{C}, 0)$ with the multiplier $\nu = \exp(-2\pi i n/m)$, where n/m is the (irreducible) ratio of the eigenvalues. This monodromy will be denoted by Δ.

The monodromy map is the modulus of analytic orbital classification of resonance saddle points, see 4.1, Paper I. Any two holomorphic maps $(\mathbb{C}, 0) \to (\mathbb{C}, 0)$ with the same multiplier ν are analytically equivalent if and only if their mth iteration are; the mth iteration of such a map has the form

$$z \mapsto z + ax^{pm+1} + \cdots, \qquad a \neq 0.$$

Then pm pairs of maps (Φ_j^-, Φ_j^+), $j = 0, \ldots, pm-1$ (the transition functions) constitute the classification modulus in this case; only the first p pairs are independent (see 4.4, Paper I). The Fourier coefficients of these independent functions constitute the tuple (1.11) which serves as the modulus for the germ v_ε.

3.1. Variation of the monodromy map. Consider the autonomous system of differential equations associated with the field v_ε:

$$\begin{cases} \dot{z} = z(-nm^{-1} + u^p(1 - \varepsilon P(z, w))(1 + \lambda u^p)^{-1}), \\ \dot{w} = w. \end{cases} \quad (3.1)$$

Denote by y the function $y = w^{1/m}$ on the m-fold covering over the domain $w \neq 0$ and pass to the new variables

$$\begin{cases} \dot{z} = z(-nm^{-1} + \tilde{u}^p(1 - \varepsilon P(z, y^m))(1 + \lambda \tilde{u}^p)^{-1}), \\ \dot{y} = m^{-1} y, \end{cases} \quad (3.2)$$

where $\tilde{u}$ stands for $z^m y^{mn}$. The system (3.2) is well defined in a neighborhood of the singular point in the space $\mathbb{C} \times \mathscr{R}^m$, where $\mathscr{R}^m$ is the covering over $(\mathbb{C}, 0) \setminus \{0\}$ given by $w = y^m$. Let $\tilde{\gamma}_0$ be the generator for the fundamental group of $\mathscr{R}^m$ (the lifting of the positive loop around the origin in $(\mathbb{C}, 0)$). It can be easily seen that the mth iteration of the monodromy Δ for (3.1) is analytically conjugated with the monodromy of (3.2) corresponding to $\tilde{\gamma}_0$.

Let us find $\Delta^{[m]}$. Carry out the substitution $(z, y) \mapsto (zy^n, y) = (x, y)$ which is biholomorphic in the neighborhood of $\tilde{\gamma}_0$. The system (3.2) then takes the form

$$\begin{cases} \dot{x} = x^{pm+1}(1 + \lambda x^{pm})^{-1}(1 - \varepsilon \tilde{P}(x, y)), \\ \dot{y} = m^{-1} y, \end{cases} \quad (3.3)$$

where $\tilde{P}(x, y) = P(xy^{-n}, y^m)$. Evidently, $\Delta^{[m]}(\xi) = x(e^{2\pi m i}; \xi, 1)$, where $x(t; \xi, 1)$ is the solution to (3.3) with $y = \exp(t/m)$ and the initial condition

$x(0;\xi,1) = \xi$. To compute the monodromy, we use the method described in 2.1. Divide both components of the field (3.3) by $1 - \varepsilon\widetilde{P}$. We obtain

$$\begin{cases} \dot{x} = x^{pm+1}(1 + \lambda x^{pm})^{-1}, \\ \dot{y} = m^{-1}y(1 + \varepsilon\widetilde{P}(x,y) + \cdots), \end{cases} \quad (3.4)$$

with the dots standing for terms of order ≥ 2 in ε. The map $\Delta^{[m]}$ is analytically equivalent to the monodromy of the system (3.4) corresponding to the loop $\widetilde{\gamma}_0$ on the invariant manifold $\{x = 0\}$. The latter monodromy will be denoted by $\widetilde{\Delta}$.

Consider the pm-fold covering $\mathscr{R}^{pm}$ over $(\mathbb{C}, \infty)$ with ramification point at infinity. On $\mathscr{R}^{pm} \setminus \{\infty\}$ construct the atlas $\{\widetilde{S}_j\}$, $j = 0, \ldots, 2pm - 1$, in the same way as in 2.1. By the following two substitutions,

$$R\colon x \mapsto \varkappa = -(pm)^{-1}x^{-pm}, \qquad Q_j\colon \varkappa_j \mapsto \zeta = \varkappa_j - \lambda(pm)^{-1}\ln \varkappa_j,$$

the system (3.4) in the domain $\widetilde{S}_j \times \mathscr{R}^n \subset \mathscr{R}^{pm} \times \mathscr{R}^n$ is transformed to

$$\begin{cases} \dot{\zeta} = 1, \\ \dot{y} = m^{-1}y(1 + \varepsilon\widehat{P}_j(\zeta, y) + \cdots); \end{cases} \quad (3.5)$$

here $\widehat{P}_j(\zeta, y) = \widetilde{P}(R^{-1} \circ Q_j^{-1}(\zeta), y)$.

The computation of the monodromy is completed by the following lemma, whose proof is analogous to that of Lemma 2.1.

LEMMA 3.1. *The monodromy of the system* (3.5) *after the substitution* $\varkappa_j \mapsto (2\pi i m)^{-1}\varkappa_j$ *takes the form*

$$\eta \mapsto \eta + 1 - \frac{\varepsilon}{2\pi mi}\int_0^{2\pi mi} \widehat{P}_j(\mu + 2\pi mi\eta, e^{\mu/m})\,d\mu + \cdots, \quad (3.6)$$

with the dots standing for terms of order ≥ 2 *in* ε.

3.2. Computation of the variations of the moduli. For the map $\widetilde{\Delta}$ given by (3.6), Lemma 2.2 also holds. Therefore to prove Theorem 2 it is sufficient to find the Fourier coefficients $c_{j,r}^{\pm}$ of the functions

$$\omega_j^{\pm} = \partial \Phi_j^{\pm}/\partial \varepsilon|_{\varepsilon=0}.$$

Fix any j, $0 \leq j \leq p - 1$. The coefficients $c_{j,r}^{\pm}$ are defined by the formula

$$c_{j,r}^{\pm} = \int_{\mp i\beta}^{\mp i\beta+1} \omega_j^{\pm}(\mu)e^{\pm 2\pi ri\mu}\,d\mu$$

for some $\beta > 0$.

Represent the function $\widetilde{P}(x, y)$ in the form of a series:

$$\widetilde{P}(x, y) = \sum_{k=2pm+1}^{\infty} \sum_{s=2pm+1}^{\infty} b_{k,s} x^k y^{ms-nk}.$$

The irreducibility of the fraction m/n implies that the parameter $q = ms - nk$ ranges over the entire set $\mathbb{Z}$ when k, s vary over $\mathbb{N}$. There the latter expression can be written in the form

$$\widetilde{P}(x, y) = \sum_{q=-\infty}^{\infty} y^q f_q(x), \tag{3.7}$$

where all the f_q are holomorphic at the origin, $f_q(x) = o(|x|^{2pm})$.

The proof of the next lemma differs from that of Lemma 2.3 only by minor technicalities; we omit it.

LEMMA 3.2.

$$c_{j,r}^+ = \frac{1}{2\pi i m} \int_{\widetilde{\gamma}_0^+} e^{\theta r/m} \widetilde{f}_r^{2j+1}(\theta) \, d\theta, \tag{3.8}$$

$$c_{j,r}^- = -\frac{1}{2\pi i m} \int_{\widetilde{\gamma}_0^-} e^{-\theta r/m} \widetilde{f}_{-r}^{2j}(\theta) \, d\theta, \tag{3.9}$$

where

$$\widetilde{f}_q^{2j} = f_q \circ R^{-1} \circ Q_{2j}^{-1}, \qquad \widetilde{f}_q^{2j+1} = f_q \circ R^{-1} \circ Q_{2j+1}^{-1},$$

and $\widetilde{\gamma}_0^\pm$ are the straight lines $(-i\infty \mp 2\pi i\beta m, i\infty \mp 2\pi i\beta m)$.

Let us find $c_{j,r}^\pm$ in explicit form. Set

$$f_q(x) = \sum_{(k,s)\in J_q} b_{k,s} x^k,$$

where $J_q = \{(k, s) : ms - nk = q, k \geq 2pm+1, s \geq 2pn+1\}$, and compute the integrals (3.8), (3.9) for each monomial x^k separately. Let us denote the first one by $c_{j,r,k}^+$, and the second one by $c_{j,r,k}^-$.

LEMMA 3.3.

$$c_{j,r,k}^\pm = e_{j,\pm r} \varepsilon_{pm}^{k(2j+1)} \frac{k(\pm r p^{-1} m^{-2})^{k/pm}}{\Gamma(1 + (\pm \lambda r + km)p^{-1}m^{-2})},$$

where $e_{j,l}$ and ε_{pm} are the constants defined in 1.3.

This lemma is proved by the same method as Lemma 2.4; the proof is omitted.

Substituting the obtained expressions into the identity

$$c_{j,r}^\pm = \sum_{(k,s)\in J_{\pm r}} b_{k,s} c_{j,r,k}^\pm,$$

and taking into account the fact that, by definition,

$$d_{j,l} \circ (d\mu) \cdot W = \begin{cases} c_{j,l}^+ & \text{if } l > 0, \\ c_{j,-l}^- & \text{if } l < 0, \end{cases}$$

we conclude the proof of Theorem 2. □

§4. Degenerate elementary singularities with nontrivial Martinet-Ramis moduli

This section is devoted to the proof of Theorem 3. Here and below definitions and notation from 1.2 and 2.1 are used; the superscript D is omitted for simplicity.

First we prove assertion 2 of the theorem. Recall that the Martinet-Ramis modulus of analytic orbital classification for vector fields of the class $D_{p,\lambda}$ is the equivalence class of tuples (1.3) with respect to the equivalence relation (1.5).

To be definite, let us assume that the support of the correction $W = \widetilde{P}(z, w)\partial/\partial w$ of an r-nondegenerate field $v = v_{p,\lambda} + W$ belongs to the set J_+ (if it is in J_-, the proof is completely analogous). Then the series $\widetilde{P}$ can be represented in the form

$$\widetilde{P}(z, w) = w \sum_{l=r}^{\infty} w^l f_l(z). \tag{4.1}$$

The orbit of the germ v under the action of the one-parameter linear transformation group $\{(z, w) \mapsto (z, \varepsilon^{-1}w) : \varepsilon \in \mathbb{C}^*\}$ contains the normal form $v_{p,\lambda}$ in its closure. Therefore the investigation of the tangent to the moduli map at the point $v_{p,\lambda}$ of this orbit permits us to clarify the analytic nature of the field v itself.

The tangent space to this orbit at the point $v_{p,\lambda}$ is determined by the coefficient f_r of the series (4.1). Indeed, the substitution $(x, y) = (z, \varepsilon^{-1}w)$ preserves the formal normal form $v_{p,\lambda}$ and sends the field v into the field $\widetilde{v}_\varepsilon = v_{p,\lambda} + \varepsilon^r \widetilde{W}_\varepsilon$, where

$$\widetilde{W}_\varepsilon = y\left(f_r(x)y^r + \sum_{l=r+1}^{\infty} \varepsilon^{l-r} y^l f_l(x)\right) \frac{\partial}{\partial y}. \tag{4.2}$$

As a result, the term having minimal degree in ε is $f_r(x)y^r$; thus it is this term which determines the derivative of the moduli map.

Since after the above transformation the coefficients of $\widetilde{W}_\varepsilon$ may only decrease as $|\varepsilon| \to 0$, without loss of generality one may assume that all the germs of the family just obtained are defined in some common neighborhood of the singular point for all small ε.

Apply Theorem 1 to the family $\{\widetilde{v}_\varepsilon\}$. The r-nondegeneracy of $\widetilde{v}_\varepsilon$ implies that there exists a j_0 such that $d_{j_0,r} \circ (d\mu) \cdot \widetilde{W}_0 \neq 0$; hence for ε small enough the component $a_{j_0,r}(\widetilde{v}_\varepsilon)$ of some tuple (1.11) representing the Martinet-Ramis modulus is also nonzero. The germ v_ε for all small ε belongs to the same equivalence class. Hence for all ε the tuples $\{a_{j,l}(\widetilde{v}_\varepsilon)\}$

represent the same equivalence class also. Thus $a_{j_0,r}(\tilde{v}_\varepsilon) \neq 0$ for all ε including $\varepsilon = 1$. The second assertion of the Theorem is proved.

Let us prove the first assertion. By the above assumption, $\operatorname{supp}\tilde{P} \subset J^+$; therefore $r > 0$ and we must prove that $a_{j,q}(v) = 0$ for all $q < r$ and $j = 0, \ldots, p - 1$. From the preceding construction it follows that to prove this it is sufficient to prove the equalities $a_{j,q}(\tilde{v}_\varepsilon) = 0$ for small ε.

Since the germ $\tilde{v}_\varepsilon$ for all $\varepsilon \neq 0$ belongs to the same analytic equivalence class, (1.5) implies

$$a_{j,q}(\tilde{v}_\varepsilon) = a_{j,q}(v)\delta^q(\varepsilon) \qquad (4.3)$$

for some $\delta = \delta(\varepsilon)$ and all j, q. From analytic dependence of the Martinet-Ramis modulus on parameters we conclude that δ depends analytically on ε. Let us estimate this function for small ε. From the r-nondegeneracy of v it follows that

$$a_{j_0,r}(\tilde{v}_\varepsilon) = O^*(|\varepsilon|^r);$$

the symbol O^* stands for the exact order indicator. Since $a_{j_0,r}(v) \neq 0$, by putting $q = r$ in (4.3) we conclude that $\delta(\varepsilon)$ is of the first order exactly.

Consider arbitrary j, q and suppose that $a_{j,q}(v) \neq 0$. Then (4.3) and the order estimate jointly imply that $a_{j,q}(\tilde{v}_\varepsilon) = O^*(|\varepsilon|^q)$. On the other hand, set

$$\tilde{v}_{\varepsilon,\mu} = v_{p,l} + \mu \widetilde{W}_\varepsilon.$$

Then $\tilde{v}_\varepsilon = \tilde{v}_{\varepsilon,\mu}|_{\mu=\varepsilon^r}$. The component $a_{j,q}(\tilde{v}_{\varepsilon,\mu})$ depends analytically on μ and ε and vanishes when $\mu = 0$. Therefore

$$a_{j,q}(\tilde{v}_{\varepsilon,\mu}) = O(\mu), \qquad a_{j,q}(\tilde{v}_\varepsilon) = O(|\varepsilon|^r).$$

The two estimates of $a_{j,q}(\tilde{v}_\varepsilon)$ imply that q must satisfy the inequality $q \geq r$. The proof of Theorem 3 is complete. □

§5. Solution of the homological equation

This section deals with Theorem 5 from 1.5. To simplify our exposition, we will prove Theorem 5 for germs of the class $D_{1,0}$: the general case is treated in an analogous manner, but the computations are more sophisticated. We omit this case in the current paper.

5.1. The Borel transform and solution of the homological equation. Consider a modification of the Borel transform for the case of series defined near infinity. Let $f(\eta) = \sum_{k=1}^\infty f_k/\eta^{k+1}$ be a formal power series. Its Borel transform is the series

$$(\mathscr{B}f)(\zeta) = \sum_{k=1}^\infty \frac{f_k \zeta^k}{k!}.$$

The following statements are direct implications of this definition.

PROPOSITION 5.1. *If the series f converges in some neighborhood of infinity, then*:

(1) $\mathscr{B}f$ *is an entire function*;
(2) $(\mathscr{B}f)(\zeta) = (2\pi i)^{-1} \int_\gamma f(\eta) e^{\zeta \eta} d\eta$, *where γ is the circle of some sufficiently large radius*;
(3) $(\mathscr{B}f')(\zeta) = -\zeta(\mathscr{B}f)(\zeta)$.

The formal transformation H conjugating the field $v = v_{p,\lambda} + W$ with its formal normal form $v_{p,\lambda}$, preserves the first coordinate. Therefore it is sufficient to consider only the equation for the second component. Denote $v_{p,\lambda} = (v_1, v_2)$ and $H = \text{id} + h$, $h = (0, \tilde{h})$. The equation (1.21) for $\tilde{h}$ takes the form

$$v_1 \frac{\partial \tilde{h}}{\partial z} + v_2 \frac{\partial \tilde{h}}{\partial w} - \tilde{h} \frac{\partial v_2}{\partial w} = P. \tag{5.1}$$

Represent the corrections P, $\tilde{h}$ in the form of power series:

$$P(z, w) = w \sum_{k=-1}^{\infty} w^k f_k(z), \tag{5.2}$$

$$\tilde{h}(z, w) = w \sum_{k=-1}^{\infty} w^k h_k(z); \tag{5.3}$$

here $h_k(z)$ is a formal power series

$$h_k(z) = \sum_{l=1}^{\infty} h_k^l z^{l+1}.$$

Substituting (5.3) into (5.1) and taking into account the fact that for $(p, \lambda) = (1, 0)$ the formal normal form is $v_{1,0} = z^2 \partial/\partial z + w \partial/\partial w$, we obtain

$$z^2 \sum_{k=-1}^{\infty} w^{k+1} h_k'(z) + \sum_{k=-1}^{\infty} (k+1) w^{k+1} h_k(z) - \sum_{k=-1}^{\infty} w^{k+1} h_k(z) = \sum_{k=-1}^{\infty} w^{k+1} f_k(z).$$

Equating the coefficients of the term w^{k+1} in this relation for each term h_k, we obtain the differential equation

$$z^2 h_k'(z) + k h_k(z) = f_k(z), \quad k = -1, 0, 1, \ldots .$$

These equations are solved as follows. Make the substitution $z \mapsto 1/\eta$ and let $g_k(\eta) = h_k(1/\eta)$, $\tilde{f}_k(\eta) = f_k(1/\eta)$, obtaining

$$g_k' - k g_k = -\tilde{f}_k.$$

Apply the Borel transformation to both sides of this equality. By virtue of Proposition 5.1 (assertion 3), one has

$$\zeta \cdot (\mathscr{B} g_k)(\zeta) + k (\mathscr{B} g_k)(\zeta) = (\mathscr{B} \tilde{f}_k)(\zeta).$$

Hence
$$(\mathscr{B} g_k)(\zeta) = (\mathscr{B} \widetilde{f_k})(\zeta)/(\zeta + k). \tag{5.4}$$

Thus we see that the homological equation is equivalent to the countable system of equations (5.4) with $k = -1, 0, \ldots$.

Prove the necessity part of Theorem 5. Suppose that the solution of the homological equation is convergent. Then each g_k is a function holomorphic at infinity. From Proposition 5.1 (assertion 1) it follows then that the functions $\mathscr{B} g_k$ are entire. On the other hand, the right-hand side of (5.4) in general has a singularity at $\zeta = -k$.

PROPOSITION 5.2. *The singularity of $\widetilde{F_k}(\zeta) = (\mathscr{B} \widetilde{f_k})(\zeta)/(\zeta + k)$ at the point $\zeta = -k$ is removable if and only if $(\mathscr{B} \widetilde{f_k})(-k) = 0$.*

The proof is evident.

Thus the homological equation may have a convergent solution only if $(\mathscr{B} \widetilde{f_k})(-k) = 0$. The necessity statement of Theorem 5 now follows from Theorem 1, since $d_{0,k} \circ (d\mu) \cdot W = (\mathscr{B} \widetilde{f_k})(-k)$ for all $k \in \mathbb{N} \cup \{-1\}$.

Let us prove the sufficiency part. If $(\mathscr{B} \widetilde{f_k})(-k) = 0$, then the right-hand side of (5.4) is an entire function. Applying the inverse Borel transformation to both sides of this equation, we get

$$g_k = \mathscr{B}^{-1}((\mathscr{B} \widetilde{f_k})(\zeta)/(\zeta + k)), \tag{5.5}$$

and finally

$$h_k(z) = g_k(1/z). \tag{5.6}$$

The following assertion completes the proof.

PROPOSITION 5.3. *The conditions of Theorem 5 being satisfied, convergence of the series P implies convergence of the series $\widetilde{h}$ of the form (5.3), (5.6).*

This proposition is proved in the next section. In the proof we present an explicit formula for the series $\widetilde{h}$ and estimate its radius of convergence.

5.2. Convergence of solutions of the homological equation. In order to find $\widetilde{h}$ we consider the monomial $F_k^l(\zeta) = \zeta^l f_k^l/l!$ and find the Taylor decomposition of the regular part of the function $F_k^l(\zeta)/(\zeta + k)$. For $l = 0$ the function F_k^l is a constant, therefore so is its regular part. For $l > 0$ we have

$$\frac{F_k^l(\zeta)}{\zeta + k} = \frac{\zeta^l - (-k)^l + (-k)^l}{l!(\zeta + k)} f_k^l = \frac{(-k)^l}{l!(\zeta + k)} f_k^l + \frac{(-k)^{l-1}}{l!} f_k^l \sum_{q=0}^{l-1} \frac{\zeta^q}{(-k)^q}. \tag{5.7}$$

Summing all the equalities (5.7) over all k, l and returning again to the variables z, w, we conclude that

$$h(z, w) = \sum_{k=-1}^{\infty} \left(\sum_{l=0}^{\infty} l! \left(\sum_{q=l+1}^{\infty} \frac{(-k)^{q-l+1}}{q!} f_k^q \right) z^{l+1} \right) w^{k+1}. \tag{5.8}$$

The formula (5.8) gives the solution of the homological equation in explicit form. Let us estimate the radius of convergence for this series.

Assume that the series $P(z, w)$ converges in the closed bidisk
$$D_{r,r} = \{|z| \leq r, |w| \leq r\}$$
and M is the maximum of its modulus in this bidisk. By virtue of the Cauchy inequalities,
$$|f_k^l| < Mr^{-(l+k+1)}.$$
Hence the series (5.8) is majorized by the convergent series
$$\sum_{k=-1}^{\infty} \left(\sum_{l=0}^{\infty} l! \left(\sum_{q=l+1}^{\infty} \frac{Mk^{q-l-1}}{q! r^{k+l+1}} \right) |z|^{l+1} \right) |w|^{k+1}.$$
Let us show that the latter series converges in the bidisk $D_{r, re^{-1/r}}$. Taking into account the fact that $l!/q! \leq 1/(q-l)!$ for $l \leq q$ and
$$\sum_{q=l+1}^{\infty} \frac{k^{q-l}}{(q-l)! r^{q-l}} \leq e^{k/r},$$
we obtain
$$|h(z, w)| \leq M \sum_{k=-1}^{\infty} \frac{e^{k/r}}{r^k} \left(\sum_{l=0}^{\infty} \frac{|z|^{l+1}}{r^{l+1}} \right) |w|^{k+1}.$$
Since
$$\sum_{l=0}^{\infty} \frac{|z|^{l+1}}{r^{l+1}} = \frac{|z|}{r} \frac{1}{1 - |z|/r},$$
we finally obtain the estimate
$$|h(z, w)| \leq \frac{M|z||w|e^{-1/r}}{(1 - |z|/r)(1 - |w|e^{1/r}/r)}.$$
Thus the series is convergent in the bidisk $D_{r, re^{-1/r}}$ and Proposition 5.3 is proved. □

§6. Proofs of the technical lemmas

6.1. Proof of Lemma 2.3. Here we compute the Fourier coefficients $c_{j,r}^+$ for the function ω_j^+; the computation for $c_{j,r}^-$ is completely analogous and we omit it here.

By virtue of Lemma 2.2 and formula (2.7), we have
$$c_{j,r}^+ = -\frac{1}{2\pi i} \int_{-i\beta}^{-i\beta+1} e^{2\pi i r \mu} \sum_{l=-\infty}^{\infty} \int_0^{2\pi i} \widetilde{P}_{2j+1}(\tau + 2\pi i(\mu + l), e^\tau) \, d\tau \, d\mu. \quad (6.1)$$

Recall that
$$\widetilde{P}_{2j+1}(\zeta, e^\tau) = P(R^{-1} \circ Q_{2j+1}^{-1}(\zeta), e^\tau),$$

where the function P is given by the series (2.8), and the transforms R and Q_{2j+1} are defined by (2.1), (2.4) respectively.

Since the series $P(z, w)$ near the origin is majorized by the function $M|z|^{p+1}$ for some $M > 0$, the transform Q_{2j+1} is close to identity, and $R^{-1}(\varkappa) \sim \varkappa^{-1/p}$, we conclude that

$$\widetilde{P}_{2j+1}(\tau + 2\pi i(\mu + l), e^{\tau})$$

is majorized by the function

$$\widetilde{M}|\tau + 2\pi i(\mu + l)|^{-1-\alpha},$$

where M, α are certain positive constants.

Therefore the series in (6.1) is absolutely converging. Substituting $\nu = \mu + l$, changing the order of integration and summation, and calculating the sum in (6.1), we get

$$c_{j,r}^+ = -\frac{1}{2\pi i}\int_0^{2\pi i}\int_{-\infty-i\beta}^{\infty-i\beta} e^{2\pi i r\nu}\widetilde{P}_{2j+1}(\tau + 2\pi i\nu, e^{\tau})\,d\tau\,d\nu.$$

Now we substitute the series (2.8) into the latter expression. After evident transformations, we get

$$c_{j,r}^+ = -\frac{1}{2\pi i}\sum_{s=-1}^{\infty}\int_0^{2\pi i}\int_{-\infty-i\beta}^{\infty-i\beta} e^{2\pi i r\nu + \tau s}\widetilde{f}_{s,2j+1}(\tau + 2\pi i\nu)\,d\tau\,d\nu,$$

where $\widetilde{f}_{s,2j+1} = f\circ R^{-1}\circ Q_{2j+1}$. The substitution $\nu = (2\pi i)^{-1}(\theta - \tau)$ changes the expression for $c_{j,r}^+$ to

$$c_{j,r}^+ = -\frac{1}{(2\pi i)^2}\sum_{s=-1}^{\infty}\int_0^{2\pi i} e^{(s-r)\tau}\int_{\gamma_\tau} e^{r\theta}\widetilde{f}_{s,2j+1}(\theta)\,d\theta\,d\tau, \tag{6.2}$$

where γ_τ is the straight line $(-i\infty + \tau + 2\pi\beta, i\infty + \tau + 2\pi\beta)$.

Since $\tau \in [0, 2\pi i]$, this line on the complex plane (modulo parametrization) does not depend on τ. The same also holds for the function under the inner integral in (6.2), therefore the multiple integral can be represented as the product of two integrals, while γ_τ can be replaced by γ_0:

$$c_{j,r}^+ = -\frac{1}{(2\pi i)^2}\sum_{s=-1}^{\infty}\int_0^{2\pi i} e^{(s-r)\tau}d\tau \cdot \int_{\gamma_0} e^{r\theta}\widetilde{f}_{s,2j+1}(\theta)\,d\theta.$$

Assertion (2.9) of Lemma 2.3 now follows from the identity

$$\int_0^{2\pi i} e^{(s-r)\tau}d\tau = \begin{cases} 0, & r \neq s, \\ 2\pi i, & r = s. \end{cases}$$

The proof of Lemma 2.3 is finished. □

6.2. Proof of Lemma 2.4. By definition,

$$c^+_{j,r,k} = -\frac{1}{2\pi i}\int_{\gamma_0} e^{r\theta}[R^{-1}\circ Q^{-1}_{2j+1}(\theta)]^k\,d\theta. \tag{6.3}$$

This coefficient depends analytically on λ, see (2.4). More precisely, for any bounded domain $\Omega \subset \mathbb{C}$ there exists a β such that the right-hand side of (6.3) is well defined for all $\lambda \in \Omega$ and the dependence on λ is analytic in Ω. After a series of subsequent substitutions and simplifications, we shall compute the integral representation (6.3) for all λ belonging to a certain open subset $\widetilde{\Omega} \subset \mathbb{C}$. The result (formula (2.11)) defines an entire function in the variable λ, independent of β. Therefore by virtue of the uniqueness theorem for analytic functions, the right-hand side of (6.3) must coincide with (2.11) for all $\lambda \in \widetilde{\Omega}$, hence for all $\lambda \in \mathbb{C}$.

Let us proceed with the computations. The substitution $\mu = Q^{-1}_{2j+1}(\theta)$ takes (6.3) into the form

$$c^+_{j,r,k} = -\frac{1}{2\pi i}\int_{\widetilde{\gamma}_0} e^{rQ_{2j+1}(\mu)}[R^{-1}(\mu)]^k (Q_{2j+1}(\mu))'\,d\mu,$$

where $\widetilde{\gamma}_0 = Q_{2j+1}(\gamma_0)$. Taking into account (2.1) and (2.4), we get

$$c^+_{j,r,k} = -\frac{1}{2\pi i}\int_{\widetilde{\gamma}_0} e^{r\mu}\left(1-\frac{\lambda}{p\mu}\right)[\mu^{1/p}]_j^{-\lambda r}[(-p\mu)^{1/p}]_j^{-k}\,d\mu, \tag{6.4}$$

where $[\mu^{1/p}]_j$ means the branch of the p-valued function defined as follows. Since μ is the parameter on $\widetilde{S}_{2j+1} \cap \widetilde{S}_{2j+2}$, by (2.1) and (2.4) we may write $\mu = \rho\exp(i\psi)$ with $-(2j+1)\pi < \psi < -2j\pi$. Put

$$[\mu^{1/p}]_j = \rho^{1/p}\exp(i\psi/p).$$

Introducing the new parameter

$$\mathbb{C} \ni \nu = \rho\exp(i\psi),\qquad -\pi < \psi < 0,$$

and putting

$$\nu^{1/p} = \rho^{1/p}\exp(i\psi/p),$$

we get

$$[\mu^{1/p}]_j = \nu^{1/p}\exp(2\pi ji/p).$$

Thus the computation of (6.4) yields

$$\begin{aligned}c^1_{j,r,k} = -p^{\,k/p}&\exp(-2\pi j\lambda ri/p)\exp(-(2j+1)k\pi i/p)\\ &\times \frac{1}{2\pi i}\int_{\widetilde{\gamma}_0} e^{r\nu}\left(1-\frac{\lambda}{p\nu}\right)\nu^{-(\lambda r+k)/p}\,d\nu.\end{aligned} \tag{6.5}$$

PROPOSITION 6.1. *Suppose that* $\operatorname{Re}\lambda > k/r$. *Then the value of the integral* (6.5) *will not be changed if we replace the curve* $\widetilde{\gamma}_0$ *by* γ_0.

The proof is postponed until 6.3.

By virtue of this proposition, in the domain $\Omega = \{\operatorname{Re}\lambda > k/r\}$, the computation of (6.5) is reduced to that of the expression

$$\mathscr{I} \equiv \int_{\gamma_0} e^{\nu r}\left(1 - \frac{\lambda}{p\nu}\right)\nu^{-(\lambda r+k)/p}\,d\nu$$
$$= \int_{\gamma_0} e^{\nu r}\nu^{-(\lambda r+k)/p}\,d\nu - \frac{\lambda}{p}\int_{\gamma_0} e^{\nu r}\nu^{-(\lambda r+k)/p-1}\,d\nu \equiv \mathscr{I}_1 - \frac{\lambda}{p}\mathscr{I}_2.$$

Integrating $\mathscr{I}_2$ by parts, we get

$$\mathscr{I}_2 = pr(\lambda r + k)^{-1}\mathscr{I}_1;$$

hence $\mathscr{I} = k(\lambda r + k)^{-1}\mathscr{I}_1$.

It is known that

$$\int_{2\pi i\beta-i\infty}^{2\pi i\beta+i\infty} z^{\alpha-1}e^{\delta z}\,dz = \frac{2\pi i}{\delta^{\alpha}\Gamma(1-\alpha)}, \qquad \delta > 0,\ \operatorname{Re}\alpha < 1,\ \beta > 0.$$

Therefore

$$\mathscr{I}_1 = \frac{2\pi i r^{(\lambda r+k)/p-1}}{\Gamma((\lambda r+k)/p)}.$$

Returning to (6.5), we obtain expression (2.11) for the coefficients $c_{j,k,r}^+$. □

6.3. Proof of Proposition 6.1. In order to prove this proposition, it is sufficient to demonstrate that

$$\int_{\widetilde{\gamma}_0} e^{\nu r}\left(1 - \frac{\lambda}{p\nu}\right)\nu^{-(\lambda r+k)/p}\,d\nu = \int_{\gamma_0} e^{\nu r}\left(1 - \frac{\lambda}{p\nu}\right)\nu^{-(\lambda r+k)/p}\,d\nu$$

provided that $\operatorname{Re}(\lambda - k/r) > 0$.

Denote by f the integrand in the above formula, and consider the closed path $\gamma_1(\sigma) \cup \gamma_2(\sigma) \cup \gamma_3(\sigma) \cup \gamma_4(\sigma)$ shown in Figure 5.

The segment $\gamma_1(\sigma)$ is symmetric with respect to the real axis, σ being its length; this length uniquely defines the entire path. Note that the integral exists for all $\sigma > 0$. Evidently,

$$\int_{\gamma_0} f\,d\nu = \lim_{\sigma\to\infty}\int_{\gamma_1(\sigma)} f\,d\nu \quad\text{and}\quad \int_{\widetilde{\gamma}_0} f\,d\nu = \lim_{\sigma\to\infty}\int_{\gamma_3(\sigma)} f\,d\nu.$$

If the parameter β determining the size and the position of the curve γ_0 is sufficiently large, then f is analytic in the domain bounded by $\gamma(\sigma) = \bigcup_j \gamma_j(\sigma)$ for all σ. On the other hand,

$$\int_{\gamma(\sigma)} f\,d\nu = \sum_j \int_{\gamma_j(\sigma)} f\,d\nu.$$

By virtue of Jordan lemma,

$$\lim_{\sigma\to\infty}\int_{\gamma_j(\sigma)} f\,d\nu = 0 \quad\text{for } j = 2,\,4.$$

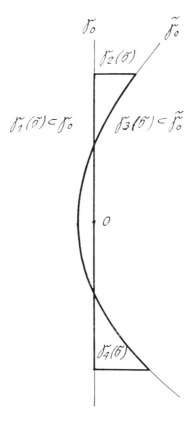

FIGURE 5

Hence

$$\lim_{\sigma \to \infty} \int_{\gamma_1(\sigma)} f\, d\nu = \lim_{\sigma \to \infty} \int_{\gamma_3(\sigma)} f\, d\nu.$$

The proposition is proved. □

REFERENCES

1. J. Martinet and J.-P. Ramis, *Problèmes de modules pour des équations différentielles non linéaires du premier ordre*, Inst. Hautes Études Sci. Publ. Math. **55** (1982), 63–164.

2. _____, *Classification analytique des équations différentielles non linéaires résonnantes du premier ordre*, Ann. Sci. École Norm. Sup. (4) **16** (1983), no. 4, 571–621.

3. P. M. Elizarov and Yu. S. Il'yashenko, *Remarks on the orbital analytic classification of germs of vector fields*, Mat. Sb. **121** (1983), 111–126; English transl. in Math. USSR-Sb. **49** (1984).

4. Yu. S. Il'yashenko, *Singular points and limit cycles of differential equations on the real and complex plane*, Preprint, Computer Center AN SSSR, Pushchino, Moscow Region, 1982. (Russian)

5. A. D. Brjuno [Bryuno], *Analytic form of differential equations*, Trudy Moskov. Mat. Obshch. **25** (1971), 119–262; **26** (1972), 199–239; English transl. in Trans. Moscow Math. Soc. **25** (1971); **26** (1972).

6. Yu. S. Il′yashenko, *In the theory of normal forms of analytic differential equations when the Bryuno condition fails divergence is the rule, convergence is the exception*, Vestnik Moskov. Univ. Ser. I Mat. Mekh. (1981), no. 2, 10–16; English transl. in Moscow Univ. Math. Bull. **36** (1982).

7. A. D. Brjuno [Bryuno], *On divergence of a real normalizing transformation*, Mat. Zametki **31** (1983), no. 3, 403–410; English transl. in Math. Notes **31** (1983).

8. J. Martinet, *Normalisation des champs de vecteurs holomorphes (d'apres A. D. Brjuno)*, Séminaire Bourbaki, 1980/1981, Lecture Notes in Math., vol. 901, Springer-Verlag, Berlin and New York, 1981, pp. 55–70.

9. P. M. Elizarov, *Orbital analytic nonequivalence of saddle resonant vector fields in* $\mathbb{C}^2$, Mat. Sb. **123** (1984), no. 4, 534–548; English transl. in Math. USSR-Sb. **51** (1985).

10. J. Horn, *Über das Verhalten der Integrale einer linearen Differenzialgleichung erster Ordnung in der Umgebung einer Unbestimmtheitsstelle*, J. Reine Angew. Math. **120** (1899), 1–26; **122** (1901), 73–83.

11. J. Ecalle, *Théorie des invariants holomorphes*, Publ. Math. d'Orsay, Université de Paris-Sud, Orsay, 1974.

12. S. M. Voronin, *Analytic classification of germs of conformal maps* $(\mathbb{C}, 0) \to (\mathbb{C}, 0)$ *with identity linear part*, Funktsional. Anal. i Prilozhen. **15** (1981), no. 1, 1–17; English transl. in Functional Anal. Appl. **15** (1981).

13. B. Malgrange, *Travaux d'Ecalle et de Martinet-Ramis sur les systèmes dynamiques*, Séminaire Bourbaki, 1981/1982, Astérisque, vol. 92–93, Soc. Math. France, Paris, 1982, pp. 59–73.

14. A. A. Shcherbakov, *On germs of maps analytically nonequivalent to their formal normal form*, Funktsional. Anal. i Prilozhen. **16** (1982), no. 2, 94–95; English transl. in Functional Anal. Appl. **16** (1982).

15. J. Ecalle, *Sur les fonctions résurgentes*, I, II, III, Publ. Math. d'Orsay, Université de Paris-Sud, Orsay, 1981, 1985.

DEPARTMENT OF MATHEMATICS, TVER STATE UNIVERSITY, TVER, RUSSIA

Translated by S. YAKOVENKO

The Darboux-Whitney Theorem and Related Questions

S. M. VORONIN

This paper is devoted to the analytic classification of singularities for various geometric problems. We consider singularities of families of plane curves, singularities of pairs of intersecting hypersurfaces in symplectic manifolds, and degenerations of tangent and symplectic structures. All these problems appear to have "hidden dynamics"; namely, germs of mappings with resonant linear part at a fixed point serve as invariants of the local classification (either formal, or smooth, or analytic). This allows us to apply methods developed in Paper I and to arrive at conceptually similar results. Namely, the problems considered admit a simple formal classification that has only a finite set of numerical invariants. The smooth classification coincides with the formal one, while the analytic classification has functional moduli.

The germ of a mapping that is given by "hidden dynamics", is called the *indicator* of the problem. As a rule, indicators not only have resonant linear parts (degeneration of finite codimension) but also possess additional symmetries (degeneration of codimension infinity). Thus the problems that may be neglected from the point of view of general theory of singularities, appear to be important in geometric applications.

V. I. Arnold studied singularities of families of plane curves at the points of the envelope. For generic families and for a smooth envelope, these singularities fall into three classes. For the first two of them, the formal classification coincides with the smooth and analytic ones. For the third class, the only coincidence is between the formal and smooth classifications. All these classifications were obtained in [4]. In [25], we studied the analytic classification for the remaining case. It appeared to have functional moduli. These results without detailed proofs are presented in Chapter I, §1.

The main part of the paper is devoted to the classification of geometric objects that were previously studied by R. Melrose and V. I. Arnold. Melrose [20] showed that the study of billiard dynamics problems can be

---

1991 *Mathematics Subject Classification*. Primary 58D27.

reduced to the study of singularities of hypersurface pairs in symplectic manifolds. V. I. Arnold [2], [3] carried out a similar reduction for the problem of avoiding an obstacle. Besides, V. I. Arnold posed a number of classification problems with natural hierarchy that are related to the problem of hypersurface pairs and obtained both formal and smooth classifications for them. When the analytic classification coincides for these problems with the formal one, the latter was also obtained by Arnold and Melrose. The present paper describes functional invariants of the analytic classification for certain cases that were left uninvestigated. Among them we mention the Darboux-Whitney problem formulated below. As a by-product of the methods developed, we prove that the smooth classification of analytic germs for the Darboux-Whitney problem coincides with the formal one.

M. E. Zhitomirskiĭ [26] studied all degenerations of generic contact and symplectic structures with "finitely defined" local smooth classification. This means that the smooth equivalence class of the form germ that defines a contact (respectively, a symplectic) structure, is defined by the finite jet of this form at a degeneration point. The problem of analytic classification of the germs of these forms leads to indicators that appear to be very similar to the indicators in the hypersurface pairs problem. The analytic classification of these degenerations involves functional moduli. In the present paper, we state the corresponding theorems but omit the proofs because of the lack of space.

The solution of each of the above-mentioned problems consists of three steps.

I. Indicator construction. Proof of the statement that the equivalence of indicators implies the equivalence of the original germs, and vice versa.

II. Analytic classification of indicators.

III. Indicator realization theorem: each germ that belongs to the functional class studied in Step II can be realized as an indicator of the original geometric problem.

The plan of the paper is the following. The first chapter is introductory. It contains formulations of all problems considered in the paper and provides an overview of known results. Besides, it contains statements of all results of the paper. Finally, in this chapter, we construct indicators for the geometric problems studied.

The second chapter is devoted to the analytic and smooth classifications of analytic indicators.

The third chapter provides a reduction of a classification problem posed by V. I. Arnold (the Darboux-Whitney theorem) to the classification problem for the corresponding indicators. A similar reduction for another classification problem of Arnold (of two folds) is elementary and is omitted here.

The second and third chapters contain the proofs of the statements of the first chapter that are related to the billiard dynamics problems and to the singularities of hypersurface pairs. For the sake of brevity, the proofs are given for indicators with two-dimensional domains. For the general case, the proofs differ in some unessential but tedious technical details.

The exact formulation of the main results may be found in 4.10 of Chapter I.

The author is grateful to V. I. Arnold and M. Ya. Zhitomirskiĭ for informing him about the classification problems studied in the paper, and to Yu. S. Il′yashenko and S. Yu. Yakovenko for valuable remarks.

CHAPTER I. CLASSIFICATION PROBLEMS AND INDICATORS

The first section of this chapter is devoted to problems with one-dimensional indicators, namely, to an envelope problem and related dynamic problems. The second section forms the main contents of the chapter. It is devoted to the obstacle avoiding problem, to its modifications, and to the construction of the corresponding indicators. The third section is devoted to degenerations of contact and symplectic structures. The main results of the paper are formulated in §4 and summarized in 4.10.

§1. Pairs of involutions, symmetry breaking, and envelopes for families of plane curves

The central result of the present section has to do with the envelope problem. We start with local dynamics problems to which it will be reduced.

1.1. Involution pairs problem. An *involution* is a mapping onto itself with square equal to the identity mapping. In this subsection, we consider germs of biholomorphic involutions $(\mathbb{C}, 0) \to (\mathbb{C}, 0)$. It is easy to prove that after an analytic change of coordinates, any involution takes the form $z \mapsto -z$. The problem is to describe the analytic equivalence classes of involution pairs. Below we consider various equivalence relations for pairs of germs of mappings, and so we write the definition of equivalence as a commutative diagram.

DEFINITION. Two pairs (I, J), $(\widetilde{I}, \widetilde{J})$ of germs of maps $(\mathbb{C}, 0) \to (\mathbb{C}, 0)$ are called *equivalent* if the following diagram

$$\begin{array}{ccccc} (\mathbb{C}, 0) & \xrightarrow{I} & (\mathbb{C}, 0) & \xrightarrow{J} & (\mathbb{C}, 0) \\ {\scriptstyle h}\downarrow & & {\scriptstyle h}\downarrow & & {\scriptstyle h}\downarrow \\ (\mathbb{C}, 0) & \xrightarrow{\widetilde{I}} & (\mathbb{C}, 0) & \xrightarrow{\widetilde{J}} & (\mathbb{C}, 0) \end{array}$$

is commutative.

The problem formulated above is a particular case of the classification problem for finitely generated groups of germs of analytic mappings considered in Paper II. However, it admits a simple independent solution that has been obtained earlier than the general result of that paper (see [25]).

Specifically, to each pair of involutions let us relate their product and call it the *pair indicator*. The linear part of an indicator is the identity mapping. The analytic equivalence of involution pairs implies the analytic equivalence of their indicators (as well as of the groups of germs generated by these involutions).

Without loss of generality, we may always assume that one (say, the first) involution of a pair (I, J) is standard: $I = I_0: z \mapsto -z$. In this subsection we confine ourselves only to such pairs. The indicator $\Delta = I_0 \circ J$ of a pair (I_0, J) is I_0-symmetric in the following sense: $I_0 \circ \Delta \circ I_0 = J \circ I_0 = \Delta^{-1}$. Let A^{I_0} be the space of all I_0-symmetric germs of conformal mappings $(\mathbb{C}, 0) \to (\mathbb{C}, 0)$ whose linear parts are the identity:

$$\Delta \in A^{I_0} \iff \Delta'(0) = 1 \text{ and } I_0 \circ \Delta \circ I_0 = \Delta^{-1}.$$

If a germ Δ is I_0-symmetric, the composition $I_0 \circ \Delta$ is an involution: $I_0 \circ \Delta \circ I_0 \circ \Delta = \Delta^{-1} \circ \Delta = \text{id}$. Hence any germ $\Delta \in A^{I_0}$ is an indicator of some pair of involutions (namely, of the pair $(I_0, I_0 \circ \Delta)$), so A^{I_0} consists exactly of all indicators of involution pairs.

Let us call two germs Δ and $\widetilde{\Delta}$ from A^{I_0} I_0-*equivalent* if there exists a germ of a holomorphism h that conjugates them and commutes with I_0:

$$\Delta \overset{I_0}{\sim} \widetilde{\Delta} \iff \exists h : (\mathbb{C}, 0) \to (\mathbb{C}, 0) : h \circ \Delta = \widetilde{\Delta} \circ h \text{ and } h \circ I_0 = I_0 \circ h.$$

It is evident that the equivalence of involution pairs implies the I_0-equivalence of their indicators, and vice versa.

Thus we have realized the two steps (the first and the third) of the plan in the Introduction, and the only thing left is to obtain the I_0-classification of germs of the class A^{I_0}. In other words, we have reduced the involution pairs problem to the so-called I_0-*symmetric conjugation problem*, which is to classify I_0-symmetric germs of mappings $(\mathbb{C}, 0) \to (\mathbb{C}, 0)$ whose linear part is identity (i.e., germs of the class A^{I_0}) with respect to the action of the group Diff^{I_0} of local holomorphisms that commute with I_0.

A solution of the I_0-symmetric conjugation problem is obtained from the solution of the ordinary conjugation problem (i.e., the classification problem for germs of conformal mappings whose linear part is identity), when one takes into account the symmetry of the germs under classification and of the conjugating holomorphisms. In Paper I we have already done this while studying germs of mappings with resonant linear part.

Let us call an involution pair (I, J) *typical* if the difference of the involutions is infinitesimal of the second order:

$$I(z) - J(z) = cz^2 + \cdots, \qquad c \neq 0.$$

A complete solution of the involution pairs problem is given in [25] (for typical pairs) and in [10] (for the general case).

Another solution of the involution pairs problem (more exactly, another version of the solution described here) utilizes the results of Paper II, §2. In fact, the commutant of the group G generated by the involutions I and J is a cyclic group generated by the indicator $I \circ J$. Therefore, the group G is solvable and one may obtain the classification of pairs of involutions from the classification theorem for solvable germ groups (Paper II, Theorem 2.4).

1.2. Symmetry breaking [6, 11]. The problem described above naturally arises in the study of envelopes (see 1.3 below).

Suppose I and g are germs of mappings $(\mathbb{C}, 0) \to (\mathbb{C}, 0)$, the first of which is an involution (a mapping into itself) and the other is a one-dimensional fold (a mapping into another space). This means that the germ g is holomorphic but not biholomorphic and in suitable charts has the form $z \mapsto z^2$. In a suitable chart, the involution I is of the form $w \mapsto -w$. In general, the charts z and w are different.

DEFINITION. Two pairs (I, g) and $(\tilde{I}, \tilde{g})$ of the form (involution, fold) are called *equivalent* if there exist germs h and H of biholomorphic mappings $(\mathbb{C}, 0) \to (\mathbb{C}, 0)$ such that the diagram

$$\begin{array}{ccccc}
(\mathbb{C}, 0) & \xrightarrow{I} & (\mathbb{C}, 0) & \xrightarrow{g} & (\mathbb{C}, 0) \\
h \downarrow & & h \downarrow & & H \downarrow \\
(\mathbb{C}, 0) & \xrightarrow{\tilde{I}} & (\mathbb{C}, 0) & \xrightarrow{\tilde{g}} & (\mathbb{C}, 0)
\end{array}$$

is commutative.

In the symmetry breaking problem it is required to classify pairs of germs of mappings $(\mathbb{C}, 0) \to (\mathbb{C}, 0)$, the first of which is an involution and the second is a one-dimensional fold.

The problem of symmetry breaking may be reduced to the classification of involution pairs in the following manner. One involution is given in the statement of the problem. The second is determined by the fold g and is constructed as a mapping that interchanges the preimages of the fold. Denote the second involution by J. Obviously, the equivalence of pairs (involution, fold) implies the equivalence of the corresponding pairs of involutions. The converse is also true and is also almost obvious. Indeed, let (I, g) and $(\tilde{I}, \tilde{g})$ be two pairs of the form (involution, fold) and suppose that the involutions J and $\tilde{J}$ are constructed from the folds g and $\tilde{g}$ as described above. Suppose the pairs (I, J) and $(\tilde{I}, \tilde{J})$ are equivalent. Then in a suitable chart z these pairs are equal: $(I, J) = (\tilde{I}, \tilde{J})$. Moreover, this chart z may be chosen so that $J(z) = -z$. Suppose the pair $(\tilde{I}, \tilde{g})$ in the chart z has the form $(I, \tilde{g})$. Then the functions g and $\tilde{g}$ are both even, i.e., $g(z) = \varphi(z^2)$ and $\tilde{g} = \psi(z^2)$, where φ and ψ are germs of biholomorphic mappings $(\mathbb{C}, 0) \to (\mathbb{C}, 0)$. Hence there exists a germ of a biholomorphism $H: (\mathbb{C}, 0) \to (\mathbb{C}, 0)$ such that $\psi = H \circ \varphi$. But then $\tilde{g} = H \circ g$, and therefore the germs (id, H)

conjugate the pairs (I, g) and $(I, \widetilde{g})$ (we refer to the commutative diagram above). We conclude that the original pairs (I, g) and $(\widetilde{I}, \widetilde{g})$ are also equivalent.

We claim that the following statement about realization is valid: every pair of involutions corresponds to a pair (involution, fold). Indeed, without loss of generality, we may assume the second involution of any pair (I, J) to be standard: $J = I_0 : z \mapsto -z$. Such a pair corresponds to the pair (I, g_0), where $g_0 : z \mapsto z^2$ is the standard one-dimensional fold.

1.3. Singularities of families of plane curves at smooth points of the envelope [1, 4, 6, 25]. We may assume that the family of plane curves with an envelope shown in Figure 1a is obtained as follows. Consider a family of parametrized curves on the plane and project this plane onto another one. Suppose the projection mapping has a singularity of the fold type (see Figure 1b).

These pictures provide a motivation to the following definition.

DEFINITION. Let us call the following diagram

$$(\mathbb{C}, 0) \xleftarrow{g} (\mathbb{C}^2, 0) \xrightarrow{F} (\mathbb{C}^2, 0)$$

where g and F are germs of holomorphic mappings, the *germ of the family of curves* on the complex plane. Two germs (g, F) and $(\widetilde{g}, \widetilde{F})$ are called *analytically equivalent* if there exist germs of biholomorphic mappings L, H, and $\widetilde{H}$ such that the diagram

$$\begin{array}{ccccc}
(\mathbb{C}, 0) & \xleftarrow{g} & (\mathbb{C}^2, 0) & \xrightarrow{F} & (\mathbb{C}^2, 0) \\
h \downarrow & & \downarrow H & & \downarrow \widetilde{H} \\
(\mathbb{C}, 0) & \xleftarrow{\widetilde{g}} & (\mathbb{C}^2, 0) & \xrightarrow{\widetilde{F}} & (\mathbb{C}^2, 0)
\end{array}$$

is commutative.

The local classification problem is formulated for a pair of mappings g and F in general position. Let us agree that this restriction (of general

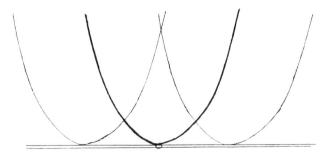

FIGURE 1a. The family of curves in a neighborhood of a generic point on an envelope.

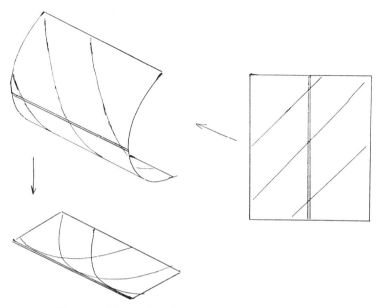

FIGURE 1b. A family of curves on the plane.

position) is imposed everywhere till the end of the present subsection. The envelopes pass near the singular points of F. For the generic case, these singularities can be either folds or pleats. In the latter case even the formal classification is boundless (it has functional moduli—see [6]). For the fold case, the following possibilities arise.

1. The fold point is not a critical point of the function g and the level curve at this point is neither tangent to the fold curve, nor to the kernel of the derivative mapping F_*.

REMARK. The critical points of the function g are isolated and in the generic case do not lie on the fold curve of F. Therefore we may assume that level curves of the function g at the fold points are smooth. Below we use this fact without explicitly mentioning it.

2. A level curve of the function g is tangent to the fold.
3. A level curve of the function g is tangent to the kernel of the derivative mapping F_*.

In the first two cases, the local formal classification of the families coincides with both smooth and analytic ones. The normal forms of the families are given by the following equations (see [1, 4, 6]):

1. $g(x, y) = x + y$, $\quad F(x, y) = (x, y^2)$,
2. $g(x, y) = y + x^2$, $\quad F(x, y) = (x, y^2)$.

The corresponding families are shown in Figures 1a, (1b), and 1c (see next page), respectively.

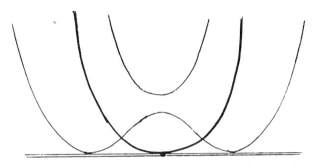

FIGURE 1c. A family of curves in a neighborhood of a tangency point of order 4.

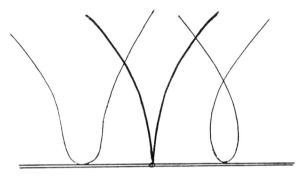

FIGURE 1d. A family of curves in a neighborhood of a cusp.

In the third case, the formal normal form of the family can be expressed as

3. $g(x, y) = x + xy + y^3$, $\quad F(x, y) = (x, y^2)$

(see Figure 1d). The analytic classification in this case involves functional moduli. Let us proceed to their description.

Let I be the involution that interchanges preimages of points under the fold mapping F. Let us define a new mapping

$$G: (\mathbb{C}^2, 0) \to (\mathbb{C}^2, 0), \quad G = (f, g), \quad f = g \circ I.$$

Simple calculations carried out in [25] (that we omit here) show that in the generic case (the only one considered here) the germ G has a singularity of the pleat type at the point studied. The set of singular points of G is a "pleat curve". It may be defined as the set of tangent points of level curves of the functions g and $f = g \circ I$. This curve is invariant under the involution I. Moreover, the function g defines a mapping of this curve with a singularity of the fold type (this is also checked by simple calculations).

Thus in the third case of the envelope problem, we have an invariantly defined "indicator" that is the germ pair (involution, fold) considered in 1.2. It is easy to check that the indicator that corresponds to the general family satisfies the following typicalness condition:

*the pair of involutions corresponding
to the pair (involution, fold) is typical.* $\quad (*)$

The invariants of analytic classification of family indicators are at the same time the invariants of the original classification problem. Namely, the following statements hold.

THEOREM 1. *Any pair (involution, fold) that satisfies the typicalness condition* $(*)$ *can be realized as an indicator of the envelope problem.*

THEOREM 2. *The analytic equivalence of families of plane curves implies the analytic equivalence of their indicators, and vice versa.*

§2. Geometry of hypersurface pairs in symplectic manifolds

The statement of the problem, the terminology, and formulations of the results are borrowed from Arnold's paper [3].

2.1. The obstacle avoiding problem. Consider an obstacle in a Ricmannian manifold V, i.e. a set bounded by a generic smooth hypersurface Γ. In the obstacle avoiding problem it is required to study singularities of the function "distance around the obstacle from an arbitrary point of the manifold V to a given generic initial point".

The phase space $M = T^*M$ has a natural symplectic structure. In terms of the cotangent fibration $\pi: T^*V \to V$ it may be defined as follows. Consider a 1-form β on M (the "action form") and suppose that its value at the point $p \in T^*V = M$ on the vector $\xi \in T_p M$ is equal to the value of the 1-form p on the vector $\pi_* \xi \in TV$. Then the integral Poincaré invariant $\omega = d\beta$ provides a natural symplectic structure on M. In the standard coordinates (q, p) on M (q is a local chart on V and $p \in \mathbb{R}^n$) we have: $\beta = p\,dq$ and $\omega = dp \wedge dq$.

Now identify the tangent and cotangent spaces to V with the help of the Riemannian metric. Let us call points of the phase space M *vectors*.

The Riemannian metric on V defines a hypersurface Γ_1 in M that is formed by vectors of unit length. The hypersurface $\Gamma \subset V$ defines a second hypersurface in M, namely, the hypersurface Γ_2 that is formed by all vectors applied at the points of Γ.

An isomorphism of two pairs (Riemannian manifold, its smooth hypersurface) defines an isomorphism of the above-defined objects (symplectic manifold, pair of hypersurfaces in it) corresponding to these pairs.

Thus the obstacle avoiding problem leads to the classification problem of (ordered) hypersurface pairs in a symplectic manifold with respect to the action of the group of symplectomorphisms (see [20]; also see [2] for the details of the relationship between these two problems). The present section is devoted to the local version of the hypersurface pairs problem.

2.2. Canonical projections.

Suppose the hypersurfaces Γ_1 and Γ_2 of a symplectic manifold M intersect transversally along a submanifold S. The symplectic structure defines a skew-scalar product on the tangent space to M. A hypersurface in the symplectic space is locally foliated into characteristics (integral curves of the field of skew-orthogonal complements of tangent planes to the hypersurface). The dimension of the characteristic space is equal to the dimension of S. For every hypersurface, let us define canonical projections of S onto the corresponding characteristic manifolds, associating to a point of S the characteristic that passes through this point.

The singularities of canonical projections of hypersurfaces constructed in 2.1 for the generic case are direct products of a plane and a standard Whitney singularity. They are locally equivalent to the singularities A_k of the projection of the hypersurface $x^{k+1} + \lambda_1 x^{k-1} + \cdots + \lambda_k = 0$ from the total space (with one-dimensional fiber) onto its base $(x, y, \lambda) \mapsto (y, \lambda)$. As usual, let us call the singularity A_1 a *fold* and the singularity A_2 a *pleat*. The critical values of canonical projections form hypersurfaces whose singularities are semicubical parabolas in the case A_1, swallowtails in the case A_2, etc. (of course, multiplied by the corresponding linear space).

REMARK. The sets of critical points of canonical projections for the two described hypersurfaces coincide. Indeed, let $P \in S$ be a critical point of one of the canonical projections (for example, of the first one). Then the skew-orthogonal complement ξ_1 of the hypersurface Γ_1 is tangent to S at P. Thus ξ_1 is skew-orthogonal to the skew-orthogonal complement ξ_2 of the hypersurface Γ_2 at the point P and is not transversal to Γ_1 because the symplectic structure is nondegenerate. We conclude that ξ_2 is tangent to Γ_1 (and hence, to the submanifold S) at the point P, so P is a critical point also for the second canonical projection.

EXAMPLE. Let Γ_1 and Γ_2 be two hypersurfaces of the symplectic manifold $M = T^*V$ constructed from a Riemannian manifold V and a hypersurface $\Gamma \subset V$ in 2.1, $S = \Gamma_1 \cap \Gamma_2$, and let (q, p) be the standard local coordinates on $M = T^*V$. On the fiber T_q^*, $q \in \Gamma$, the manifold V cuts out a surface S_q (a level set of the positive definite quadratic form that defines the Riemannian metric, an "indicatrix") that bounds a convex domain. The characteristics of Γ_2 passing through the points of S_q, lie entirely in the same fiber and form a pencil of parallel lines in it (with direction vector equal to the (co)normal to Γ at the point q) (see Figure 2b on page 151). We conclude that, in particular, a fold A_1 is a single singularity of the second canonical projection. In the Euclidean case, the characteristics of Γ_2 that pass through intersection points of the "fiber" $\{p = \text{const}\}$ and the manifold S, lie in the same "fiber" and form a pencil of parallel lines in it with the direction vector p (see Figure 2a on page 151). The critical points of (both!) canonical projections are the points $(q, p) \in S$ such that $p \in T_q\Gamma$ (and only those points).

Throughout what follows we confine ourselves to hypersurface pairs where the second canonical projection is either nondegenerate or is a fold (as in the Riemannian case considered above).

2.3. Problem hierarchy. The normalization of a hypersurface pair implies the normalization of all objects that depend solely on this pair, in particular, of the pair formed by the first hypersurface and the trace of the second one on it.

The characteristic space on Γ_1 (that we denote by M_1) has a natural symplectic structure inherited from the symplectic structure of the original symplectic manifold.

The reduction to normal form of the pair (hypersurface Γ_1, submanifold $S \subset \Gamma_1$) by a symplectomorphism f of the ambient manifold implies the reduction to normal form of the set of critical values of the canonical projection $\pi_1: S \to M_1$ by the corresponding symplectic diffeomorphism of the space of characteristics of the hypersurface Γ_1 onto a similar space for $f\Gamma_1$.

In accordance with this, the problem of hypersurface pairs on a symplectic manifold will be studied on three levels. The subject for the zero level is hypersurface pairs; the subject for the first one are pairs (hypersurface, submanifold of codimension 1), while the subject for the second one is hypersurfaces with singularities (semicubical parabolas and swallowtails multiplied by a linear space). The classification at all levels is performed with respect to the action of the group of (local) symplectomorphisms. For each level, we shall study smooth, formal, and analytic classifications.

The interest in higher levels is motivated by the two following facts. Firstly, for some cases it is possible to solve the problem for these levels even when a zero level solution appears to be tedious. On the other hand, for the obstacle avoiding problem, these levels produce some information of importance for the original problem (see [3]).

2.4. Nondegenerate case. Local Darboux coordinates $(q, p) \in (\mathbb{R}^{2n}, 0)$ will be used below. In these coordinates, the symplectic structure has the standard form $\omega = dp \wedge dq$.

In the generic case, the canonical projections are nondegenerate at any generic point of the submanifold $S = \Gamma_1 \cap \Gamma_2$ and the pair Γ_1, Γ_2 is reduced to the normal form $(q_0 = 0, p_0 = 0)$ by an analytic (formal, smooth) symplectomorphism [3].

The corresponding normal form on the first level is the pair $(q_0 = 0, q_0 = p_0 = 0)$. We do not study the problem on the second level since both projections are nondegenerate.

2.5. Two folds. Melrose involutions. Indicators of hypersurface pairs. In the case when the singularities of both canonical projections are folds, the smooth and formal classifications of hypersurface pairs for the generic case were obtained by Melrose [20]. Namely, any hypersurface pair can be reduced to the normal form $(q_0 = 0, q_0 = p_0^2 + p_1)$ in the Darboux coordinates by a local (either smooth or formal) symplectomorphism.

In particular, this implies the possibility of formal and smooth normalization on higher levels. Namely, when the singularity of the canonical projection is a fold, the pair (hypersurface, submanifold of codimension 1) is reduced to the normal form $(q_0 = 0, q_0 = p_1 + p_0^2)$ by a local symplectomorphism (the first level, see [3]). For the second level, the corresponding statement claims that it is possible to normalize a smooth hypersurface by symplectomorphisms. This is equivalent to a well-known result about the normalization of a symplectic structure by diffeomorphisms that preserve a smooth hypersurface [7]. One may interpret this very statement as a theorem about the simultaneous reduction of a projection with "fold" singularity to the Whitney normal form and of a symplectic structure in the image to the Darboux normal form (the so-called Darboux-Whitney theorem).

The analytic classification coincides with the formal one for the higher levels [3] and involves functional moduli (and will be obtained below) for the zero one.

Following the plan in the Introduction, we begin our study of the analytic classification of hypersurface pairs by constructing their indicators.

A mapping of the "fold" type defines an involution of the neighborhood of the set of critical points that is identical on this set and interchanges the preimages of noncritical values of the mapping. The set of fixed points of an involution is called its *mirror*. The canonical projections for the case of "two folds" define a pair of involutions that are called *Melrose involutions*. Note that the mirrors of Melrose involutions coincide (see Remark in 2.2). Let us call the composition of Melrose involutions (i.e., the Birkhoff billiard transformations [9]) the *indicator of the hypersurface pair*.

2.6. Properties of indicators of hypersurface pairs for the case of two folds. The indicator of hypersurface pairs has the following properties: it is real analytic and can be represented as the composition of a pair of involutions with common mirror (and therefore has a hypersurface of fixed points). Besides, as will follow from the lemma proved below, the indicator (and the involutions that generate it) preserves the restriction $\alpha = \omega|_S$ of the symplectic structure ω to the hypersurface intersection S.

LEMMA. *Let S be a submanifold of codimension 1 of a smooth hypersurface Γ of a symplectic manifold M, let ω be a symplectic structure on M, and suppose there is a mapping I that interchanges the points of the intersection of S and the characteristics of Γ. Then I preserves the form $\alpha = \omega|_S$: $I^*\alpha = \alpha$.*

PROOF. For any area element $\sigma \subset S$ that is transversal to the characteristics, consider the "cylinder" C formed by characteristic segments cut out by S and passing through σ (see Figure 3 on page 152).

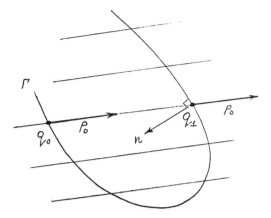

FIGURE 2a. Characteristics on the "fiber" $\{p = p_0\}$. The first Melrose involution $I\colon (p_0, q_0) \mapsto (p_0, q_1)$.

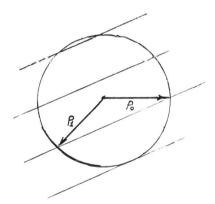

FIGURE 2b. Characteristics on the "fiber" $\{q = q_1\}$. The second Melrose involution $J\colon (p_0, q_1) \mapsto (p_1, q_1)$.

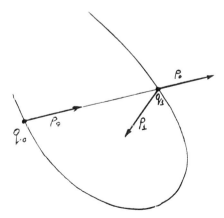

FIGURE 2c. The Birkhoff billiard transformation (indicator) $\Delta\colon (p_0, q_0) \mapsto (p_1, q_1)$.

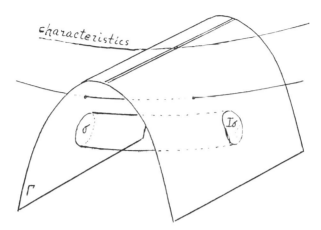

FIGURE 3. A shift along characteristics preserves the induced almost symplectic structure.

Since ω is closed, we have $\int_{\partial C} \omega = 0$. Since the lateral surface B of the cylinder C is formed by characteristics, we also have $\int_B \omega = 0$. Thus the integrals of ω along the "bases" σ and $I(\sigma)$ of the cylinder are equal. Together with the arbitrary choice of σ, this implies the lemma. □

The 2-form $\alpha = \omega|_S$ gives an "almost" symplectic structure: it is closed but degenerates at all fixed points of the indicator.

EXAMPLE 1. Consider the example in 2.2, where V is a domain in Euclidean space $\mathbb{R}^n$. For this case, the Melrose involutions $J\colon (p_0, q_0) \mapsto (p_0, q_1)$, $I\colon (p_0, q_1) \mapsto (p_1, q_1)$, and the Birkhoff billiard transformation $\Delta = I \circ J\colon (p_0, q_0) \mapsto (p_1, q_1)$ are shown in Figures 2a, 2b, and 2c, respectively (on page 151).

EXAMPLE 2. Let (p, q) be coordinates in $\mathbb{R}^{2n}$, where $p = (p_0, p_1, p')$, $q = (q_0, q_1, q')$ and $p_0, p_1, q_0, q_1 \in \mathbb{R}$, while $p', q' \in \mathbb{R}^{n-2}$, $n \geqslant 2$. Let $\omega = dp \wedge dq$ be the standard symplectic structure in $\mathbb{R}^{2n}$, (Γ_1, Γ_2) be the Melrose normal form, $S = \Gamma_1 \cap \Gamma_2$, and $\alpha = \omega|_S$. Let J_0 (I_0) be the first (second) Melrose involution and $\Delta_0 = I_0 \circ J_0$ be the indicator of the pair Γ_1, Γ_0. Then $\Gamma_1 = \{q_0 = 0\}$, $\Gamma_2 = \{q_0 = p_0^2 + p_1\}$, and $S = \{q_0 = p_0^2 + p_1 = 0\}$, so that p_0, q_1, p', and q are the coordinates in S. In these coordinates,

$$I_0\colon (p_0, q_1, p', q') \mapsto (-p_0, q_1, -2p_0, p', q'),$$
$$J_0\colon (p_0, q_1, p', q') \mapsto (-p_0, q_1, p', q'),$$
$$\Delta_0\colon (p_0, q_1, p', q') \mapsto (p_0, q_1 + 2p_0, p', q'),$$
$$\alpha = -2p_0 dp_0 \wedge dq_0 + dp' \wedge dq'.$$

The mappings I_0, J_0, and Δ_0 preserve the 2-form α that degenerates on the set $\{p_0 = 0\} \subset S$ of fixed points of these mappings.

As we noted in 2.5, for the first level and the generic case, it is possible to perform the analytic normalization. Therefore below we assume that the first hypersurface and the intersection of the hypersurfaces (and hence also the first canonical projection, the first Melrose involution, and the almost symplectic structure α) of the pairs being classified are the same as in the example above.

Observe that the restriction of the symplectomorphism that transforms one hypersurface pair into another to the intersection of the hypersurfaces conjugates the Melrose involutions that correspond to these pairs, and preserves the almost symplectic structure α. In particular, the indicator of any hypersurface pair studied here can be reduced to the normal form from Example 2 by a formal real change of coordinates that preserves the almost symplectic structure α and the first Melrose involution J_0.

The almost symplectic structure α degenerates at points of the mirror of Melrose involutions. Its kernel (i.e., the set of vectors that are skew-orthogonal to the whole tangent space) is two-dimensional at these points and intersects the tangent space to the mirror along a one-dimensional space. Let us call any basis vector ξ of this space a *characteristic vector*. For the linear mapping Δ_0 in Example 2, all vectors tangent to the mirror $\{p_0 = 0\}$ are its eigenvectors (with eigenvalue 1), while all vectors transversal to the mirror are adjoint to the characteristic vector ξ. Since there exists a formal symplectomorphism that conjugates the indicator Δ to the linear mapping Δ_0 and preserves α, the same properties hold for the linear part of the germ Δ. Thus the 1-jet of the germ Δ at zero is of the form

$$(p_0, q_1, p', q') \mapsto (p_0, q_1 + kp_0, p', q'), \qquad k \neq 0,$$

in the local chart (p_0, q_1, p', q'). Critical points of the projection that are close to folds are themselves fold points. Repeating the above argument for them, we conclude that the 1-jet of the germ Δ has the form indicated at all points of the mirror.

2.7. First classification problem. In 2.5, we considered the classification of intersections of two hypersurfaces in a symplectic manifold in the case when both singularities of the projections of the corresponding hypersurfaces are folds. The involution that corresponds to each fold interchanges the preimages of the points of this fold. The pair of involutions corresponding to the two folds is said to be the *indicator of the original hypersurface pair*.

In 2.6, we established the following properties of the indicator Δ: there exists a chart (x, y, z), $x \in \mathbb{R}$, $y \in \mathbb{R}$, $z \in \mathbb{R}^{2n-2}$, such that

1) the germ Δ is real analytic and is the product of two involutions with common mirror of fixed points $\{x = 0\}$, one of which is standard and has the form $I: (x, y, z) \mapsto (-x, y, z)$;

2) the 1-jet of the germ Δ at mirror points has the form

$$(x, y, z) \mapsto (x, y + kx, z), \qquad k \neq 0$$

(generally speaking, the coefficient k depends on the mirror point);

3) the germ Δ preserves the standard almost symplectic structure:

$$\Delta^*\alpha = \alpha, \quad \text{where } \alpha = -x\,dx \wedge dy + dz_1 \wedge dz_2 + \cdots + dz_{2n-3} \wedge dz_{2n-2}.$$

REMARK 1. The chart (x, y, z) used above is obtained from the chart constructed in 2.6 by a linear change of coordinates.

REMARK 2. As it was shown in 1.1, the product $\Delta = I \circ J$ of the involutions I and J is symmetric with respect to each of them in the following sense: $I \circ \Delta \circ I = \Delta^{-1}$, $J \circ \Delta \circ I = \Delta^{-1}$.

Denote by $\mathscr{B}_{1,\mathbb{R}}^{I,\alpha}$ the space of germs of mappings Δ with properties 1)–3). The subscript 1 indicates the lowest terms of the Taylor expansion of the germ correction term (i.e., of the difference between the given germ and the identity). The superscripts I and α are to remind of properties 1) and 3). The subscript $\mathbb{R}$ shows that we deal with the real case.

As we noted in 2.6 above, an analytic symplectomorphism of the ambient space induces a real-analytic diffeomorphism on the intersection of hypersurfaces that commutes with the involution I and preserves the 2-form α. Denote by $\text{Diff}_{\mathbb{R}}^{I,\alpha}$ the space of germs of these diffeomorphisms. We are interested in the classification of indicators with respect to the action of substitutions of this class. Below we solve a more general problem, namely, we classify germs from the class $\mathscr{B}_{1,\mathbb{R}}^{I,\alpha}$ that contains indicators of hypersurface pairs. Chapter II is partly devoted to this problem.

FIRST CLASSIFICATION PROBLEM. *Classify equivalence classes of germs in $\mathscr{B}_{1,\mathbb{R}}^{I,\alpha}$ with respect to the action of $\text{Diff}_{\mathbb{R}}^{J,\alpha}$.*

This problem is a particular case of a general classification problem stated and discussed in §4. In the same section, we state the results obtained (in particular, an analytic classification theorem of hypersurface pairs for the case of "two folds").

Finally, we would like to quote V. I. Arnold. In [2] he writes that for the case of two folds "reduction to Melrose normal form is possible in the C^∞-case or on the formal series level but not in the analytic case. The pairs of involutions encountered in other problems also have a simple formal classification without functional moduli and have functional moduli in the analytic situation. To the best of my knowledge, functional moduli of the analytic classification for the problem of hypersurface pairs in symplectic manifolds are not presented explicitly (this is also true for the analytic classification of canonical pairs of involutions or of their products)." The functional moduli mentioned in the above quotation (and for all problems mentioned above) will be presented in the present paper. As to "other" problems, they were already considered above in §1.

2.8. Pleats and folds. Suppose the singularity of one canonical projection is a pleat (or A_k, $k \geq 2$) and the singularity of the other one is a fold. Then it is impossible to construct a unique normal form on the zero level, since already the formal classification involves functional moduli [21]. Still on the first and second levels, the normalization for the pleat case ($k = 2$) is possible in both formal and analytic cases [2, 3]. Namely, if the canonical projection has a pleat singularity, the pair (hypersurface, submanifold of codimension 1) in the generic case can be reduced (in Darboux coordinates) to the form ($q_0 = 0$, $q_0 = p_0^3 + p_1 p_2 + q_1 = 0$) by a local symplectomorphism. In accordance with this, at the second level we shall see that a hypersurface with singularity "semicubical cuspidal edge" in the generic case can be reduced to the normal form $L_0 = \{p_0^3 = q_0^2\}$ by a local symplectomorphism. Or equivalently, a generic symplectic structure can be reduced to Darboux normal form by local coordinate changes that preserve a standard hypersurface with singularities L_0 (the Darboux-Whitney theorem for the singularity A_2), see [23].

2.9. Singularities A_k, $k \geq 3$. The formal Darboux-Whitney theorem for A_3. Suppose the canonical projection has an A_k singularity. Then for the first and second levels, the formal classification is trivial if $k = 3$ and involves moduli if $k > 3$. Namely, the pair (hypersurface, submanifold of codimension 1) with canonical projection having the singularity A_3 in the generic case can be reduced to the form

$$(q_0 = 0, \; q_0 = p_0^4 + p_1 p_0^2 + q_1 p_0 + p_2 = 0)$$

in Darboux coordinates [3] by a formal symplectomorphism. The set of critical values of this projection (the extended swallowtail) also admits a corresponding normalization (second level!).

DEFINITION. Let us call a hypersurface with singularities $L \subset \mathbb{R}^3$ consisting of all points $(A, B, C) \in \mathbb{R}^3$ a *swallowtail* if the polynomial $x^4 + Ax^2 + Bx + C$ has a multiple root. The points that correspond to polynomials with two multiple roots form a line of self-intersection, while the points that correspond to polynomials with a root of multiplicity three form the cuspidal edge of the swallowtail (see Figure 4 on next page).

The product $\Lambda = L \times \mathbb{R}^{2n-1} \subset \mathbb{R}^{2n+2}$ is called the *extended swallowtail*. In the same manner, we define the self-intersection "line" and the cuspidal "edge". Let us call the germ H of the coordinate change in $(\mathbb{R}^{2n+2}, 0)$ that preserves a germ of the extended swallowtail at zero (i.e., such that $H(\Lambda) = \Lambda$), a Λ-*diffeomorphism*.

Following V. I. Arnold, let us formulate the statement about normalization on the second level for the singularity A_3 as the so-called Darboux-Whitney theorem.

DARBOUX-WHITNEY THEOREM (Formal version, Arnold, [3]). *A generic symplectic structure (i.e., nondegenerate on the space $B = C = 0$) can be reduced to the normal form*

$$\omega_0 = dA \wedge dD + dC \wedge dB + dE \wedge dF + \cdots$$

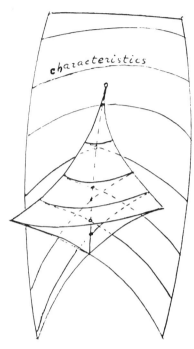

FIGURE 4. Swallowtail L. Projections of the characteristics of Λ on L.

(here $(A, B, C, D, E, F, \ldots)$ are the coordinates in $\mathbb{R}^{2n+2}$ and A, B, and C were used in the definition of the swallowtail).

The smooth (C^∞) and analytic versions of this theorem are studied in the third chapter of the present paper. It turns out that the Darboux-Whitney theorem holds also in the smooth case but not in the analytic one: the classification of generic symplectic structures with respect to the action of the group of Λ-diffeomorphisms involves functional moduli.

Let us call the classification problem of generic symplectic structures with respect to the action of the group of Λ-diffeomorphisms the *Darboux-Whitney problem*. This problem is solved below following the approach outlined in the Introduction. Namely, for any generic symplectic structure, we construct its invariant, which is a germ of a mapping with a hypersurface of fixed points. The Darboux-Whitney problem reduces to the corresponding classification problem for indicators.

The indicators in the Darboux-Whitney problem are constructed in 2.10. The indicator classification problem (the second classification problem) is stated in 2.12. The reduction of the Darboux-Whitney problem to the second classification problem is carried out in Chapter III (this problem is outlined in §1 of Chapter III). The solution to the second classification problem is discussed in §4. We also refer to §4 for the statement of all results concerning the Darboux-Whitney problem that are derived in this paper.

2.10. Construction of the indicators for the Darboux-Whitney problem.

The indicator for the Darboux-Whitney problem is (as above for the hypersurface pairs problem) the product of two involutions with common mirror. For example, these involutions may be defined as follows (Arnold's construction, private communication).

Let e be the extended self-intersection line of an extended swallowtail Λ and suppose the subset $\gamma \subset e$ consists of all points corresponding to the polynomial with a root of multiplicity 4. The intersection of a sufficiently narrow neighborhood of the set $e \setminus \gamma$ and of Λ consists of two transversally intersecting leaves (smooth surfaces). The set of vectors tangent to Λ at the points of $e \setminus \gamma$ consists of a pair of hyperplanes tangent to these leaves and of one hyperplane at the points of γ. Consider the space $\hat{e}$ whose points are the pairs (point of e; hyperplane formed by vectors tangent to Λ at this point). Note that we may define the natural projection $\hat{e} \to e$ by omitting the second component of the pair. The points of $e \setminus \gamma$ have two-point preimages in $\hat{e}$ while the points of γ have one-point preimages. The first Arnold involution interchanges the points of the two-point preimages of this projection. The second one is constructed from the generic symplectic structure as follows. The characteristics of the surface $\Lambda = L \times \mathbb{R}^{2n-1}$ leave the set $e \setminus \gamma$ along one leaf and return along the other one. Their projections on the swallowtail L (along the second factor) are shown in Figure 4. The second Arnold involution interchanges the points of $\hat{e}$ that correspond to the intersection points of characteristics with $e \setminus \gamma$ and is the identity over the points of γ.

Let us give a "coordinate" description of the same construction. This construction relies on certain corollaries of genericness that are proved in Chapter III, §1.

For example, let us choose the following parametrization of the extended swallowtail. Let $P(x) = x^4 + Ax^2 + Bx + C$. Let us express B and C from the system that defines the swallowtail $L \subset \mathbb{R}^3$ in terms of the variables A and x (which therefore become parameters on L). This leads to the following parametrization of the extended swallowtail Λ.

DEFINITION. Let $A, B, C, D, t, x, y \in \mathbb{R}$ and $z, Z \in \mathbb{R}^{2n-1}$. Let us call the mapping given by $A = t$, $B = -2tx - 4x^3$, $C = 3x^4 + x^2 t$, $D = y$, and $Z = z$ the *standard parametrization* $\varphi \colon \mathbb{R}^{2n+1} \to \mathbb{R}^{2n+2}$, $\varphi \colon (t, x, y, z) \mapsto (A, B, C, D, Z)$.

Observe that the standard parametrization φ is injective outside the preimage $\tilde{e}$ of the extended self-intersection line and its restriction to this preimage given by the equation $t = -2x^2$ is a fold. Denote by I the involution on $\tilde{e}$ that interchanges points of preimages under the mapping φ, $I \colon (-2x^2, x) \mapsto (-2x^2, -x)$. The involution I is the identity on the set $\tilde{\gamma}$: $t = x = 0$ of points of the space of parameters that correspond to polynomials with a root of multiplicity four.

Let us construct a second involution on $\tilde{e}$ with the same mirror. Let ω be a generic symplectic structure in $(\mathbb{R}^{2n+2}, 0)$. Let us transfer it to the parameter space using the parametrization and put $\tilde{\omega} = \varphi^*\omega$. From the genericness condition, we infer that the kernel of $\tilde{\omega}$ is one-dimensional at each point of $(\mathbb{R}^{2n-1}, 0)$. Observe that the kernel of the tangent map $(\varphi|_{\tilde{e}})_*$ is nontrivial at the points of the fold curve. In particular, this implies that the kernel of $\tilde{\omega}$ at the points of $\tilde{\gamma}$ is tangent to $\tilde{e}$. The parameter space is foliated into characteristics that are the integral curves of the kernel field of $\tilde{\omega}$. To each point of $\tilde{e}$ let us associate the characteristic passing through this point. We obtain a projection of $\tilde{e}$ onto the characteristic space of dimension equal to the dimension of $\tilde{e}$. As was noted above, characteristics are tangent to $\tilde{e}$ at the points of $\tilde{\gamma}$, so for this projection all the points of $\tilde{\gamma}$ are singular. For the generic case, this projection is a fold. Thus the involution that interchanges points with same projection is well defined and $\tilde{\gamma}$ is its mirror.

DEFINITION. A product of two involutions (the first on the left, the second on the right) constructed above with the help of the parametrization from a generic symplectic structure is called the *indicator* of this structure.

2.11. Properties of indicators of generic symplectic structures. The indicator Δ_ω of a symplectic structure ω is a product of two involutions with common mirror of fixed points. Each of these involutions preserves the almost symplectic structure $\alpha = \varphi^*\omega|_{\tilde{e}}$ (this is evident for the first involution and follows from Lemma 2.6 for the second). Thus the indicator Δ_ω also preserves the form α: $A^*_\omega \alpha = \alpha$.

EXAMPLE. For the normal form ω_0 from the formal Darboux-Whitney theorem, the preimage $\tilde{e}$ of an extended self-intersection curve in coordinates $(t, x, y, z) \in \mathbb{R}^{2n+1}$ has the form $\tilde{e} = \{t + 2x^2 = 0\}$. Thus (x, y, z) are the coordinates on $\tilde{e}$. In these coordinates the objects that correspond to ω_0 have the following form:

first involution $I_0 \colon (x, y, z) \mapsto (-x, y, z)$,
second involution $J \colon (x, y, z) \mapsto (-x, y + (32/15)x^5, z)$,
almost symplectic structure:

$$\alpha = -4x\, dx \wedge dy + dz_1 \wedge dz_2 + \cdots + dz_{2n-3} \wedge dz_{2n-2}.$$

Both involutions and the indicator preserve α: $I_0^*\alpha = J^*\alpha = \delta^*\alpha = \alpha$. Both involutions (and the indicator) have the common mirror $\tilde{\gamma} = \{x = 0\} \subset \tilde{e}$. The almost symplectic structure degenerates on the set $\tilde{\gamma}$ of fixed points of the indicator. These results can be obtained by simple calculations that we omit here.

The restriction of the parametrization φ to $\tilde{e}$ is a fold. The restriction δ of the symplectic structure ω to the extended self-intersection line $\varphi(e)$ is nondegenerate due to the genericness condition from the formal Darboux-Whitney theorem. Then the analytic version of the normalization statement at the second level in the case of a fold (see 2.5) allows us to normalize both the mapping $\varphi|_{\tilde{e}}$ and the symplectic structure δ. Hence without loss of generality, throughout what follows we may assume that the almost symplectic structure $\alpha = (\varphi|_{\tilde{e}})^*\delta$ and the first of the two involutions constructed above are standard (namely, are the same as in 2.7). The additional structures I_0 and α from the example are reduced to the standard form from 2.7 by linear substitutions.

The Λ-diffeomorphism maps the extended self-intersection line e onto itself. Hence its restriction to e can be lifted to a diffeomorphism of the double covering $\tilde{e}$ over e and commutes with the mapping that transposes points of the leaves of this covering (i.e., with the first of the two constructed involutions). If the Λ-diffeomorphism maps one symplectic structure onto another, it also maps characteristics on Λ of the first structure onto the characteristics of the second. Therefore the corresponding lifting conjugates (on $\tilde{e}$) the "second" involutions of the original symplectic structures (and hence their indicators). Besides, this lifting preserves the standard almost symplectic structure α from the example (this structure coincides with the restrictions to $\tilde{e}$ of both symplectic structures). From this and the formal Darboux-Whitney theorem, it follows in particular that the indicator of generic symplectic structure can be reduced to the normal form Δ from the example by a formal diffeomorphism that commutes with the standard involution I_0 and preserves the 2-form α. Concluding this subsection, we note that having fixed the standard almost symplectic structure α, we have thus selected two coordinates satisfying the following conditions: the form α degenerates at the points of the surface $\{x = 0\}$; the vector $\partial/\partial y$ belongs to the kernel of α at the points of this surface. Taking this statement into account, we conclude that the 5-jet of the indicator at zero is of the form

$$(x, y, z) \mapsto (x, y + kx^5, z), \qquad k \neq 0.$$

Repeating the above argument for points of the surface $\{x = 0\}$ that are close to zero, one may see that at all these points the 5-jet of the indicator has the same form (k may depend on the point).

2.12. Second classification problem. As was shown in the two preceding subsections, the indicator Δ of a generic symplectic structure has the following property: there exists a chart (x, y, z) in $\mathbb{R}^{2n}$, $x, y \in \mathbb{R}$, $z \in \mathbb{R}^{2n-2}$ such that:

1) the germ Δ is real-analytic and can be represented as the product of two involutions with common mirror (the set of fixed points). One of these involutions is standard:

$$I: (x, y, z) \mapsto (-x, y, z);$$

2) the 5-jet of the germ Δ at the mirror points has the form

$$(x, y, z) \mapsto (x, y + kx^5, z), \qquad k \neq 0;$$

3) the germ Δ preserves the standard almost symplectic structure

$$x = -x\,dx \wedge dy + dz_1 \wedge dz_2 + \cdots + dz_{2n-3} \wedge dz_{2n-2}.$$

As in 2.7, let us denote by $\mathscr{B}_{5,\mathbb{R}}^{I,\alpha}$ the space of germs Δ with properties 1)–3) (here the subscript 5 indicates the lowest term of the germ's discrepancy).

Suppose two generic symplectic structures in $(\mathbb{R}^{2n}, 0)$ are Λ-equivalent (this means that there exists a local real-analytic diffeomorphism that preserves the swallowtail Λ and maps one structure onto the other). As shown in 2.4 above, under these conditions there exists a local real-analytic diffeomorphism that conjugates the indicators of these structures, commutes with the involution I, and preserves the almost symplectic structure α. Recall that in 2.7 we denoted by $\operatorname{Diff}_{\mathbb{R}}^{I,\alpha}$ the class of germs of diffeomorphisms that have these properties. We conclude that the Darboux-Whitney problem leads to the following classification problem.

SECOND CLASSIFICATION PROBLEM. *Give the classification of germs from $\mathscr{B}_{5,\mathbb{R}}^{I,\alpha}$ with respect to the group action of $\operatorname{Diff}_{\mathbb{R}}^{I,\alpha}$.*

This problem will be considered in 4.1 together with the first classification problem.

2.13. Complex case. The whole argument of the previous 12 subsections can be extended in a natural way to the complex case. In particular, the complex versions of the classification problem for hypersurface pairs in symplectic manifolds lead to the complex versions of the first and second classification problems. These versions are obtained from the problems in the real case by substituting $\mathbb{C}$ for $\mathbb{R}$ and "holomorphic" for "real-analytic" and will be also treated in §4 (see 4.1).

§3. Degeneracies of symplectic and contact structures

The study of singularities of the Pfaff equations H_T and of singularities of closed 2-forms [7, 26] for codimension 3 leads in a natural way to the problem of orbital classification of germs of real-analytic vector fields with singular points forming a submanifold of codimension 2, which have simple normal form. M. Ya. Zhitomirskiĭ proved that for such germs of vector fields the orbital formal classification does not coincide with the orbital analytic one and suggested the study of the latter to the author.

It turned out that for all these problems one can construct in a natural way indicators similar to those considered in the previous section. Their study allows us to obtain the analytic classification for the original problems.

The proofs of results presented in this section will appear in a separate paper. Here we only outline them.

The germs of vector fields mentioned above in the simplest cases are orbitally formally equivalent to one of the following two germs: either to

$$v_- = -x\frac{\partial}{\partial x} + z\frac{\partial}{\partial z} + xz\frac{\partial}{\partial y}$$

or to

$$v_+ = z\frac{\partial}{\partial x} - x\frac{\partial}{\partial z} + (x^2 + z^2)\frac{\partial}{\partial y}.$$

Denote by $\mathscr{V}_\pm$ the formal orbital equivalence class of the germ $v_\pm$. The following two subsections contain the formal orbital analytic classification of germs from $\mathscr{V}_\pm$. Their connection with the degeneracies of symplectic and contact structures is described in 3.3 and 3.4.

3.1. Germs of vector fields of the class $\mathscr{V}_-$. Together with every germ v of a real-analytic vector field let us consider its complexification $^\mathbb{C}v$. It is clear that the orbital analytic equivalence of germs implies the analytic orbital equivalence of their complexifications. The converse is also true if the holomorphism that maps phase portraits of the complexified germs onto each other commutes with the operator of complex conjugation.

At first, consider the normal form of v_-. Let us use the coordinates (x, y, z, u) in $\mathbb{C}^n$, where $x, y, z \in \mathbb{C}$ and $u \in \mathbb{C}^{n-3}$. All the points of the surface $\gamma = \{x = z = 0\}$ are singular points of the germ

$$^\mathbb{C}v_- = -x\frac{\partial}{\partial x} + z\frac{\partial}{\partial z} + xz\frac{\partial}{\partial y}.$$

Consider the "separatrix" $S = \{x = y = u = 0\}$ of the germ $^\mathbb{C}v_-$ and the transversal $\Gamma = \{z = 1\}$ to it. Let $\Delta_0: \Gamma \to \Gamma$ be the monodromy transformation of the germ $^\mathbb{C}v_-$ that corresponds to a single turn in positive direction of the punctured point $z = 0$ on the separatrix S. In the coordinates (x, y, u) on Γ this mapping has the form

$$\Delta_0: (x, y, u) \mapsto (x, y + 2\pi i x, u).$$

All the points of the intersection γ of the transversal Γ with the invariant manifold $G = \{x = 0\}$ of the germ v_- are fixed points of the germ Δ_0.

Now let v be an arbitrary germ from $\mathscr{V}_-$. Assume that e_v is an eigenvector of the linearization of v at the origin that corresponds to a negative eigenvalue. By the Bibikov theorem on normal forms on invariant manifolds (see [8]), there exists a separatrix S_v (a 1-dimensional analytic invariant manifold) of the germ v that is tangent to e_v at the origin. The singular point 0 divides the separatrix S_v into two phase curves of the germ v; each of these curves is called a *singular phase curve* of the germ.

REMARK. Two singular curves of a germ of class $\mathscr{V}_-$ cannot be distinguished from the intrinsic point of view. However, the monodromy transformations defined in terms of transversals to distinct singular phase curves of the germ, cannot in general be conjugated by a diffeomorphism from Diff^{σ_0}.

To avoid this difficulty (although it is not one of principle) and to simplify formulations of results, we shall replace the class $\mathscr{V}_-$ by $\dot{\mathscr{V}}_-$.

A germ of the class $\mathscr{V}_-$ is said to be *distinguished* if one of its two singular phase curves is distinguished. Denote by $\dot{\mathscr{V}}_-$ the class of all distinguished germs. Two distinguished germs are said to be *orbitally analytically equivalent* if there exists an analytic change of coordinates that conjugates their phase portraits (takes one of them to the other) and takes one of the distinguished singular phase curves to the other.

Assume that v is a germ from $\dot{\mathscr{V}}_-$ and S_v^+ is its distinguished singular phase curve. Denote by $^{\mathbb{C}}v$ the complexification of v and by S_v the phase curve (separatrix) of $^{\mathbb{C}}v$ that contains S_v^+. The separatrix S_v is biholomorphically equivalent to a punctured disk and is invariant with respect to the complex conjugation σ: $\sigma S_v = S_v$. Let Γ be the germ of the transversal to the separatrix S_v at a point of S_v^+ such that $\sigma \Gamma = \Gamma$. The monodromy transformation $\Delta_v : \Gamma \to \Gamma$ corresponding to a single turn of the punctured point of S_v in the positive direction is said to be the *indicator* of the germ $v \in \dot{\mathscr{V}}_-$.

Let us study the properties of the indicator Δ_v of a germ $v \in \dot{\mathscr{V}}_-$.

By the Bibikov theorem, there exists a smooth invariant hypersurface G_v of the germ $^{\mathbb{C}}v$ that contains both the set γ_v of its singular points and the separatrix S_v. The indicator is the identity on the intersection of the hypersurface and the transversal Γ:

$$\Delta_v = \mathrm{id} \quad \text{on} \quad G_v \cap \Gamma.$$

Since the complex conjugation σ reverses the direction of rotation of the punctured point on the separatrix and maps the phase portrait of the germ $^{\mathbb{C}}v$ onto itself, the indicator Δ_v is symmetric with respect to the restriction $\sigma_0 = \sigma|_\Gamma$ in the sense that $\sigma_0 \Delta_v \sigma_0 = \Delta_v^{-1}$. Finally, simple calculations show that the antisymmetry σ_0 is standard in a suitable local chart (x, y, u):

$$\sigma : (x, y, u) \mapsto (\overline{x}, \overline{y}, \overline{u}),$$

the set of fixed points γ_v of the indicator Δ_v is given by $\{x = 0\}$, and the 1-jet of the germ Δ_v at any point of γ_v has the form

$$(x, y, u) \mapsto (x, y + kx, u), \qquad k \neq 0.$$

Denote by $\mathscr{B}_1^{\sigma_0}$ the class of germs that possesses the indicator properties stated above and let Diff^{σ_0} be the class of germs of holomorphisms that commute with σ_0. Utilizing the methods developed in Paper I, one can prove the following theorem.

THEOREM. *Two germs from $\dot{\mathscr{V}}_-$ are orbitally analytically equivalent iff their indicators are conjugate with respect to the germ of a holomorphism from Diff^{σ_0}. Every germ from $\mathscr{B}_1^{\sigma_0}$ is realized as an indicator of some germ from $\dot{\mathscr{V}}_-$.*

This theorem reduces the orbital analytic classification problem of germs of class $\mathscr{V}_-$ to the following problem.

THIRD CLASSIFICATION PROBLEM. *Give the analytic classification of the germs from $\mathscr{B}_1^{\sigma_0}$ with respect to the group action of Diff^{σ_0}.*

The third classification problem will be considered in 4.1 in the framework of a more general approach.

The constructions of this subsection can be extended naturally to the complex case: one must only distinguish one of the two separatrices of the complex germ instead of the singular phase curve of the real analytic germ.

3.2. Germs of vector fields of class $\mathscr{V}_+$. Let (x, y, z, u) be coordinates in $\mathbb{R}^n$, where $x, y, z \in \mathbb{R}$ and $u \in \mathbb{R}^{n-3}$. Denote by γ_v the set of singular points of the germ $v \in \mathscr{V}_+$. Since by definition all the germs from $\mathscr{V}_+$ are formally orbitally equivalent to the germ

$$v_+ = z\frac{\partial}{\partial x} - x\frac{\partial}{\partial z} + (x^2 + z^2)\frac{\partial}{\partial y},$$

we conclude that γ_v is a smooth surface of codimension 2. It can be shown that the phase curves of the germ v are near to helices wound around γ_v (see Figure 5 on next page).

Consider an arbitrary smooth hypersurface Γ in $(\mathbb{R}^n, 0)$ that contains γ^v. Define the succession map $e_v: \Gamma \to \Gamma$ by associating to the intersection point of Γ with an arbitrary phase curve of the germ v its next intersection point with Γ. Let us call the iterative square $\Delta_v = e_v \circ e_v$ of the succession map e_v the *indicator* of the germ v. Note that on γ_v the indicator is identical.

EXAMPLE. For the formal normal form v_+, the surface γ_{v_+} can be expressed as $\{x = z = 0\}$. Let $\Gamma = \{z = 0\}$ and (x, y, u) be the coordinates on Γ. Then

$$e_{v_+}: (x, y, u) \mapsto (-x, y + \pi x^2, u),$$

$$\Delta_{v_+}: (x, z, u) \mapsto (x, y + 2\pi x^2, u).$$

As in 3.1, the formal orbital equivalence of the germs v and v_+ implies that in a suitable chart in which the set γ_v of fixed points of the germ v coincides with the plane $\{x = 0\}$, the 2-jet of the indicator Δ_v at the points of γ_v can be expressed as

$$(x, y, u) \mapsto (x, y + kx^2, u), \quad k \neq 0.$$

Denote by $\mathscr{B}_{2,\mathbb{R}}^{[2]}$ the class of germs of real-analytic mappings that have hyperplane of fixed points $\{x = 0\}$, may be represented as the iterative square of mappings with linear part $(x, y, u) \mapsto (-x, y, u)$, and satisfy the condition on 2-jets stated above. By definition, the indicator of the germ $v \in \mathscr{V}_+$ belongs to the class $\mathscr{B}_{2,\mathbb{R}}^{[2]}$. As above, the following theorem is valid.

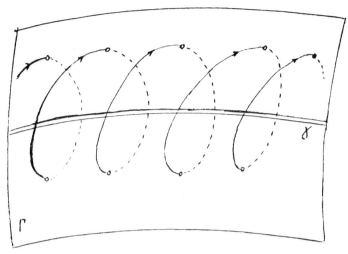

FIGURE 5. Phase curves of the germ $v \in \mathscr{V}_+$.

THEOREM. *Any germ from $\mathscr{B}_{2,\mathbb{R}}^{[2]}$ can be realized as the indicator of a germ from $\mathscr{V}_+$. Two germs from $\mathscr{V}_+$ are orbitally analytically equivalent iff their indicators are analytically conjugate.*

This theorem reduces the orbital analytic classification problem for germs of class $\mathscr{V}_+$ to the following problem.

FOURTH CLASSIFICATION PROBLEM. *Give the classification of germs of class $\mathscr{B}_{2,\mathbb{R}}^{[2]}$ with respect to the action of the group of germs of real-analytic diffeomorphisms $\mathrm{Diff}_\mathbb{R}$.*

This problem will be also considered in 4.1.

Now let us explain the origin of the classes of germs $\mathscr{V}_-$ and $\mathscr{V}_+$.

3.3. Degenerations of almost symplectic structures. A closed but not necessarily nondegenerate 2-form ω is called an *almost symplectic structure* on the even-dimensional smooth manifold M.

In the generic case, the almost symplectic structure ω is nondegenerate at the generic points of M (i.e., is a symplectic structure) and in suitable local coordinates coincides with the Darboux normal form $\omega_0 = dp \wedge dq$ [7]. The degeneration of ω occurs at the points of the smooth hypersurface $L \subset M$ (first degeneration hypersurface). The kernel of ω is two-dimensional at a generic point of L and is transversal to L. In the neighborhood of such a point, the form ω reduces to the normal form ω_1 [10], where

$$\omega_1 = p_1 dp_1 \wedge dq_1 + dp_2 \wedge dq_2 + \cdots + dp_n \wedge dq_n.$$

However the two-dimensional kernel of ω is tangent to L at the points of a smooth submanifold $\gamma \subset L$ of codimension 2. Let us call the surface γ the *second degeneration surface*. At a generic point of the second degeneration surface γ, the kernel of ω is transversal to γ and in the neighborhood of such a point ω can be reduced to one of the two following normal forms $\omega_\pm$ [7]:

$$\omega_\pm = d(x - z^2/2) \wedge dy + d(xz \pm ty - z^3/3) \wedge dt + dp_1 \wedge dq_1 + \cdots + dp_{n-2} \wedge dq_{n-2}.$$

The almost symplectic structure ω can be reduced to the normal forms ω_0 and ω_1 by both formal and smooth or analytic coordinate changes. The reduction to the normal form $\omega_\pm$ was obtained previously only for the formal and smooth cases; the analytic case will be studied in this subsection.

Let us call the germs of almost symplectic structures with real-analytic coefficients *analytically (formally) equivalent* if there exists a germ of a real-analytic coordinate change (formal change) that transforms the germ of one structure into the germ of the other one. Let $\Omega_\pm$ be a formal equivalence class of the 2-form $\omega_\pm$.

The kernel field of the form $\omega \in \Omega_\pm$ cuts out an analytic direction field on the first degeneration surface L_ω outside the second degeneration submanifold γ_ω. This direction field can be given by the analytic vector field v_ω that will be called the *characteristic vector field* of the form ω. The characteristic vector field vanishes on the second degeneration submanifold: $v_\omega = 0$ on γ_ω.

EXAMPLE. For the normal form $\omega_\pm$, the first degeneration surface L has the form $\{x = 0\}$ while the second one has the form $\gamma = \{x = z = t = 0\}$. In the coordinates $(y, z, t, \ldots)$ on L, the characteristic vector field $v_{\omega_\pm}$ can be expressed as

$$v_{\omega_\pm} = -z^2 \frac{\partial}{\partial y} + z \frac{\partial}{\partial t} \mp t \frac{\partial}{\partial z}.$$

In particular, the field $v_{\omega_\pm}$ can be reduced to the normal form $v_\pm$ by a polynomial change of coordinates (see the beginning of this section, as well as 3.1 and 3.2).

Let us describe a specific way of finding the characteristic vector field v_ω of the germ $\omega \in \Omega_\pm$. Let L_ω be the manifold of the first degeneracy for ω. Choose an orientation on ω, and let Q be the volume form on L_ω (that is, a nondegenerate $(2n-1)$-form on L_ω, $2n-1 = \dim L_\omega$) that defines this orientation. Define the field v_ω by

$$Q(v_\omega, \cdot) = (\omega|_{L_\omega})^{n-1}(\cdot).$$

Let $\omega, \tilde{\omega} \in \Omega_\pm$, and suppose that H takes ω to $\tilde{\omega}$: $H^*\tilde{\omega} = \omega$. Then obviously $H(L_\omega) = L_{\tilde{\omega}}$. Moreover, suppose that the mapping $h = H|_{L_\omega}$ takes the orientation defining v_ω to that defining $v_{\tilde{\omega}}$. Then h establishes the orbital equivalence of the germs v_ω and $v_{\tilde{\omega}}$. This fact (together with the example above) implies, in particular, that $v_\omega \in \mathscr{V}_\pm$ for any germ $\omega \in \Omega_\pm$.

To simplify further presentation (compare with the Remark in 3.1), we change the setting of the initial problem slightly.

DEFINITION. A *distinguished almost symplectic structure* is an almost symplectic structure from $\Omega_\pm$ with a distinguished orientation of its manifold of first degeneracy (and a distinguished singular phase curve, see 3.1, of its characteristic vector field in the case of Ω_-). Denote by $\dot{\Omega}_\pm$ the class of distinguished almost symplectic structures. The *characteristic vector field* $v_\omega \in \mathscr{V}_+$ ($v_\omega \in \mathscr{V}_-$) of a distinguished almost symplectic structure $\omega \in \dot{\Omega}_+$ ($\omega \in \dot{\Omega}_-$) is defined by its distinguished orientation. Two distinguished almost symplectic structures are said to be *equivalent* if there exists a change of coordinates that takes both the structure and all of its distinguished elements to the other structure and its corresponding distinguished elements. The *indicator* of a distinguished almost symplectic structure is the indicator of its characteristic vector field.

Let Γ_ω be the transversal to the characteristics (phase curves of the field v_ω) that was used in the definitions of indicators of the characteristic vector fields v_ω in 3.1 and 3.2. Denote by α_ω the restriction of ω to Γ_ω. Then by virtue of the lemma in 2.6, the indicator Δ_ω preserves the 2-form α_ω: $\Delta_\omega^* \alpha_\omega = \alpha_\omega$. For the 2-form α_ω (in the case $\omega \in \dot{\Omega}_+$), the submanifold of fixed points of the germ Δ_ω is the first degeneration surface and therefore α_ω can be reduced to the normal form ω_1 by a suitable choice of (analytic!) chart on the transversal Γ_ω. In the case $\omega \in \dot{\Omega}_-$ the 2-form α_ω is nondegenerate; it can be reduced to Darboux normal form by an analytic transformation respecting the submanifold of fixed points of the indicator Δ_ω. In the charts (x, y, u) that were used to write out the indicator in 3.1 and 3.2, these normal forms can be expressed as follows:

$$\alpha_+ = -x\,dx \wedge dy + du_1 \wedge du_2 + \cdots + du_{2n-3} \wedge du_{2n-2},$$
$$\alpha_- = dx \wedge dy + du_1 \wedge du_2 + \cdots + du_{2n-3} \wedge du_{2n-2}.$$

Denote by $\mathscr{B}_1^{\alpha_-, \sigma_0}$ (by $\mathscr{B}_{2,\mathbb{R}}^{[2], \alpha_+}$) the class of germs from the space $\mathscr{B}_1^{\sigma_0}$ ($\mathscr{B}_{2,\mathbb{R}}^{[2]}$) defined in 3.1 (3.2) that preserve the 2-form α_-: $\Delta^* \alpha_- = \alpha_-$ (respectively, that preserve α_+: $\Delta^* \alpha_+ = \alpha_+$). The indicators of almost symplectic structures from $\dot{\Omega}_-$ ($\dot{\Omega}_+$) belong to this class.

Similarly one can check the following statement: a diffeomorphism that maps one symplectic structure onto another induces a diffeomorphism of the transversals that conjugates its indicators and preserves the 2-form α_- (respectively, α_+). Denote by $\text{Diff}^{\alpha_-, \sigma_0}$ ($\text{Diff}_\mathbb{R}^{\alpha_+}$) the class of diffeomorphisms from Diff^{σ_0} ($\text{Diff}_\mathbb{R}$) that preserves α_- (respectively, α_+). Then the following theorem is valid.

THEOREM. *The indicator of a germ from $\dot{\Omega}_-$ (from $\dot{\Omega}_+$) belongs to the class $\mathscr{B}_1^{\alpha_-, \sigma_0}$ ($\mathscr{B}_{2,\mathbb{R}}^{[2], \alpha_+}$). Any germ from $\dot{\Omega}_-$ (from $\dot{\Omega}_+$) is an indicator of a germ from $\mathscr{B}_1^{\alpha_-, \sigma_0}$ ($\mathscr{B}_{2,\mathbb{R}}^{[2], \alpha_+}$). Two germs from $\dot{\Omega}_-$ (from $\dot{\Omega}_+$) are analytically equivalent iff their indicators can be conjugated by the germ of a diffeomorphism from $\text{Diff}^{\alpha_-, \sigma_0}$ ($\text{Diff}_\mathbb{R}^{\alpha_+}$).*

We observe that the analytic classification problem for germs of class $\dot{\Omega}_-$ (class $\dot{\Omega}_+$) reduces to the following problem.

SYMPLECTIC VERSION OF THE THIRD AND FOURTH CLASSIFICATION PROBLEMS. *Give the germ classification of the class $\mathscr{B}_1^{\alpha_-,\sigma_0}$ (of the class $\mathscr{B}_{2,\mathbb{R}}^{[2],\alpha_+}$) with respect to the group action of* $\operatorname{Diff}^{\alpha,\sigma_0}$ *(respectively,* $\operatorname{Diff}_\mathbb{R}^\alpha$*).*

This problem will be considered in 4.1.

3.4. Degenerations of almost contact structures. Suppose that a hyperplane that depends smoothly on the application point is defined in the tangent space at every point of a smooth manifold. In this case let us say that a field of hyperplanes (an almost contact structure) is defined on this manifold. The *almost contact structure* is locally determined by a nonvanishing 1-form ω; $\omega|_x = 0$ is the equation of the hyperplane of the field (of the contact element) at x. An almost contact structure on a $(2n+1)$-dimensional manifold is called the *contact structure* if the form $\omega \wedge (d\omega)^n$ is nondegenerate (it is evident that the definition is independent of the choice of the defining 1-form ω). Let us call the restriction to the contact element $\{\omega = 0\}$ of the differential $d\omega$ of the 1-form ω that defines the almost contact structure the *induced almost symplectic structure on the contact element*. The kernel of the induced almost symplectic structure is independent of the choice of the 1-form ω that defines the almost contact structure.

In the generic case, the almost contact structure is nondegenerate (is a contact structure) at a generic point of the manifold and can be reduced to Darboux normal form [7]:

$$\omega = du + x_1 dy_1 + \cdots + x_n dy_n$$

by an analytic (formal, smooth) substitution.

The degeneration of the generic almost contact structure occurs at points of the smooth hypersurface L (first degeneration surface). At the generic points of L, the kernel of an induced almost symplectic structure is two-dimensional and transversal to L. In the neighborhood of such a point the almost contact structure can be reduced to Martinet normal form [18]

$$\omega_1 = u\,du + (1+x_1)\,dy_1 + x_2 dy_2 + \cdots + x_n dy_n$$

by a local analytic (formal, smooth) substitution. However, the two-dimensional kernel of the induced almost symplectic structure is tangent to L at the points of the smooth submanifold $\gamma \subset L$ of codimension 2 (in L). Let us call the surface γ the *second degeneration surface* of the almost contact structure. At generic points of γ, the kernel of the induced almost symplectic structure is transversal to the second degeneration surface and in the neighborhood of such points the almost contact structure can be reduced to

Zhitomirskiĭ normal form [15]

$$\omega_\pm = dy_1 + x_2 dy_2 + \cdots + x_n dy_n + x_1(y_2 + x_1^2 \pm u^2)\, du$$

by a local smooth (or formal) substitution. In this subsection, we intend to study the possibility of reduction to Zhitomirskiĭ normal form by analytic substitutions.

DEFINITION. Let us call two almost contact structures *analytically (formally) equivalent* if one of them can be transformed to the other by a local analytic diffeomorphism (a formal substitution). Denote by $Q_\pm$ the classes of formal equivalence of almost contact structures with representatives $\omega_\pm$.

Let ω be an almost contact structure from $Q_\pm$ and let L_ω and γ_ω be its first and second degeneration surfaces, respectively. The kernel of the almost symplectic structure induced by ω on the contact element and applied at the point $x \in L_\omega \setminus \gamma_\omega$ cuts out a one-dimensional direction that is called *characteristic* on $T_x L_\omega$. The field of characteristic directions is defined on L_ω outside of γ_ω and can be defined by the analytic vector field v_ω (that we also call characteristic). The characteristic vector field v_ω vanishes on the second degeneration surface γ_ω.

EXAMPLE. For the normal form $\omega_\pm$, the first degeneration surface $L_{\omega_\pm}$ can be expressed as $y_2 + 3x_1^2 \pm u^2 = 0$, so $(u, x_1, y_1, x_2, x_3, y_3, \ldots, x_n, y_n)$ are the coordinates on $L_{\omega_\pm}$. In these coordinates, the second degeneration surface $\gamma_{\omega_\pm}$ has the form $\gamma_{\omega_\pm} = \{x_1 = u = 0\} \subset L_{\omega_\pm}$. The characteristic vector field is given by

$$v_{\omega_\pm} = -3x_1 \frac{\partial}{\partial u} \pm u \frac{\partial}{\partial x_1} - 3x_1^2 \frac{\partial}{\partial x_2} - 6x_1^4 \frac{\partial}{\partial y_1};$$

on $\gamma_\pm$ we have $v_{\omega_\pm} = 0$ and it can be reduced to the form $v_\pm$ from 3.1 and 3.2 by a polynomial substitution.

As in 3.3, the definitions of the classes $Q_\pm$ imply $v_\omega \in \mathscr{V}_\pm$ if $\omega \in Q_\pm$.

As in the previous subsections, to simplify formulations, we endow each almost contact structure $\omega \in Q_\pm$ with a distinguished element by choosing one of the two possible directions for going around its characteristics (besides, in the case of Q_-, we distinguish one of the two singular phase curves of the corresponding field v_ω). Denote by $\dot{Q}_\pm$ class of *distinguished almost contact structures* thus obtained. The characteristic vector field $v_\omega \in \dot{\mathscr{V}}_+$ ($v_\omega \in \dot{\mathscr{V}}_-$) of the germ $\omega \in \dot{Q}_+$ ($\omega \in \dot{Q}_-$), as well as the equivalence of germs in $\dot{Q}_\pm$, are defined in a natural way.

DEFINITION. The indicator of the characteristic field of the distinguished almost symplectic contact structure $\omega \in \dot{Q}_\pm$ is called an *indicator* of this structure. We denote it by Δ_ω.

Repeating the argument of the previous subsection, we arrive to the following contact version of the theorem stated in 3.3.

Let
$$\beta_+ = du_1 + x^2 dy - 2xy\, dx + u_2 du_3 + \cdots + u_{2n-3} du_{2n-2},$$
$$\beta_- = du_1 + x\, dy - y\, dx + u_2 du_3 + \cdots + u_{2n-3} du_{2n-2}.$$

Note that β_+ is equivalent to the Martinet model, while β_-, to the Darboux one.

Define the classes $\mathscr{B}_1^{\beta_- \sigma_0}$, $\mathscr{B}_{1,\mathbb{R}}^{[2]\beta_+}$, $\mathrm{Diff}^{\beta_- \sigma_0}$, $\mathrm{Diff}_{\mathbb{R}}^{\beta_+}$ by replacing the invariance conditions for $\alpha_\pm$ by the invariance of the almost contact structures $\beta_\pm$ in the corresponding definitions in 3.3. Then changing $\dot{\Omega}_\pm$ to $\dot{Q}_\pm$ and the almost symplectic structure $\alpha_\pm$ to the almost contact structure $\beta_\pm$ in the claim of Theorem in 3.3, we obtain a valid theorem. This means that the analytic classification problem for almost contact structures of the class $\dot{Q}_-$ ($\dot{Q}_+$) can be reduced to the contact version of the third (fourth) classification problem. Its statement is *mutatis mutandis* the statement of the corresponding symplectic problem.

§4. General classification problem with an outline of the solution. Statement of the results

In this section we consider a problem for which the classification problems in §§2, 3 are particular cases. It is formulated in 4.1. The following eight subsections are devoted to a detailed outline of its solution. The results of this paper are summarized in 4.10.

4.1. Standard structures and the general classification problem. All the problems in §§2 and 3 (namely the problems one to four, the symplectic and contact versions of the third and fourth, as well as their complex versions) are related to each other and allow a unified formulation and solution.

DEFINITION. Let $\mathscr{B}_q$ be the class of germs of all holomorphic mappings $(\mathbb{C}^n, 0) \to (\mathbb{C}^n, 0)$ that are the identity on the hypersurface $\{x = 0\}$ and have a q-jet of the form
$$(x, y, z) \mapsto (x, y + kx^q, z), \qquad k \neq 0,$$
at the points of this hypersurface (here (x, y, z) are the coordinates in $\mathbb{C}^n$, $x, y \in \mathbb{C}$, $z \in \mathbb{C}^{n-2}$).

Let Diff be the group of germs of holomorphic transformations $(\mathbb{C}^n, 0) \to (\mathbb{C}^n, 0)$.

The germs and transformations considered above, used to define the equivalence of germs, possess additional properties of symmetry. Let us summarize them.

Let $(x, y, z) \in \mathbb{C}^n$, $x \in \mathbb{C}$, $y \in \mathbb{C}$, $z \in \mathbb{C}^{n-2}$. Suppose that as above $I: (x, y, z) \mapsto (-x, y, z)$ is the involution and $\sigma: (x, y, z) \mapsto (\overline{x}, \overline{y}, \overline{z})$ is the antiholomorphic involution. If n is even, then let
$$\alpha_+ = x\, dx \wedge dy + dz_1 \wedge dz_2 + \cdots$$

be the Martinet model of almost symplectic structure and let
$$\alpha_- = dx \wedge dy + dz_1 \wedge dz_2 + \cdots$$
be the Darboux model. If n is odd and $q \in \mathbb{N}$, then let
$$\beta_+^q = dz_1 + x^2 dy - qxy\, dx + z_2 dz_3 + \cdots$$
and
$$\beta_-^q = dz_1 + x\, dy - qy\, dx + z_2 dz_3 + \cdots$$
(the superscript q is defined by the corresponding class $\mathscr{B}_q$). Note that β_+^q is equivalent to the Martinet model of the almost contact structure and β_-^q is equivalent to the Darboux model.

Let us call these forms "*geometric structures*". In the next definition "structure" is one of the following objects: "symmetry", "antisymmetry", "geometric structure".

DEFINITION. A germ $F \in \mathscr{B}_q$ is said to *agree* with the structure s (a germ $H \in \text{Diff}$ preserves the structure s) if:

s is a symmetry: for q odd $IFI = F^{-1}$ ($IHI = H$) and for q even F is an iteration square of a germ with linear part I (no restrictions on H);

s is an antisymmetry: either $\sigma F \sigma = F$ or $\sigma F \sigma = F^{-1}$ ($\sigma H \sigma = H$);

s is a geometric structure α: for n even $F^*\alpha = \alpha$ ($H^*\alpha = \alpha$) and for n odd the germ F (H) maps an almost contact structure defined by the form α onto itself.

Let us call these structures ("symmetry", "antisymmetry" and "geometric structure") *elementary structures* and an arbitrary collection of them a *standard structure*. A germ $F \in \mathscr{B}_q$ is said to *agree* with the standard structure s (the germ $H \in \text{Diff}$ *preserves* s) if the germ F agrees with each of the elementary structures of the set s (H preserves each of these elementary structures).

Denote by $\mathscr{B}_q^s$ (Diff^s) the set of germs $F \in \mathscr{B}_q$ ($H \in \text{Diff}$) that agree with the structure s (that preserve s).

GENERAL CLASSIFICATION PROBLEM FOR THE CLASS $\mathscr{B}_q^s$. *Give the classification of germs from $\mathscr{B}_q^s$ with respect to the group action of* Diff^s.

Thus, for example, the case $q = 1$ ($q = 5$) and the complete set of elementary structures (the geometric structure is α_+) correspond to the first (second) classification problem, the case $q = 2$ and the "antisymmetry" structure correspond to the third problem, and the case $q = 2$ and the "symmetry, antisymmetry" structure correspond to the fourth one. The symplectic and contact versions of the third (fourth) problem are obtained by adding the geometric structure with "minus" ("plus") and n even (respectively, odd). The complex versions of the first three problems as well as of the symplectic and contact versions of the third are obtained by forgetting the "antisymmetry" structure.

REMARK 1. A justification for the somewhat perplexing name for the matching condition for a germ and the "symmetry" structure may be the following: this condition implies the symmetry (the existence of an involution) of the quotient space of the neighborhood of zero by the germ action.

REMARK 2. Note that the matching condition for a germ and an antisymmetry (a symmetry) implies the solvability of the group generated by this germ and the involution σ (or I for q odd).

REMARK 3. Below we impose a single restriction on the nonelementary structures s: if s contains the structures "symmetry" and "geometric structure", then the latter is necessarily a Martinet geometric structure. This restriction is motivated by certain consistency conditions for the structures of s, see 6.5, Chapter II.

The general classification problem is solved as follows. First one solves a so-called conjugation problem for the class $\mathscr{B}_q$ (i.e., one constructs the analytic classification of germs with respect to the group action of Diff) using the methods described in Paper I. Next one takes into account additional structures as it was done for the problem of taking roots in 2.7 of Paper I and in 2.6 of Paper II.

Note that, for simplicity, in proving all results, we confine ourselves to the case of $n = 2$ and q odd.

Besides, for convenience of notation, the structure index s is always written in deciphered form: we include the symbol $\alpha_\pm$ or $\beta_\pm^q$ into the set s for "geometric structure", the symbol I for "symmetry" and q odd, and the symbol [2] (representability in the form of iteration square) for "symmetry" and q even. Finally, we write σ instead of "antisymmetry". We write σ^+ or σ^- when it is necessary to state explicitly which of the two ways, $\sigma F \sigma = F$ or $\sigma F \sigma = F^{-1}$, of the agreement with this structure is considered. (This does not cause trouble, since the group action of Diff^σ does not take us out of any of the classes $\mathscr{B}_q^{\sigma^\pm}$.)

4.2. Plan of construction of functional moduli. We construct functional moduli for the conjugation problem in the same manner as it was done in the one-dimensional case. Namely, we normalize an arbitrary germ in a domain that is not a neighborhood of zero (sectorial domain) and establish the asymptotic proximity of the formal and holomorphic sectorial normalizing mappings. We construct an atlas in which the germ coincides with its formal normal form (the normalizing atlas) by covering the punctured neighborhood of zero with such sectorial mappings. The normalizing atlas is defined ambiguously, namely, multiplication of any of its charts by a mapping (its own mapping for each chart) that commutes with the normal form (i.e., the action of the "automorphism group of the normal form"), preserves the normalization property. The transition functions of the normalizing atlases form cocycles (the sets consisting of mappings that are asymptotically close to the identity and commute with the formal normal form). Next we show that any such set may be realized as the set of transition functions of some normalizing atlas. Finally, we factorize the space of these sets with respect to the action of the normal form automorphism group. We arrive at the moduli space. The principal tools are the successive approximation method used for normalizing germs and almost complex structures used for the realization of moduli.

We solve the general classification problem in a similar way. The only additional thing that we must require is the preservation of the additional structure s both by charts of the normalizing atlases and by the mappings that transpose cocycles and automorphism groups of the normal form.

4.3. Formal classification of germs of class $\mathscr{B}_1$. Recall that the germs of class $\mathscr{B}_q$ were defined in 4.1.

THEOREM (on the formal normalization of germs of class $\mathscr{B}_1$). *Any germ of class $\mathscr{B}_1$ is formally equivalent to the germ*

$$F_1: (x, y, z) \mapsto (x, x+y, z).$$

This theorem can be strengthened. First let us give some definitions.

DEFINITION 1. A formal Taylor series (in x) is called *semiformal*, if its coefficients holomorphically depend on y and z in the same domain. A formal change of coordinates is called *semiformal* if its components are semiformal series.

DEFINITION 2. The semiformal coordinate change

$$G: (x, y, z) \mapsto (xk(x, z), yk(x, z) + c(x, z), \varphi(x, z))$$

with coefficients affine in y is called a *semiformal affine coordinate change*.

REMARK. Any semiformal affine change commutes with F_1.

A coordinate change H with the property $H(x, 0, z) \equiv (x, 0, z)$ is called *normed*.

THEOREM. *For any germ $F \in \mathscr{B}_1$ there exists a unique semiformal normed change H_0 that conjugates F and F_1. Any semiformal normalizing change (i.e., transposing F and F_1) can be represented as the product of a semiformal normed change and a semiformal affine change.*

The formal normalization theorem is proved in §1 of Chapter II using the standard successive approximations method.

4.4. Sectorial normalization. Normalizing cochain. Automorphism group of a normal form. Let us call an arbitrary domain in $\mathbb{C}^n$ of the form $\{\alpha < \arg x < \beta, |x| < \varepsilon\} \times \{|y| < \varepsilon\} \times \{|z| < \varepsilon\}$ a *sectorial domain*. The pair α, β is called its *direction*, the difference $\beta - \alpha$ is called the *aperture*, and ε is called its *radius* (in Chapters II and III we use a somewhat different terminology).

Let us call a holomorphic injective mapping $H: U \to \mathbb{C}^n$ a *sectorial normalizing change for a germ* $\mathbb{F}$ *on the sectorial domain* $S \subset U$ if H conjugates a certain representative F of this germ on S with the normal form F_1, $H \circ F = F_1 \circ H$ on S. A semiformal mapping Φ is called *asymptotic on S for H* if

$$H(x, y, z) = \Phi_n(x, y, z) + o(x^n) \quad \text{as } x \to 0$$

for any partial sum Φ_n of the series Φ, where $(x, y, z) \in S$.

THEOREM (on sectorial normalization). *For any germ $\mathbb{F}$ and any sectorial domain S with given direction, aperture less than π, and sufficiently small radius, there exists a sectorial normed normalizing change for $\mathbb{F}$ on S and the normed semiformal normalizing change is asymptotic for it on S.*

The sectorial normalization theorem is proved in Chapter II, §§2, 3 using the finite order operator technique [1]. Namely, in §2 the homological equation is studied and in §3 the convergence of successive approximations is proved.

4.5. Normalizing cochain. Automorphisms of the normal form. Functional invariants. $\mathscr{B}_1$ **Theorem.** The functional invariants of the germs of $\mathscr{B}_1$ are constructed in Chapter II, §4 using an approach that is standard for the Stokes phenomenon. Namely, a small neighborhood of zero with deleted fixed points of a germ $F \in \mathscr{B}_1$ is covered by sectorial domains S_j. In accordance with the sectorial normalization theorem, normalizing mappings H_j with common asymptotic series are constructed in each of these domains. These mappings form a normalizing cochain $\{H_j\}$ of the germ F. The normalizing cochain is defined ambiguously: for two normalizing cochains $H = \{H_j\}$ and $\widetilde{H} = \{\widetilde{H}_j\}$ of the germ F (after a suitable renumbering of covering domains and, possibly, after their refinements and narrowings) one can find an automorphism $\Phi = \{\Phi_j\}$ of the normal form F_1 (i.e., a normalizing cochain for F_1) such that

$$\Phi \circ H = \widetilde{H} \circ \Phi \quad (\text{i.e., } \Phi_j \circ H_j = \widetilde{H}_j \circ \Phi_j \text{ for all } j) \tag{1}$$

(Chapter II, §4, Lemma 10).

One may consider the compositions $H_{j+1,j} = H_{j+1} \circ H_j^{-1}$ of sectorial normalizing mappings on the intersection of neighboring sectorial domains S_j and S_{j+1} (note that the numbering is cyclic). These compositions form a set of mappings that is called the *coboundary* $\delta\{H_j\}$ of the cochain $\{H_j\}$ below. The mappings $H_{j+1,j}$ of the coboundary $\delta\{H_j\}$ possess the following properties: they commute with the formal normal form F_1 and the identity mapping is asymptotic for them. These two properties define a class of sets of holomorphic mappings, that are called *cocycles* below. Let us take the relation $\Phi \circ \delta = \delta \circ \Phi$ between the coboundaries $\delta = \delta H$ and $\widetilde{\delta} = \delta \widetilde{H}$ that follows from (1) for the definition of the equivalence of cocycles δ and

$\tilde{\delta}$. Let us also call the equivalence classes of cocycles δ and $\tilde{\delta}$ *functional invariants* of germs of the class $\mathscr{B}_1$.

$\mathscr{B}_1$ THEOREM. *The space of functional invariants of germs of class $\mathscr{B}_1$ is the moduli space in the conjugation problem for the class $\mathscr{B}_1$. A functional class of equivalent cocycles that contains the coboundary of the normalizing cochain of a germ is the germ modulus. The modulus analytically depends on the germ: certain representatives of moduli of the family of analytic germs form an analytic family.*

REMARK. A more precise, detailed, and tedious description of functional invariants can be found in Chapter II, §4.

Since each equivalence class of cocycles is rather large, it is by far not obvious that the space of functional invariants constructed above is nontrivial (contrary to the analogous one-dimensional problem in Paper I). This question is discussed in the two following subsections.

4.6. The moduli space is infinite-dimensional. The quotient space $\hat{S}$ of a sectorial domain S of aperture not less that π with respect to the action of F_1 is not holomorphically convex. Its holomorphic hull is the direct product of the projection of S on the x-axis, of the punctured plane $\mathbb{C}_*$, and of a polydisc in $\mathbb{C}^{n-1}$ (Chapter II, Lemma 14). From this it follows, in particular, that any bounded function, holomorphic in $\hat{S}$ (and hence any bounded function holomorphic in S that is constant on the orbits of F_1) is constant on the lines $\{x = \text{const}, z = \text{const}\}$. Hence any holomorphic change that commutes with F_1 in a sectorial domain of aperture greater than π is affine in the sense of Definition 2 in 4.3. In particular, for special cocycles all whose mappings except one are the identity, the automorphism group consists only of affine mappings. This allows us to construct numerous examples of nonequivalent special cocycles.

EXAMPLE (see Chapter II, 5.4, Remark 2). Let $n = 2$ and suppose that all functions $p_k(x)$ of the set $p = \{p_k\}_{k=1}^\infty$ are holomorphic in the disk $\{|x| < 1\}$ and their modulus is less than or equal to one. Finally, let $p_0 = p_1 \equiv 1$. Let

$$c_p \colon (x, y) \mapsto \left(x, y + \ln\left(\sum_{k=0}^\infty p_k(x) \exp(k(2\pi i y - 1)/x) \right) \right)$$

be a sectorial mapping that is holomorphic in the domain $S = \{|\arg x| < \pi/4, |x| < 1, |y| < 1\}$. This mapping commutes with F_1 and the identity mapping is asymptotic for it in S. Then all cocycles δ_p with a single nonidentity mapping c_p are pairwise nonequivalent.

4.7. Some observations about the quotient space of the punctured neighborhood of zero with respect to the normal form action. The functional invariants in the analytic classification problem for germs of class $\mathscr{B}_1$ are much simpler than their analogs in Paper I (in the analytic classification problem of one-dimensional mappings whose linear part is identity). Moreover, it seems doubtful that the examples in Chapter II, §2 of the present paper that prove the "almost exactness" of all results of the paper can be simplified. Below we attempt to explain (though only intuitively) the reasons for this difference, comparing the quotient spaces of the punctured neighborhoods of zero with respect to the normal form action.

The presence of the Stokes phenomenon in the (orbital) analytic classification problem for the class of formally (orbitally) equivalent germs of (vector field) mappings is equivalent to the nontriviality of the cohomology set, i.e., of classes of homologous gluings (in terms of §4 of Chapter II of this paper), or of classes of equivalent cocycles (in terms of Paper I), or of the space of orbits (leafs) of the formal normal form (described in the Introduction). This cohomology set is closely related to the ordinary cohomology group of the complex manifold "tangent" to it (cf. [11]). For the cohomology group of the complex manifold $\hat{S}$ to be nontrivial it is necessary that $\hat{S}$ possess "non-Stein property". The nature of this non-Stein property is different for the two problems mentioned above. Namely, for the quotient space of the punctured neighborhood of zero with respect to the action of the mapping $f_0: x \to x/(1-x)$ considered in Paper I, the holomorphic separability condition is violated: topologically nonseparable points are also analytically nonseparable. (Recall that this quotient space is non-Hausdorff and is obtained from two punctured spheres by identifying neighborhoods of the deleted points). The condition of holomorphic convexity is violated for the quotient space $\hat{S}$ of the "punctured" neighborhood of zero $S = \{0 < |x| < \varepsilon, |y| < 2\varepsilon\}$ with respect to the action of the normal form F_1. Indeed, $\hat{S}$ is biholomirphically equivalent to the image of S under the projection $\pi: (x, y) \mapsto (x, \exp(2\pi i y/x))$. The image $\pi(S)$ contains the Hartogs domain $u\{(x, t) : 0 < |x| < \varepsilon, \exp(-|x|^{-1}) < |t| < \exp|x|^{-1}\}$. Any superharmonic majorant of the function $|x|^{-1}$ is equal to $+\infty$ and any superharmonic majorant of $-|x|^{-1}$ is equal to $-\infty$ (Chapter II, 5.1, Lemma 19). Observe that the Hartogs theorem on analytic continuation [24] implies that the holomorphic hull of u has the form $\{0 < |x| < \varepsilon\} \times \mathbb{C}_*$. Hence the holomorphic convexity condition is violated for the quotient space $\hat{S}$.

Thus the Stokes phenomenon in our two problems has two different "parents" and therefore the difference in the results must cause regret rather than surprise.

4.8. General classification problem for the class $\mathscr{B}_1^s$. $\mathscr{B}_1^s$ Theorem. The general classification problem for the class $\mathscr{B}_1^s$ (for any of the standard structures in 4.1) is solved according to the following plan. It is possible to choose the normalizing cochain H of the germ $F \in \mathscr{B}_1^s$ that agrees with the structure s, in such a way that it preserves s (in this case we call it *s-normalizing*); then its coboundary δH also preserves s. In accordance with this, let us add the structure preservation condition to the definition of cocycles (and of their equivalence) and call their *s*-equivalence classes *functional invariants of germs of class* $\mathscr{B}_1^s$.

$\mathscr{B}_1^s$ THEOREM. *The space of functional invariants of germs of class $\mathscr{B}_1^s$ is a moduli space in the general classification problem for the class $\mathscr{B}_1^s$. The functional invariant that contains the coboundary of the s-normalizing cochain of a germ matched to s is the germ modulus. The modulus analytically depends on the germ.*

The $\mathscr{B}_1^s$ Theorem is proved in Chapter II, §6 at first for each of the elementary structures and then for nonelementary structures. In the same place one may find examples that show that the moduli space for $\mathscr{B}_1^s$ is infinite-dimensional. Finally, let us state a corollary of the $\mathscr{B}_1^s$ Theorem (see Chapter II, 6.7, Corollary 1).

COROLLARY. *All real-analytic germs of class $\mathscr{B}_1$ are smoothly equivalent.*

4.9. $\mathscr{B}_q$ and $\mathscr{B}_q^s$ Theorems. The solution to the general classification problem obtained in §§1–6 of Chapter II for the class $\mathscr{B}_1$ may be slightly changed to fit the case of arbitrary q (see §7). Namely, one must replace everywhere the normal form F_1 by $F_q: (x, y) \mapsto (x, y + x^q)$, the condition "aperture less than π" by the condition "aperture less than π/q", etc. For example, the formulations of the $\mathscr{B}_q$ (and $\mathscr{B}_q^s$) Theorems that give a solution to the general classification problem in $\mathscr{B}_q$ and $\mathscr{B}_q^s$ are obtained from the $\mathscr{B}_1$ ($\mathscr{B}_1^s$) Theorems by changing the index 1 to q. In the same manner, one may obtain examples that show that the moduli spaces for each of these problems are infinite-dimensional.

4.10. Resumé of the results (geometric problems). Here we state the results that were obtained for the geometric problems in §§2, 3. The structures I, [2], $\alpha_\pm$, $\beta_\pm^q$, $\sigma^\pm$, and σ are defined in 4.1.

THEOREM. *For any of the problems mentioned in Table 1 (for the real or complex case), the space of element indicators with formal normal forms shown in the table coincides with the space $\mathscr{B}_q^s$. Table 1 also shows the value of q and the structure type. The equivalence of any two elements implies the equivalence of their indicators, and vice versa. The moduli space for each of these problems is the corresponding moduli space from the $\mathscr{B}_q^s$ Theorem (see 4.8, 4.9). The modulus of an element analytically depends on the element.*

This statement along with the table contains 5 classification theorems. The proof of any of them consists of two parts, namely, the "analytic" and "geometric" ones. The first part is devoted to the analytic classification of indicators (of germs of the spaces $\mathscr{B}_q^s$). The second part contains a reduction of the "geometric" classification problems studied to indicator classification.

TABLE 1

Geometric problem	Ground field	Formal normal form		Corresponding classification problem	Indicator class		
						q	structure s
Hypersurface pair in a symplectic manifold	$\mathbb{R}$	$(q_0 = 0, q_0 = p_3^2 + p_1)$		First		1	I, σ^+, α_+
	$\mathbb{C}$						I, α_+
Darboux-Whitney problem	$\mathbb{R}$	$\omega_0 = dA \wedge dD + dC \wedge dB + dE \wedge dF + \cdots$		Third		5	I, σ^+, α_+
	$\mathbb{C}$						I, α_+
Germs of vector fields that correspond to the degenerations of symplectic and contact structures	$\mathbb{R}$	$v^- = y\partial/\partial y - x\partial/\partial x + xy /\partial t$		Third		1	σ^-
		$v^+ = y\partial/\partial y - x\partial/\partial x + (x^2+y^2)/\partial t$		Fourth		2	σ^+, [2]
	$\mathbb{C}$	v_-		Third; complex version		1	no additional structures
Degenerations of closed 2-forms	$\mathbb{R}$	$\omega_\pm = d(x - z^2/2) \wedge dy$ $+ d(xz \pm ty - z^3/3) \wedge dt$ $+ du_1 \wedge du_2 + \cdots$	ω_-	Third; symplectic version		1	α_-, σ^-
			ω_+	Fourth; symplectic version		2	α_-, σ^+, [2]
	$\mathbb{C}$	ω_-		Third; complex symplectic version		1	α
Degenerations of contact structure	$\mathbb{R}$	$\omega_\pm = du_1 + x_2 t y_2 + \cdots + x_n dy_n$ $+ x_1(y_2 + x_1^2 \pm u^2) du$	ω_-	Third; contact version		1	β_-^1, σ^-
			ω_+	Fourth; contact version		2	β_+^2, σ^+, [2]
	$\mathbb{C}$	ω_-		Third; complex contact version		1	β_-^1

This reduction is the implication "indicator equivalence" $\Longrightarrow$ "germ equivalence". The converse follows directly from the construction. Besides, the "geometric" part contains the indicator realization theorem. Chapter II is devoted to the classification of germs for $\mathscr{B}_q^s$. The case $q = 1$, $n = 2$ is given the most detailed study; the study of the case $q > 1$, $n = 2$ is outlined in Chapter II, §7. Thus Chapter II contains the most complicated "analytic" part of the solution of *all five problems* mentioned above.

In this paper we focused mostly on the Darboux-Whitney problem. Therefore the geometric part is developed in detail only for this part. The theorem is stated separately in Chapter III, 1.6, 1.7. The geometric investigation of the hypersurface pair problem for the "two folds" case is much simpler than that of the Darboux-Whitney problem and is omitted here.

We have also performed a complete geometric investigation of the three last problems mentioned in the table, though it is omitted here because of the lack of space. We intend to publish it elsewhere.

As a by-product of the methods developed above, we prove that the smooth classification of indicators of the space $\mathscr{B}_q^s$ for $q = 5$ and $s = (\sigma, I, \alpha)$ coincides with the formal one and the classified germs are real-analytic. This gives a smooth version of the Darboux-Whitney theorem, which is also proved in the present paper (see Chapter III, 1.6).

4.11. Other possible applications. As to Paper I, the results obtained above can be applied to the study of "resonance" mappings with a certain iteration power belonging to the class $\mathscr{B}_q$ (cf. also [14, 18, 19]).

Similar to Paper I, one can use the language of functional invariants to solve problems of embedding into a flow, of extracting an "iteration" root, etc. (see [14]).

One more application is discussed in the following example.

EXAMPLE. A mapping is called *seminormalized* if it maps every straight line $x = \text{const}$ onto itself. The following statement is valid.

The germ $F \in \mathscr{B}_q$ is seminormalizable by holomorphic coordinate changes iff its functional invariant contains a cocycle, all mappings of which are seminormalized. Since it is easy to construct a cocycle that is not seminormalizable by the action of the automorphism group of the normal form, it follows that the seminormalization in $\mathscr{B}_q$ is, as a rule, impossible.

CHAPTER II. GERMS OF HOLOMORPHIC MAPS WITH A LINE OF FIXED POINTS

In this chapter the classification of indicators from Chapter I, §2 is given.

§1. Formal classification of germs of class $\mathscr{B}_1$

1.1. Germs of semiformal maps. Let U be an arbitrary neighborhood of the origin in $\mathbb{C}$, and $O(U)$ be the ring of functions holomorphic in U. Denote by $\mathfrak{F}_U = O(U)[[x]]$ the class of formal power series in the variable

x with coefficients from $O(U)$. For another neighborhood $V \subset U$, the restriction of a series $\alpha \in \mathfrak{F}_U$ to V is the series from $\mathfrak{F}_V$ denoted by $\alpha|_V$ with the coefficients obtained by restriction of those of α to V.

Two series with coinciding restrictions are *equivalent*. The corresponding equivalence classes are called *germs of semiformal series*.

The set of all germs of semiformal series is denoted by $\mathfrak{F}$. Let $\mathfrak{F}_0 = \{\varphi \in \mathfrak{F} : \varphi(0, 0) = 0\}$. The *order* $\operatorname{ord}\varphi$ of a germ $\varphi \in \mathfrak{F}$ is the number of its first nonzero coefficient. We write $\varphi = o(x^n)$ if $\operatorname{ord}\varphi > n$, and $\varphi = O(x^n)$ if $\operatorname{ord}\varphi \geq n$.

A sequence $\alpha_n \in \mathfrak{F}$ *converges* to a germ $\alpha \in \mathfrak{F}$, $\alpha_n \to \alpha$, if $\operatorname{ord}(\alpha_n - \alpha) \to \infty$ for $n \to \infty$.

DEFINITION. Let $\varphi \in \mathfrak{F}$, $a, b \in \mathfrak{F}_0$,

$$\varphi(x, y) = \sum_{k=0}^{\infty} c_k(y) x^k, \qquad (1.1)$$
$$b(x, t) = \beta_0(t) + \beta_1(x, t), \qquad a, \beta_1 = O(x).$$

By definition we put

$$\varphi(a(x, t), b(x, t)) = \sum_{k=0}^{\infty} \left[\sum_{n=0}^{\infty} \frac{1}{n!} \left. \frac{d^n c_k(y)}{dy^n} \right|_{y=\beta_0(t)} (\beta_1(x, t))^n \right] a^k(x, t). \quad (1.2)$$

This expression is well defined since by (1.1) the coefficient at x^m for any $m \geq 0$ is obtained by summing a finite number (not exceeding $m + 1$) of functions analytic in a certain neighborhood of the origin. The definition is natural, since for convergent series φ, a, b the series (1.2) converges to the composition $\varphi(a, b)$. The composition $\varphi(a, b)$ depends continuously on the germs φ, a and b: $\varphi_n \to \varphi$, $a_n \to a$, $b_n \to b$ imply $\varphi_n(a_n, b_n) \to \varphi(a, b)$.

Pairs $(\varphi, \psi) \in \mathfrak{F}^2$ are called *germs of semiformal maps*. The order of a pair $\operatorname{ord}(\varphi, \psi)$ is the number $\min(\operatorname{ord}\varphi, \operatorname{ord}\psi + 1)$. Composition of semiformal maps is defined via the composition law for series from $\mathfrak{F}$.

The following assertion is easily proved using the successive approximation method.

LEMMA 1. *Suppose*
$$H(x, y) = (x\alpha(y) + o(x^n), \varphi(y) + O(x^n)) \in \mathfrak{F}_0^2, \qquad n \geq 1, \quad \alpha(0)\varphi'(0) \neq 0.$$
Then H is invertible: the germ of the inverse map $H^{-1} \in \mathfrak{F}_0^2$ has the form
$$H^{-1}(x, y) = \left(\frac{x}{\alpha \circ \varphi^{-1}(y)} + o(x^n), \varphi^{-1}(y) + O(x^n) \right).$$

1.2. Theorem on formal classification of germs of class $\mathscr{B}_1$. Let $\mathscr{B}_1$ be the class of germs of holomorphic maps $(\mathbb{C}^2, 0) \to (\mathbb{C}^2, 0)$ of the form
$$F(x, y) = (x + \varphi_1(x, y), y + x + \varphi_2(x, y)), \qquad (1.3)$$
$$\varphi_1 = x^2 \psi_1, \qquad \varphi_2 = x \psi_2,$$

where ψ_i are analytic in $(\mathbb{C}^2, 0)$ and $\psi_2(0, 0) \neq -1$. Identifying the germ F with its semiformal series (the Hartogs series), we may put $\mathscr{B}_1 \subset \mathfrak{F}_0^2$.

The germ
$$F_0: (x, y) \mapsto (x, y + x)$$
belongs to $\mathscr{B}_1$ and plays an important role in our subsequent exposition. We call it the *standard germ*. For any other germ $F \in \mathscr{B}_1$, the difference $F - F_0$ is called the *discrepancy* of the germ F. The discrepancy always has order no less than 2.

Two germs $F, G \in \mathfrak{F}_0^2$ are *formally equivalent*, if the germ of an invertible semiformal map $H \in \mathfrak{F}_0^2$ conjugating them exists: $H \circ F = G \circ H$. If the germ G is standard, then the semiformal map H satisfying
$$H \circ F = F_0 \circ H \tag{1.4}$$
is called the *formal normalizing transformation* for F. The germ H is *normed*, if
$$H(x, 0) = (x, 0). \tag{1.5}$$

THEOREM 1. *All germs of class $\mathscr{B}_1$ are formally equivalent to the standard germ F_0. The normed formal normalizing transformation is unique.*

Considering partial sums of the normalizing series, we get the following corollary from Theorem 1:

COROLLARY. *For any $F \in \mathscr{B}_1$ and any $n \in \mathbb{N}$ there exists a germ of an analytic (and even polynomial in x) transformation H such that the order of discrepancy of the germ $F_1 = H \circ F \circ H^{-1}$, analytically equivalent to F, is no less than n.* □

REMARK (on preliminary normalization). Without loss of generality one may assume the order of discrepancy of the germ F in Theorem 1 to be no less than 3. Indeed, let φ_1, φ_2 from (1.3) be of the form
$$\varphi_1(x, y) = x^2 a(y) + O(x^3), \qquad \varphi_2(x, y) = xb(y) + O(x^2).$$
Define the analytic germs $k(y)$, $\varphi(y)$, $c(y)$ by the relations
$$c' = a/(b+1), \quad k = \exp(-c), \quad \varphi' = k/(b+1), \quad c(0) = \varphi(0) = 0,$$
and let
$$H(x, y) = (xk(y), \varphi(y)).$$
Then the analytic germ H is invertible and normed, and the germ $F_1 = H \circ F \circ H^{-1}$ equivalent to F has discrepancy of at least third order. Actually we have given an independent proof of the corollary for $n = 3$.

1.3. Homological and auxiliary equations. Let us look for the normalizing transformation H for the germ F in the form
$$H(x, y) = (x + h_1(x, y), y + h_2(x, y)).$$

Assuming φ_j, h_j small and neglecting second order terms with respect to φ_j, h_j in (1.4), we get the so-called *homological equation*

$$F_0 \circ H - H \circ F_0 = F - F_0, \qquad (1.6)$$

or, in component form,

$$\begin{cases} h_1(x, y) - h_1(x, y + x) = \varphi_1(x, y), \\ h_2(x, y) - h_2(x, y + x) = \varphi_2(x, y) - h_1(x, y). \end{cases} \qquad (1.7)$$

The map H being normed, we have in addition

$$h_1(x, 0) = h_2(x, 0) = 0.$$

DEFINITION. The normed solution H of equation (1.6) is called the *correction map*, and the difference $H - \mathrm{id}$ is the *correction associated with the discrepancy* $F - F_0$.

Along with the homological equations (1.6), (1.7), we consider the *auxiliary equation*

$$h(x, y) - h(x, y + x) = \varphi(x, y), \qquad (1.8)$$

whose solution is subject to the initial condition

$$h(x, 0) = 0. \qquad (1.9)$$

LEMMA 2. *Let* $\varphi \in \mathfrak{F}$. *Then*:
1. *Equation* (1.8) *admits a solution* $h \in \mathfrak{F}$ *if and only if* $\mathrm{ord}\,\varphi > 0$.
2. *The normed solution* $h \in \mathfrak{F}$ *is unique, and* $\mathrm{ord}\,h = \mathrm{ord}\,\varphi - 1$.

PROOF. Expanding $h(x, y + x)$ in accordance with the definition of superposition of series and comparing the coefficients of the same powers of x, we immediately obtain both assertions of the lemma. □

Applying Lemma 1 twice, we get the following statement.

COROLLARY. *The semiformal correction exists and is unique for any discrepancy of order greater than* 1. *The order of the correction is less than the order of the discrepancy by* 1.

REMARK. In particular, the correction map is invertible if the order of the discrepancy is 3 or more (see Lemma 1).

1.4. Existence of the formal normalizing series. The following statement is evident both in the formal and in the analytic case.

LEMMA 3 (on subsequent discrepancies). *Let* $h = (h_1, h_2)$ *be the correction associated with the discrepancy* $(\varphi_1, \varphi_2) = F - F_0$. *Assume the germ* $H = \mathrm{id} + h$ *invertible and denote* H^{-1} *by* $\mathrm{id} + g$, $g = (g_1, g_2)$. *Then the subsequent discrepancy* $(\alpha_1, \alpha_2) = H \circ F \circ H^{-1} - F_0$ *is given by the formulas*

$$\begin{cases} \alpha_1(x, y) = h_1(x_1 + \Delta_1, y_1 + \Delta_2) - h_1(x_1, y_1), \\ \alpha_2(x, y) = h_2(x_1 + \Delta_1, y_1 + \Delta_2) - h_2(x_1, y_1), \end{cases} \qquad (1.10)$$

where

$$x_1 = x + g_1, \quad y_1 = y + x + g_1 + g_2, \quad g_i = g_i(x, y), \quad i = 1, 2,$$
$$\Delta_1 = \varphi_1 \circ H^{-1}(x, y), \quad \Delta_2 = \varphi_2 \circ H^{-1}(x, y).$$

This lemma together with the order estimates from the Corollary and Remark in 1.3 and Lemma 1 imply the following statement.

COROLLARY. *If the order of the discrepancy is equal to $n \geq 3$, then the subsequent discrepancy is well defined and has order no less than $2n - 2$.*

Now we construct the formal normalizing transformation for a germ $F \in \mathscr{B}_1$ by successive approximations. Namely, define by induction the sequences $\{G_n\}$, $\{H_n\}$ and $\{\mathscr{H}_n\}$ from the initial conditions $G_1 = F$, $\mathscr{H}_0 = \mathrm{id}$, and for $n \geq 1$ let H_n be the correction map associated with the discrepancy $G_n - F_0$, and put $\mathscr{H}_n = H_n \circ \mathscr{H}_{n-1}$, $G_{n+1} = H_n \circ G_n \circ H_n^{-1}$. Evidently, these definitions imply

$$\mathscr{H}_n \circ F = G_{n+1} \circ \mathscr{H}_n. \tag{1.11}$$

From the Corollary to Lemma 3 and the Remark in 1.2, we conclude that G_n converges to F_0 and $\mathscr{H}_n$ converges to a certain normed germ $H \in \mathfrak{F}_0^2$, which is invertible by Lemma 1. Passing to the limit in (1.11), we conclude that H is the required normed formal normalizing transformation for F.

1.5. Uniqueness of the normed formal normalizing transformation. Note that for $F = F_0$ the functional equation (1.4) coincides with the homological equation (1.6). But two solutions of (1.4) differ by a germ commuting with F_0. Hence the uniqueness of the normed formal normalizing transformation follows from the uniqueness of the normed solution of the homological equation established in the Corollary in 1.3.

DEFINITION. We call *affine* any semiformal germ of the form

$$G(x, y) = (\varphi(x), y\varphi(x)/x + k(x)). \tag{1.12}$$

The above uniqueness statement has the following consequence.

COROLLARY. *Let H be an arbitrary normalizing transformation, and H_0 the normed one for the germ $F \in \mathscr{B}_1$. Then there exists a unique affine invertible germ $G \in \mathfrak{F}_0^2$ such that $H = G \circ H_0$.*

Indeed, for G take the germ (1.12) with $(\varphi(x), k(x)) = H(x, 0)$.

§2. Sectorial solutions to the homological equation

The most important step in the proof of Theorem 1 was to establish the formal solvability of the homological equation (1.6), which was implied by the formal solvability of the auxiliary equation (1.8). But later we shall show that in general the auxiliary equation does not have solutions holomorphic in $(\mathbb{C}^2, 0)$, see 2.7. Nevertheless one can find solutions of the auxiliary and homological equations in domains of a certain special form, the so-called *sectorial domains*, which do not contain any entire neighborhood of the origin in $\mathbb{C}^2$. The formal solutions found previously serve as asymptotic series for those sectorial ones.

2.1. Sectorial solutions of the auxiliary equation.
Let
$$V = V_{\varepsilon_1} = \{x \in \mathbb{C} : \alpha < \arg x < \beta, \, |x| < \varepsilon_1\}$$
be the sector with the angle (aperture) $\gamma = \gamma(V) = \beta - \alpha < \pi$, and
$$R_\varepsilon = \{t_1 e^{i\alpha} + t_2 e^{i\beta} : t_j \in \mathbb{R}, \, |t_j| < \varepsilon, \, j = 1, 2\}$$
be the rhombus with sides parallel to the edges of the sector V. Denote by W an arbitrary neighborhood of the origin. Domains of the form $V \times W$ will be called *sectorial domains*, while the products of the form $V \times R_\varepsilon$ with $\gamma(V) < \pi$ will be referred to as *standard sectorial domains*.

A function h analytic in $S \cup F_0(S)$ is the *sectorial solution* of the equation (1.8) in the sectorial domain S, if (1.8) holds for all the points $(x, y) \in S$. The solution is *normed*, if the condition (1.9) holds for the points $(x, 0) \in S$.

LEMMA 4. *Let the function φ be holomorphic in a standard sectorial domain $S = V \times R_\varepsilon$ of aperture $\gamma < \pi$, and continuous in $\bar{V} \times \bar{R}_\varepsilon$. Then equation (1.8) admits a normed sectorial solution in S.*

PROOF. We look for the solution h to (1.8) in the form of a series
$$h(x, y) = \sum_{k=0}^{\infty} \varphi(x, y + kx) \quad \text{or} \quad h(x, y) = -\sum_{k=-1}^{-\infty} \varphi(x, y + kx).$$
The following two problems then arise. The function φ may not be defined at some points $(x, y + kx)$ (especially for large $|k|$). Next, the series may diverge. The first difficulty can be overcome by splitting the function φ into terms with larger domains of definition. The second can be avoided by the choice of an appropriate kernel in the Cauchy type integral realizing the above splitting.

More exactly, let $\Gamma_\pm$ be the parts of the boundary Γ of the rhombus R_ε such that $y \mp x \notin \bar{R}_\varepsilon$ for all $y \in \Gamma_\pm$, $x \in V$; then $\Gamma = \Gamma_+ \cup \Gamma_-$. Let $y_0 \in \Gamma_+ \cap \Gamma_-$ be a vertex of the rhombus. Put
$$\varphi_\pm(x, y) = \frac{1}{2\pi i} \int_{\Gamma_\pm} \varphi(x, t) \mathscr{E}(t, y_0, y) \, dt, \tag{2.1}$$
where
$$\mathscr{E}(t, y_0, y) = \frac{1}{t - y} - \frac{1}{y_0 - y}.$$
Then the $\varphi_\pm$ are holomorphic in $V \times (\mathbb{C} \setminus \Gamma_\pm)$, and $\varphi_+ + \varphi_- = \varphi$ on $V \times R_\varepsilon$. Note that $\mathscr{E}(t, y_0, y) = O(y^{-2})$ as $y \to \infty$, therefore the series
$$\sum_{k=0}^{\infty} \varphi(x, t) \mathscr{E}(t, y_0, y + kx) \tag{2.2}$$

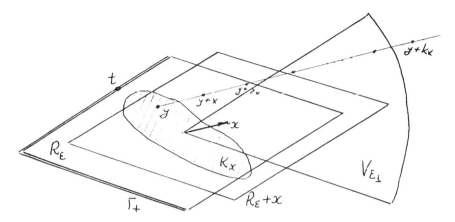

FIGURE 6. The rhombus R_ε.

converges uniformly in $t \in \Gamma_+$, $(x, y) \in K$ for an arbitrary compact $K \subset U = S \cup F_0(S)$, see Figure 6. Hence the function

$$h_+(x, y) = \frac{1}{2\pi i} \int_{\Gamma_+} \sum_{k=0}^{\infty} \varphi(x, t)\mathscr{E}(t, y_0, y + kx)\, dt \qquad (2.3)$$

is holomorphic in U, the series $\sum_{k=0}^{\infty} \varphi_+(x, y + kx)$ converges in U and

$$h_+(x, y) = \sum_{k=0}^{\infty} \varphi_+(x, y + kx) \quad \text{on } U. \qquad (2.4)$$

In the same way we construct the function

$$h_-(x, y) = -\frac{1}{2\pi i} \int_{\Gamma_-} \sum_{k=-1}^{-\infty} \varphi(x, t)\mathscr{E}(t, y_0, y + kx)\, dt$$

analytic in U, and the convergent series $\sum_{k=-1}^{-\infty} \varphi_-(x, y + kx)$ satisfying the relation

$$h_-(x, y) = -\sum_{k=-1}^{-\infty} \varphi_-(x, y + kx) \quad \text{on } U. \qquad (2.5)$$

Now by putting $h = h_+ + h_-$ we obtain a function holomorphic in U for which relation (1.8) holds in the entire domain S. Replacing the function h if necessary by the difference $h(x, y) - h(x, 0)$, we obtain the normed sectorial solution. □

REMARK 1. Using the fact that for $x \in V$ the sum $\varphi_+(x, y) + \varphi_-(x, y)$ is equal to $\varphi(x, y)$ if $y \in R_\varepsilon$, and to zero otherwise (if $y \notin \overline{R}_\varepsilon$), we conclude from (2.1), (2.4) and (2.5) that

$$h(x, y) = \sum_{k \geq 0,\, R_\varepsilon} \varphi(x, y + kx) - \sum_{k \in \mathbb{Z}} \int_{\Gamma_-} \varphi(x, t)\mathscr{E}(x, y_0, y + kx)\, dt,$$

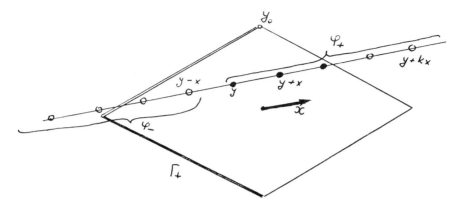

FIGURE 7. Constructing the solutions $h_\pm$.

where the symbol R_ε in the subscript means that the summation is only over those k for which the point $(x, y + kx)$ belongs to R_ε, see Figure 7.

Using the uniform convergence of the series (2.2) and the well-known formula

$$\sum_{-\infty}^{\infty} \frac{1}{t+k} = \pi \cot \pi t,$$

we obtain

$$h(x, y) = \sum_{k \geq 0, R_\varepsilon} \varphi(x, y + kx)$$
$$- \frac{1}{2ix} \int_{\Gamma_-} \varphi(x, t) \left[\cot \frac{\pi(t-y)}{x} - \cot \frac{\pi(y_0 - y)}{x} \right] dt \quad (2.6)$$

if $y + kx \notin \partial R_\varepsilon$ for all $k \in \mathbb{Z}$. This function differs from the following one:

$$\tilde{h}(x, y) = \sum_{k \geq 0, R_\varepsilon} \varphi(x, y + kx) - \frac{1}{2ix} \int_{\Gamma_-} \varphi(x, t) \cot \frac{\pi(t-y)}{x} dt \quad (2.7)$$

only by a term which is analytic and x-periodic in R_ε. Therefore $\tilde{h}$ is another solution to problem (1.8). By the Cauchy formula one can write

$$\tilde{h}(x, y) = - \sum_{k < 0, R_\varepsilon} \varphi(x, y + kx) + \frac{1}{2ix} \int_{\Gamma_+} \varphi(x, t) \cot \frac{\pi(t-y)}{x} dt. \quad (2.7')$$

These formulas are remarkable by the fact that the sum of the two terms discontinuous in R_ε is analytically extendable to all of R_ε: all the singularities

$$\bigcup_{k \in \mathbb{Z}} \{ y \in R_\varepsilon : y + kx \in \partial R_\varepsilon \}$$

are removable. In general, it is rather amusing that the attempt which at first glance seemed hopeless finally turned out to be almost completely realizable.

REMARK 2. Let $\sigma: (x, y) \mapsto (\bar{x}, \bar{y})$ be complex conjugation. If the sectorial domain S mentioned in Lemma 4 is σ-symmetric ($\sigma(S) = S$), and the right-hand side φ of equation (1.8) is σ-invariant ($\varphi \circ \sigma = \bar{\varphi}$), then (2.7) and (2.7') provide σ-invariant solutions for (1.8): $\tilde{h} \circ \sigma = \bar{\tilde{h}}$. Thus, equation (1.8) with a real right-hand side possesses a real solution.

2.2. Estimates of the kernel. The kernel of the integral operator $\varphi \mapsto h_+$ from (2.3) has the form

$$\frac{1}{2\pi i}\sum_{k\geq 0}\mathscr{E}(t, y_0, y+kx) = \frac{1}{2\pi i x}G(a, b), \qquad (2.8)$$

where

$$G(a, b) = \sum_{k\geq 0}\left[\frac{1}{k+a} - \frac{1}{k+b}\right], \qquad a = \frac{y-y_0}{x}, \quad b = \frac{y-t}{x}.$$

The following simple properties of the function $G(a, b)$ are evident:

$$G(a, b) = \sum_{k=0}^{n-1}\left[\frac{1}{k+a} - \frac{1}{k+b}\right] + G(a+n, b+n), \qquad (2.9)$$

$$G(a, b) = G(a, z) + G(z, b). \qquad (2.10)$$

LEMMA 5. *Let* $\Omega^\delta = \{z \in \mathbb{C}: |\arg z| < \pi - \delta\}$. *Then*

$$|G(a, b)| \leq C\delta^{-2}|a-b|(\min\{|a|, |b|, |ab|\})^{-1}, \qquad a, b \in \Omega^\delta, \qquad (2.11)$$

for a certain constant C *and all* δ, $0 < \delta < \pi/4$.

PROOF. The estimate (2.11) can be easily obtained for points $a, b \in \Omega_1 = \{z: |\arg z| < \delta\}$ by comparing the sum with the improper integral. Using (2.9) and replacing each term of the sum by the term with the maximal modulus, we deduce from estimate (2.11) on Ω_1 an analogous estimate on $\overline{\Omega}_2$, $\Omega_2 = \Omega^\delta \setminus \Omega_1$. Finally, choosing $z \in \partial\Omega_1 \cap \partial\Omega_2$ in such a way that $\operatorname{Re} z = \operatorname{Re} a$, $\operatorname{Im} z \cdot \operatorname{Im} b > 0$, we conclude from (2.10) and the above estimates for G in $\overline{\Omega}_1$ and $\overline{\Omega}_2$ that (2.11) holds also for $a \in \overline{\Omega}_1$, $b \in \overline{\Omega}_2$. □

2.3. Estimate of the sectorial solution to the auxiliary equation. In this section we prove a proposition which can be interpreted in the following way: the operator solving the auxiliary equation in the sectorial domain (Lemma 4) is of the first order in the sense of [1, § 28].

LEMMA 6. *Let* h *be the normed sectorial solution from Lemma 4 to equation* (1.8) *in the standard sectorial domain* $V_{\varepsilon_1} \times R_\varepsilon$ *of aperture* $\gamma < \pi$ *and* $0 < \varepsilon_1 \leq \varepsilon < 1/2$. *Then for any* $\varepsilon' < \varepsilon$

$$|h(x, y)| \leq C_0 \delta^{-3}\varepsilon |x|^{-1}\max_{t \in R_\varepsilon}|\varphi(x, t)|, \qquad \forall (x, y) \in U_{\varepsilon'}, \qquad (2.12)$$

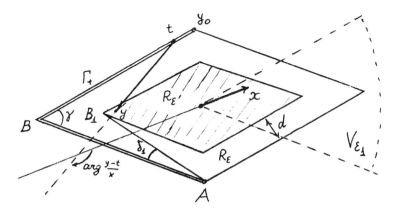

FIGURE 8. Domains in Lemma 6.

where $U_{\varepsilon'} = S_{\varepsilon'} \cup F_0(S_{\varepsilon'})$, $S_{\varepsilon'} = V_{\varepsilon_1} \times R_{\varepsilon'}$, $\delta = (\varepsilon - \varepsilon')/\varepsilon$ and the constant C_0 depends only on γ: $C_0 \sim (\sin\gamma)^{-3}$.

PROOF. From (2.3) and (2.8) we conclude that

$$|h_+(x,y)| \leq |\Gamma_+|(2\pi|x|)^{-1} \max_{t \in \Gamma_+}(|\varphi(x,t)||G((y-y_0)/x,(y-t)/x)|), \quad (2.13)$$

where $|\Gamma_+| = 4\varepsilon$ is the length of Γ_+. We shall estimate the factor $G(\cdot,\cdot)$ in (2.13) by using Lemma 5. To do this, we shall prove the above inequalities for $\delta_1 = C\delta$, $C = O(\sin\gamma)$. The above estimates are purely geometrical (see Figure 8). For any $t \in \Gamma_+$, $x \in V_{\varepsilon_1}$, $y \in R_{\varepsilon'}$, we have

$$|\arg((y-t)/x)| > \delta_0, \qquad \delta_0 = \angle B_1 AB.$$

But $\sin\gamma_0 > d/\operatorname{diam} R_\varepsilon$, where d is the distance between the rhombus $R_{\varepsilon'}$ and the curve Γ_+. Therefore $\delta_0 > C' \sin\gamma \cdot \delta$ for some constant C'. Thus

$$\left|\arg\frac{y-y_0}{x}\right| < \pi - \delta_1, \quad \left|\arg\frac{y-t}{x}\right| < \pi - \delta_1, \quad \frac{y-y_0}{x} \in \Omega^{\delta_1}, \quad \frac{y-t}{x} \in \Omega^{\delta_1}.$$

Applying Lemma 5 and estimating the fractions $|x/(y-y_0)|$, $|x/(y-t)|$ from above by $|x|d^{-1}$, from (2.13) we obtain the estimate (2.12) for h_+ on $S_{\varepsilon'}$. Proceeding in the same way, we get a similar estimate for h_- and hence for the sum $\tilde{h} = h_+ + h_-$.

Therefore (2.12) is valid for the normed solution $\tilde{h}(x,y) - \tilde{h}(x,0)$ for $(x,y) \in S_{\varepsilon'}$. Finally, expressing $h(x, y+x)$ from (1.8), we obtain the estimate (2.12) on $S_{\varepsilon'} \cup F_0(S_{\varepsilon'})$. □

REMARK 1. If in the conditions of Lemma 6 one has $\varphi(x,y) = o(x^{n+1})$ as $x \to 0$, $x \in V$ uniformly in $y \in R_\varepsilon$ and $n \geq 0$, then $h(x,y) = o(x^n)$ as $x \to 0$, $x \in V$ uniformly in $y \in R_{\varepsilon'}$. In particular, h is bounded.

REMARK 2. The claims of Lemma 6 hold true for the solutions of the auxiliary equation (1.8) defined by (2.7), (2.7') (see Remark 1 in 2.1). This fact can be verified by comparing (2.7), (2.7') with the solution (2.6) obtained in Lemma 4.

2.4. Sectorial solutions for the homogeneous auxiliary equation. Since the auxiliary equation (1.8) is linear, its general solution is the sum of the particular solution constructed in 2.1 and the general solution to the homogeneous equation

$$h(x, y) - h(x, y + x) = 0. \tag{2.14}$$

LEMMA 7. *For any neighborhood W of the origin there exists a neighborhood $U \subset W$ of the origin such that for any bounded normed sectorial solution of the homogeneous equation* (2.14) *in the sectorial domain $S = V \times W$ one has $h(x, y) = o(x^n)$ for any $n \in \mathbb{N}$ uniformly in $y \in U$, where $x \to 0$, $x \in V$.*

PROOF. Identify the quotient space $\widehat{S} = S/F_0$ with the image $\pi_0(S)$ of the sectorial domain S under the map $\pi_0 \colon (x, y) \mapsto (x, \exp(2\pi i y/x))$. Since h is constant along any orbit F_0 (see (2.14)), the bounded holomorphic function $\widehat{h}$, $\widehat{h} \colon \widehat{S} \to \mathbb{C}$, with the property $h = \widehat{h} \circ \pi_0$ is well defined. Let $\varepsilon > 0$ be such that the square $W_0 = \{y \in \mathbb{C} \colon |\operatorname{Im} y| < 2\varepsilon, |\operatorname{Re} y| < 2\varepsilon\}$ lies in W, and put $\widetilde{V} = \{x \in V \colon |x| < \varepsilon/2\}$. Then the domain $\widehat{S}$ contains a Hartogs domain

$$\widehat{S}_0 = \{(x, t) \colon x \in \widetilde{V}, \exp(-2\pi\varepsilon/|x|) < |t| < \exp(2\pi\varepsilon/|x|)\}.$$

The function $\widehat{h}$, which is holomorphic in $\widehat{S}_0$, can be expanded in a Hartogs-Laurent series (see [24]) convergent in $\widehat{S}_0$:

$$\widehat{h}(x, t) = \sum_{k \in \mathbb{Z}} c_k(x) t^k. \tag{2.15}$$

Let $M = \sup_{\widehat{S}_0} |\widehat{h}|$; the Cauchy inequalities yield

$$|c_k(x)| \leqslant M \exp(-2\pi\varepsilon|k|/|x|). \tag{2.16}$$

Since h is normed, one can represent $c_0(x)$ via other coefficients: $c_0(x) = -\sum_{k \neq 0} c_k(x)$. Applying inequalities (2.16) and bounding the sum of the geometric progression by twice its first term (since the ratio of the progression is less than $1/2$), one obtains $|c_0(x)| \leqslant 4M \exp(-2\pi\varepsilon/|x|)$ for $x \in \widetilde{V}$. Hence (2.15), (2.16) imply

$$|h(x, y)| = |\widehat{h}(x, \exp(2\pi i y/x))| \leqslant 6M\lambda/(1 - \lambda)$$

with $\lambda = \exp(-2\pi(|y| - \varepsilon)/|x|)$, thus the statement of the lemma is proved for any disk $U = \{|y| < \varepsilon'\}$ with $\varepsilon' < \varepsilon$. □

REMARK 1. Evidently, there exist nontrivial normed bounded sectorial solutions for (2.14). For instance, such a solution for the sectorial domain

$S = \{|x| < \varepsilon, |\arg x| < \alpha\} \times \{|y| < \varepsilon\}$ is given by

$$h(x, y) = [\exp(2\pi i y/x) - 1]\exp(-2\pi\varepsilon/(x\cos\alpha))$$

(of course, provided that $\alpha < \pi/2$).

REMARK 2. Observe that, as it was shown in the proof of Lemma 7, the quotient space of an arbitrary sectorial domain $V \times W$ by the action of the mapping F_0 contains a domain

$$\{x \in V, \exp(-c/|x|) < |t| < \exp(c/|x|)\}$$

for some $c > 0$.

The claim of Lemma 7 can be significantly strengthened for sectorial domains of large aperture. Namely, the following proposition is true.

LEMMA 7'. *Bounded normed sectorial solutions of the homogeneous equation in a sectorial domain whose aperture exceeds π vanish identically.*

Lemma 7' is proved in 5.2.

2.5. Asymptotic series.

DEFINITION. Let f be a holomorphic function in a sectorial domain $V \times W$. The germ of a semiformal power series $\widehat{f} \in \mathfrak{F}$ (see 1.1) is said to be
- *strictly asymptotic for f in $V \times W$* if there exists a representative $\widetilde{f} \in \mathfrak{F}_W$ of the germ $\widehat{f}$ such that for any n the partial sum $\widetilde{f}_n$ of the semiformal power series $\widetilde{f}$ satisfies the relation $f(x, y) - \widetilde{f}_n(x, y) = o(x^n)$ uniformly in $y \in W$, where $x \to 0$, $x \in V$;
- *asymptotic for f in $V \times W$* if it is strictly asymptotic for f in $V \times U$, where $U \subset W$ is a neighborhood of the origin.

REMARK 1. The statement of Lemma 7 can be reformulated now as follows: the null series is asymptotic for any bounded normed sectorial solution of the homogeneous equation (2.14).

DEFINITION. A germ $\widehat{h} \in \mathfrak{F}$ is said to be a *formal solution* of (1.8) if the difference $\widehat{h}(x, y) - \widehat{h}(x, y + x)$ is an asymptotic germ for φ (for the right-hand side of the equation).

LEMMA 8. *A normed formal solution of the auxiliary equation (1.8) is asymptotic for any bounded normed sectorial solution of the same equation.*

PROOF. Let h be a bounded normed sectorial solution of (1.8) (with φ as the right-hand side) defined in a sectorial domain $V \times W$. Suppose that $\widehat{\varphi}$ is an asymptotic germ for φ, $\widehat{h}$ is a normed semiformal solution of (1.8) (with $\widehat{\varphi}$ as the right-hand side); the latter solution exists by Lemma 2 of 1.3. Let h_n and φ_n be the nth partial sums of the series $\widehat{h}$ and $\widehat{\varphi}$, respectively. Then

$$\varphi - \varphi_n = o(x^n) \quad \text{and} \quad h_n - h_n \circ F_0 = \varphi_n + o(x^n)$$

if $x \to 0$, $x \in V$, uniformly in y from some neighborhood U of the origin. Let $\Delta_n = h - h_n$; then (1.8) yields

$$\Delta_n - \Delta_n \circ F_0 = o(x^n)$$

if $x \to 0$, $x \in V$, uniformly in $y \in U \cap W$. Since Δ_n is normed, Remark 1 from 2.3 and Lemma 7 imply that $\Delta_n = o(x^n)$ if $x \to 0$, $x \in V$, uniformly in y from some neighborhood W' of the origin. This means that the series $\widehat{h}$ is strictly asymptotic for h in $V \times W'$ (and asymptotic in $V \times W$). □

2.6. A theorem on sectorial solutions of the homological equation. In this section we make use of results obtained for the auxiliary equation (1.8) in our study of the homological equation (1.7). Namely, we show that the latter equation possesses a solution in any sectorial domain of aperture $< \pi$; the operator that provides solutions for the homological equation is, in a sense (see the remark at the end of the section), a first-order operator; finally, a formal solution for the homological equation is asymptotic for the sectorial one.

DEFINITION. A map $H: \widetilde{S} \to \widehat{\mathbb{C}}^2$, $\widetilde{S} = S \cup F_0(S)$, is called a *sectorial correction map* (and the difference $H - \mathrm{id}$ is called a *sectorial correction*) for the discrepancy $F - F_0$ in the sectorial domain S if H is holomorphic in the domain $\widetilde{S}$ and (1.6) is true at all points of S. The germ of a semiformal map $(\widehat{\varphi}_1, \widehat{\varphi}_2) \in \mathscr{F}^2$ is said to be *(strictly) asymptotic* in S for a holomorphic map (φ_1, φ_2) if $\widehat{\varphi}_j$ is a (strictly) asymptotic germ for φ_j, $j = 1, 2$, in S. The map $H : S \to \mathbb{C}^2$, $H = (H_1, H_2)$, is called *sectorially bounded* if both functions $H_1(x, y)/x$ and $H_2(x, y)$ are bounded in the sectorial domain S.

THEOREM 2. *Suppose the discrepancy $\varphi = F - F_0$ is holomorphic in a standard sectorial domain $S = V \times R_\varepsilon$ of aperture $\gamma < \pi$, and let $\widehat{\varphi} \in \mathscr{F}^2$ be an asymptotic germ for φ in S, $\mathrm{ord}\, \widehat{\varphi} > 1$. Then:*

1. *For any ε', $0 < \varepsilon' < \varepsilon$, a normed sectorial correction map $H = \mathrm{id} + h$ associated with the discrepancy φ exists and is defined in $S' \cup F_0(S')$, where S' is the standard sectorial domain: $S' = V \times R_{\varepsilon'}$.*

2. *There exists a constant $c_1 = c_1(\gamma) = \mathrm{const} \cdot (\sin \gamma)^{-6}$ such that for $\varepsilon' < \mathrm{diam}\, V$ the sectorial correction $h = (h_1, h_2)$ in the domain $S' \cup F_0(S')$ satisfies the inequalities*

$$|h_1(x, y)| \leq c_1 \delta^{-3} \varepsilon |x|^{-1} \max_{t \in R_\varepsilon} |\varphi_1(x, t)|,$$

$$|h_2(x, y)| \leq c_1 \left(\delta^{-6} \varepsilon^2 |x|^{-2} \max_{t \in R_\varepsilon} |\varphi_1(x, t)| + \delta^{-3} \varepsilon |x|^{-1} \max_{t \in R_\varepsilon} |\varphi_2(x, t)| \right),$$

where $\delta = (\varepsilon - \varepsilon')/\varepsilon$.

3. *There exists a unique normed formal correction $\widehat{h}$ for the discrepancy $\widehat{\varphi}$; $\widehat{h}$ is an asymptotic germ for any sectorially bounded normed correction h in S'.*

PROOF. Put $\delta_0 = \varepsilon - \varepsilon'$, $\varepsilon_j = \varepsilon - j\delta_0/3$, $S_j = V \times R_{\varepsilon_j}$, $j = 1, 2, 3$ (and hence $S_3 = S'$). Using Lemma 4, let us find a solution h_1 for the first equation of (1.7) in S_1, and then a solution h_2 for the second equation in S_2. Next, apply Lemma 6 twice to obtain the required bounds associated with the discrepancy h. The last assertion follows from the Corollary in 1.3 and Lemma 8. □

REMARK 1. Remark 2 in 2.1 and Remark 2 in 2.3 imply that for a real-analytic discrepancy φ (and a σ-symmetric sectorial domain S), the correction map H in Theorem 2 can be regarded as a real-analytic as well.

REMARK 2. Unlike the solutions of the auxiliary equation from Lemma 4, those of the homological equation obtained above depend on ε'. There is a separate solution h of (1.7) for each $\varepsilon' < \varepsilon$, therefore the operator that provides solutions for the homological equation is not literally a first-order operator (see [1]); this means that only the corresponding estimates are fulfilled.

2.7. Examples. In this section we present a number of examples showing that the assertions of Lemma 4 (and thus of Theorem 2) are, in a sense, "almost exact".

EXAMPLE 1 (of a functional equation (1.8) that has no holomorphic solutions). Consider the equation

$$h(x, y) - h(x, y + x) = x/(y - a), \qquad a \neq 0. \tag{2.17}$$

One can easily check that the following functions are solutions of (2.17):

$$h_\pm(x, y) = \pm \sum_k {}^{(\pm)} \left(\frac{x}{a - kx} - \frac{x}{a - y - kx} \right);$$

the superscripts + and − mean that we sum over all integer $k \geq 0$ and $k < 0$, respectively. Note that $h_+(x, y) = G((a - y)/x, -a/x)$ (see 2.2), and by Lemma 5, $h_+(x, y)$ is bounded in the above sectorial domain of aperture $2\pi - 2\varepsilon$ (the fact that h_- is bounded is proved in the same way). Since the functions $h_\pm$ have poles in any neighborhood of the origin (for any arbitrarily small fixed $y \neq 0$), the uniqueness theorem for normed solutions of (1.8) in sectorial domains of aperture $> \pi$ (Lemma $7'$) implies (once more) the absence of holomorphic solutions for (2.17).

REMARK. Taking derivatives of (2.17) with respect to y, one obtains sectorial solutions h_+ of (1.8) for any right-hand side of the form $cx(y - a)^{-n}$, $n \in \mathbb{N}$. By normalizing these solutions, one can see that the $h_\pm$ have (for a fixed $y \neq 0$) infinitely many poles of order n on the ray $\arg(\pm x) = \arg a$. Moreover, for any polynomial $P(y)$, there exists a polynomial $h(x, y)$ satisfying equation (1.8) with $xP(y)$ as its right-hand side; for example, it can be reconstructed from its values at the points $(x, y + kx)$, $0 \leq k \leq \deg P$, with the help of the Lagrange interpolation formula. Now the uniqueness theorem for sectorial solutions (Lemma $7'$) implies

PROPOSITION. *Suppose $R(y)$ is a rational function with singularities at the points y_k, $k = 1, 2, \ldots, n$, $\arg y_k = \psi_k$, the points $\pm e^{i\psi_k}$ divide the unit circle into m parts ($m \leqslant 2k$), and δ is the maximal length of these parts. Then equation (1.8) with $xR(y)$ as its right-hand side has no bounded sectorial solutions in any sectorial domain of aperture $> \pi + \delta$.*

The case $\delta = \pi$ must be considered separately; this leads to the following

COROLLARY. *Equation (1.8) with rational right-hand side $xR(y)$ possesses a holomorphic solution if and only if R is a polynomial.*

EXAMPLE 2 (equation (1.8) having no sectorial solutions in any sectorial domain of aperture $> \pi$). Let

$$\varphi(x, y) = \sum_{k=0}^{\infty} \frac{xy^{2^k-1}}{1 - y^{2^k}}.$$

Let us show that equation (1.8) with $\varphi(x, y)$ as its right-hand side does not have holomorphic bounded sectorial solutions in any sectorial domain of aperture $> \pi$. Indeed, let h be such a normed solution in a sectorial domain S of aperture $\pi + \delta$, $\delta > 0$. Let $\lambda = \exp(2\pi i/2^n)$, $n > -\log_2 \delta + 3$, and put $\varphi_1(x, y) = \varphi(x, y) - \varphi(\lambda x, \lambda y)$, $h_1(x, y) = h(x, y) - h(\lambda x, \lambda y)$. Then h_1 is the normed bounded sectorial solution of the equation $h_1(x, y) - h_1(x, y+x) = \varphi_1(x, y)$ with right-hand side

$$\varphi_1(x, y) = \sum_{k=0}^{n-1} xy^{2^k-1} \left(\frac{\lambda^{2^k}}{1 - (\lambda y)^{2^k}} - \frac{1}{1 - y^{2^k}} \right)$$

in the sectorial domain $S_1 = S \cap \lambda^{-1} S$, in contradiction to the Proposition.

§3. The sectorial normalization theorem

3.1. Formulation of the result and an outline of the proof. A biholomorphic map $H: U \to H(U)$ defined in some domain $U \subset \mathbb{C}^2$ will be called the *sectorial normalizing transformation in a sectorial domain S for a germ $\mathscr{F}$ of the class $\mathscr{B}_1$*, if:

(1) for a certain representative F of the germ $\mathscr{F}$ in the sectorial domain S the superposition $H \circ F \circ H^{-1}$ coinciding with F_0 is defined;
(2) $H = (H_1, H_2)$ is S-bounded in the following sense: the two functions $H_1(x, y)/x$, $H_2(x, y)$ are bounded on S.

Recall that the map $H: S \to \mathbb{C}^2$ is called *normed* if $H|_{\{y=0\} \cap S} = \mathrm{id}$. We shall say that the germ $\mathscr{F} \in \mathscr{B}_1$ admits *sectorial normalization on the sectorial domain S*, if there exists a normed sectorial normalizing transformation for it, defined in S.

THEOREM 3. *Each germ of the class $\mathscr{B}_1$ admits sectorial normalization in any sectorial domain of angle less than π and sufficiently small diameter.*

The normed formal normalizing transformation is the asymptotic series for any normed sectorial normalizing transformation. A real-analytic germ of the class $\mathscr{B}_1$ admits a sectorial normalization, which is real on the real plane.

The two last assertions of Theorem 3 imply the following

COROLLARY. *Any real-analytic germ of the class $\mathscr{B}_1$ is smoothly equivalent to the formal normal form F_0.*

REMARK. It may be derived from Remark 1 following Lemma 7 that the sectorial normalizing transformation (even the normed one) is not unique.

OUTLINE OF THE PROOF. The proof follows the standard pattern described in [1]. A normed transformation in the sectorial domain S for the germ $\mathscr{F} \in \mathscr{B}_1$ will be constructed using successive approximations (in the same way as in the proof of Theorem 1 in 1.4). Namely, we define by induction the sequences $\{G_n\}$, $\{H_n\}$ and $\{\mathscr{H}_n\}$ such that $G_1 = F$ is the representative of the germ $\mathscr{F}$, $\mathscr{H}_0 = \mathrm{id}$, and, for $n \geqslant 1$,

$$\mathscr{H}_n = H_n \circ \mathscr{H}_{n-1}, \qquad G_{n+1} = H_n \circ G_n \circ H_n^{-1},$$

where H_n is the correction map associated with the discrepancy $G_n - F_0$ provided by Theorem 2. After checking that the convergence $G_n \to F_0$, $\mathscr{H}_n \to H$ is uniform on S, we get the required transformation. The convergence is analyzed using the technique of operators of finite order, see [1]. A detailed exposition of this proof is given below.

3.2. One step of the successive approximation method. Let $S = V \times R_\varepsilon$ be a standard sectorial domain of the angle $\gamma < \pi$, where

$$V = V_{\varepsilon_0} = \{x \in \mathbb{C} : \alpha < \arg x < \alpha + \gamma, \ |x| < \varepsilon_0\}.$$

By the δ-*narrowing* of the domain S we shall mean the standard sectorial domain $S^\delta = V^\delta \times R_\varepsilon^\delta$, where the sector

$$V^\delta = \{x \in V : \alpha + \delta < \arg x < \alpha + \gamma - \delta, \ |x| < \varepsilon_0 e^{-\delta} \cos(\gamma/2)/\cos(\gamma/2 - \delta)\}$$

is the δ-narrowing of the sector V and R_ε^δ is the rhombus inscribed in $R_{\varepsilon \exp(-\delta)}$ with edges parallel to the sides of the sector V^δ, see Figures 9a and 9b on next page. Evidently, $(S^\delta)^\delta = S^{2\delta}$.

REMARK. The ratio of the parameters ε_0 (the radius of the sector) and ε (the half-length of the side of the rhombus), which define the size of the domain $S = V_{\varepsilon_0} \times R_\varepsilon$, remains the same for the initial domain S and for its narrowing S^δ. This property guarantees the applicability of Lemma 9 in the proof of Theorem 3. This remark motivates the choice of the radius in the definition of V^δ.

LEMMA 9 (on subsequent discrepancies). *Let $S = V \times R_\varepsilon$ be the standard sectorial domain of angle γ with $0 < \gamma_0 < \gamma < \pi - \gamma_0$, $\operatorname{diam} V \leqslant \varepsilon < 10^{-2}$.*

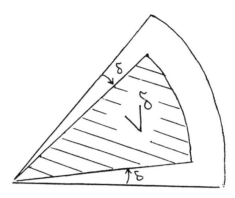

FIGURE 9a. Narrowing of the sector V^δ.

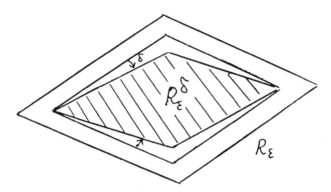

FIGURE 9b. The restriction of the rhombus R^δ.

Th n there exist constants $M_0 = M_0(\gamma_0)$, $N_0 = N_0(\gamma_0)$ such that for any positive $\delta < 1/2$ and for all $N > N_0$, $n \geqslant 2$ and for any F holomorphic in S with discrepancy $\varphi = F - F_0 = (\varphi_1, \varphi_2)$ satisfying the conditions

$$|\varphi_1(x,y)| < \delta^N |x|^{n+1}, \qquad |\varphi_2(x,y)| \leqslant \delta^N \varepsilon |x|^n \qquad (3.1)$$

in S we can find a sectorial domain S' and a sectorial correction map H in S' such that the superposition $\widetilde{F} = H \circ F \circ H^{-1}$ is defined and holomorphic in S^δ, and, moreover, the discrepancy $\widetilde{F} - F_0 = (\psi_1, \psi_2)$ satisfies

$$|\psi_1(x,y)| \leqslant \delta^{2N-M_0} |x|^{n+2}, \qquad |\psi_2(x,y)| \leqslant \delta^{2N-M_0} \varepsilon |x|^{n+1} \qquad (3.2)$$

for all $(x,y) \in S^\delta$; in particular, $H^{-1}(S^\delta) \subset S$.

In other words, the operator $(\varphi_1, \varphi_2) \mapsto (\psi_1, \psi_2)$ is of the second order in the sense of [1].

PROOF. Let $\delta_1 = \delta/4$, $S_0 = S$, $S_j = (S_{j-1})^{\delta_1}$, $j = 1, 2, 3, 4$, so that $S_4 = S^\delta$. Let $h = (h_1, h_2)$ be the correction associated with the discrepancy

φ in S_1 given by Theorem 2; the estimates in this theorem and (3.1) imply

$$|h_1(x,y)| \leq \delta^{N-M_1}\varepsilon|x|^n, \qquad |h_2(x,y)| \leq \delta^{N-M_1}\varepsilon^2|x|^{n-1} \qquad (3.3)$$

for $(x,y) \in \widetilde{S}_1 = S_1 \cup F_0(S_1)$ and for some $M_1 = M_1(\gamma_0)$. In particular,

$$|x/2| \leq |x + h_1(x,y)| \leq |2x| \quad \text{for } (x,y) \in \widetilde{S}_1 \qquad (3.4)$$

if N is sufficiently large. Estimating the derivatives of h_j, $j = 1, 2$, by the Cauchy integral formula and taking (3.3) into consideration, we obtain

$$|h'_{jx}(x,y)| \leq \delta^{N-M_2}\varepsilon^j|x|^{n-j}2^n, \qquad |h'_{jy}(x,y)| \leq \delta^{N-M_2}\varepsilon^{j-1}|x|^{n-j+1} \qquad (3.5)$$

for $(x,y) \in \widetilde{S}_2 = S_2 \cup F_0(S_2)$, $j = 1, 2$, and a certain constant $M_2 = M_2(\gamma_0)$.

If N is sufficiently large, then the partial derivatives h'_{jx}, h'_{jy} are small, and the convexity of S_2 implies the injectivity of the map $H = \text{id} + h$ in S_2. Putting $H(0,y) = y$, we get the continuity of H in $\overline{S}_2$. Note that for N sufficiently large it follows from (3.3) that for all $t \in [0,1]$ we have $S_3 \cap (\text{id}+th)(\partial S_2) = \varnothing$. Therefore, for any point $A \in S_3$, the topological degree of the map $H = \text{id} + h$ with respect to the point A and the domain S_2 is the same as that for the map id homotopic to H (and thus equals 1, since $S_3 \subset S_2$), see [15]. Hence A belongs to the image of the domain S_2 under the map H, and therefore $S_3 \subset H(S_2)$.

Hence the inverse H^{-1} is well defined on S_3, $H^{-1}(S_3) \subset S_2$ and $H^{-1} = \text{id} + g$ is holomorphic in S_3. Let $g = (g_1, g_2)$, $(\tilde{x}, \tilde{y}) \in S_3$, $(x,y) = H^{-1}(\tilde{x}, \tilde{y})$. Then $h_j(x,y) + g_j(\tilde{x}, \tilde{y}) = 0$. Using (3.3) and (3.4), we get

$$|g_1(\tilde{x}, \tilde{y})| \leq \delta^{N-M_3}\varepsilon(2|\tilde{x}|)^n, \qquad |g_2(\tilde{x}, \tilde{y})| \leq \delta^{N-M_3}\varepsilon^2(2|\tilde{x}|)^{n-1}, \qquad (3.6)$$

valid on S_3, and

$$|\tilde{x}/2| \leq |\tilde{x} + g_1(\tilde{x}, \tilde{y})| \leq |2\tilde{x}| \quad \text{for } (\tilde{x}, \tilde{y}) \in S_3 \qquad (3.7)$$

if N is sufficiently large.

It follows from (3.1)–(3.7) that the superposition $H \circ F \circ H^{-1}$ is defined on S_4 for large N. Using formula (1.10) for the subsequent discrepancy, and using (3.1), (3.5)–(3.7), we obtain

$$|\psi_j(x,y)| \leq 2\delta^{2N-M_2}\varepsilon^j \cdot 32^n|x|^{2n+1-j}, \qquad j = 1, 2, \quad (x,y) \in S^\delta,$$

whence comes the assertion of the lemma, since $|x| < \varepsilon < 1/100$. □

3.3. Proof of Theorem 3. Let $\gamma \in \mathbb{R}$, $0 < \gamma < \pi$; choose γ' so that $\gamma < \gamma' < \pi$. Let G_1 be a representative of the germ $\mathscr{F} \in \mathscr{B}_1$. Because of the Corollary to Theorem 1 in 1.2, one may assume that the discrepancy $\varphi = (\varphi_1, \varphi_2) = G_1 - F_0$ is of order ≥ 4. Therefore we can find real numbers $\varepsilon_0 < \varepsilon < 1/100$ such that for $|x| < \varepsilon_0$, $|y| \leq 2\varepsilon$ the functions $\varphi_j(x,y)$ are defined and satisfy

$$|\varphi_1(x,y)| < \delta_1^{N_1}|x|^3, \qquad |\varphi_2(x,y)| < \delta_1^{N_1}|x|^2\varepsilon;$$

the numbers δ_1, N_1 depending only on γ, γ' will be specified afterwards.

Define the sequence $\{\delta_n\}$ by the condition $\delta_{n+1} = \delta_n^{3/2}$, and choose $\delta_1 < 1/10$ in such a way that

$$\Delta' = \sum_{n=1}^{\infty} \delta_n < (\gamma - \gamma')/4 = \Delta.$$

Put $\gamma_0 = (1/2)\min\{\pi - \gamma', \gamma\}$ and suppose that $N_0 = N_0(\gamma_0)$, $M_0 = M_0(\gamma_0)$ are the constants given by Lemma 9.

Suppose $\alpha, \beta \in \mathbb{R}$ are any numbers with $\beta - \alpha = \gamma$, and

$$V = \{x \in \mathbb{C} : \alpha - \Delta < \arg x < \beta + \Delta, \ |x| < \varepsilon_0\}$$

is a sector of aperture γ', $S_1 = V \times R_\varepsilon$ is the standard sectorial domain. Finally put $S' = S_1^{\Delta'}$, $\tilde{S} = (S')^{\Delta - \Delta'}$, $S = (\tilde{S})^{\Delta}$. The normalizing transformation will be constructed in the domain S.

Put $S_{n+1} = S_n^{\delta_n}$, $N_{n+1} = (5/4)N_n$. Then

$$\delta_{n+1}^{N_{n+1}} = \delta_n^{15 N_n/8} \geq \delta_n^{2N_n - M_0}, \quad \text{if } N_n \geq 8M_0.$$

Therefore for $N_1 \geq 8M_0$ the sequences $\{G_n\}$, $\{H_n\}$ and $\{\mathscr{H}_n\}$ from 3.1 are well defined by virtue of Lemma 9. Using the estimate (3.3) for the correction $(h_1^k, h_2^k) = H_k - \mathrm{id}$, we get

$$|h_j^k(x, y)| \leq \delta_k^{N_k - M_0} \varepsilon^j |x|^{k+2-j} \tag{3.8}$$

for $(x, y) \in S_k \cup F_0(S_k)$, $k \geq 1$, $j = 1, 2$.

Let $\{x_n, y_n\} = \mathscr{H}_n(x_0, y_0)$; then $(x_n, y_n) = H_n(x_{n-1}, y_{n-1})$. From the estimates (3.8) it follows that if $N_1 \geq M_0 + 1$ and $(x_{n-1}, y_{n-1}) \in S_n$, then

$$|x_n - x_{n-1}| < \delta_n, \quad |\arg x_n - \arg x_{n-1}| < \delta_n, \quad |y_n - y_{n-1}| < \delta_n. \tag{3.9}$$

Now let $(x_0, y_0) \in S$. Since $\sum \delta_n = \Delta' < \Delta$, $S = (\tilde{S})^{\Delta}$, by induction we see that $(x_n, y_n) \in \tilde{S}$ for all $n \in \mathbb{N}$. Thus the estimates (3.9) are valid for all $n \in \mathbb{N}$. Hence, the series $\sum(\mathscr{H}_n - \mathscr{H}_{n-1})$ converges uniformly on S. Therefore the sequence $\{\mathscr{H}_n\}$ converges uniformly on S to a certain holomorphic map H. In the same way as in the proof of Lemma 9 (estimating derivatives of H) we prove the invertibility of H provided that N_1 is sufficiently large. Finally, passing to the limit in the identities $\mathscr{H}_n \circ \mathscr{H}_n^{-1} = \mathrm{id}$, $G_{n+1} \circ \mathscr{H}_n = \mathscr{H}_n \circ G_n$, we check that H is the required holomorphic normalizing transformation. The first assertion of Theorem 3 is proved.

REMARK. From the estimates (3.8) it follows that $\mathscr{H} - \mathscr{H}_n = o(x^n)$ for $x \to 0$, $(x, y) \in S$ uniformly in y. Moreover, $\mathscr{H} - \mathrm{id} = o(x^n)$ for $x \to 0$, $(x, y) \in S$ also uniformly in y provided the order of the discrepancy $F - F_0$ is greater than $n + 2$.

3.4. Asymptotic series. First let us point out that a formal series which is the finite superposition of asymptotic series, will be the asymptotic expansion for the superposition of maps holomorphic in a sectorial domain provided that the superposition is well defined. Comparing the proof of the formal (Theorem 1) and the sectorial (Theorem 3) assertions on the normalization, from Theorem 3 and the Remark in 3.3 we see that the sectorial normalization constructed in the current section has the formal normalizing series as its asymptotic expansion.

Now let $\widetilde{H}$ be another arbitrary normed normalizing transformation, defined in the sectorial domain S. Then for the normalizing transformation H constructed above we see that $H_0 = \widetilde{H} \circ H^{-1}$ is well defined on $H(S) = \widetilde{S}$ and commutes with F_0. Therefore H_0 satisfies the homogeneous homological equation, and by virtue of Theorem 2, the germ of the identity map will be the asymptotic expansion for H_0 in any sectorial domain $S' \subset S$, whence we get the second assertion of Theorem 3.

The third assertion of Theorem 3 follows from Remark 1 in 2.6.

In the same way as in 1.5, we get

COROLLARY. *For any sectorial normalizing transformation $\widetilde{H}$ defined in S there exists a holomorphic in $\widetilde{S} = H(S)$ map G affine in the sense of (1.12) such that the zero germ $(0, 0)$ is the asymptotic germ for $\widetilde{H} - G \circ H$ (here H is the normed normalizing transformation given by Theorem 3).*

§4. Theorem on functional invariants for the class $\mathscr{B}_1$

4.1. The normalizing atlas. Theorem 3 allows us to normalize the germ $F \in \mathscr{B}_1$ in any sectorial domain of aperture less than π. Covering a neighborhood of the origin by domains of such form, we get an object called the normalizing atlas.

DEFINITION 1. A *cut neighborhood* of the origin is a neighborhood $(\mathbb{C}^2, 0)$ with the line $\{x = 0\}$ deleted.

A finite collection of domains $\omega = \{\Omega_j\}_{j \in J}$ (where $J = J(w)$ is a finite subset of $\mathbb{N}$) will be called a *regular covering of the domain* Ω, if

1) $\Omega = \bigcup_{j \in J} \Omega_j$ is a cut neighborhood of the origin;
2) each domain Ω_j contains (and is contained in) some sectorial domain;
3) each domain Ω_j has a nonempty intersection with exactly two of the other domains from ω.

A regular covering consisting of sectorial domains will be called a *sectorial covering*.

If ω is a regular covering, then $J'(\omega)$ will denote the set of pairs of indices of the domains having nonempty intersections. The collection ω' is the finite set of nonempty pairwise intersections: $\omega' = \{\Omega_i \cap \Omega_j\}_{(i,j) \in J'(\omega)}$.

The regular covering $\widetilde{\omega} = \{\widetilde{\Omega}_j\}$ will be called a *narrowing* of a regular covering ω, if $\widetilde{\Omega}_j \subset \Omega_j$ for all $j \in J$.

The regular covering $\widetilde{\omega} = \{\widetilde{\Omega}_j\}|_{j \in \widetilde{J}}$ will be called a *refinement* of a regular covering $\omega = \{\Omega_j\}|_{j \in J}$, if for a certain surjective map $e \colon \widetilde{J} \to J$ each Ω_j is the union of all $\widetilde{\Omega}_{\widetilde{j}} \in \widetilde{\omega}$ such that $e(\widetilde{j}) = j$, $j \in J$.

If the regular covering $\widetilde{\omega}$ is the narrowing (refinement) of a regular covering ω, then the corresponding collection of intersections $\widetilde{\omega}'$ will be called the *narrowing* (respectively, *refinement*) of the collection ω'. The surjection $e \colon \widetilde{J} \to J$ induces the injective map $e' \colon J'(\omega) \to \widetilde{J}'(\widetilde{\omega})$, $e'(j, k) = (\widetilde{j}, \widetilde{k})$ provided that $j = e(\widetilde{j})$, $k = e(\widetilde{k})$.

Let all the domains constituting a sectorial covering $\omega = \{\Omega_j\}$ have angles smaller than π and sufficiently small diameter. Using Theorem 3, for the given representative F of some germ $\mathscr{F} \in \mathscr{B}_1$ we construct a normalizing map H_j in each domain so that the formal normalizing transformation of the germ $\mathscr{F}$ is an asymptotic germ for H_j. Replacing the covering ω by its narrowing if necessary, we may always assume that the following holds:

NA-1. H_j is the sectorial normalizing transformation for the representative F of the germ $\mathscr{F}$ for all $j \in J$.

NA-2. There exists a germ Φ of the formal normalizing transformation for F which is strictly asymptotic for all H_j in Ω_j.

DEFINITION 2. A pair $\tau = (\omega, H)$, $\omega = \{\Omega_j\}$, $H = \{H_j\}$, $j \in J$, consisting of a sectorial covering and a collection of maps satisfying NA-1, NA-2, will be called a *normalizing atlas* for F (and $\mathscr{F}$) on the cut neighborhood of the origin $\Omega = \bigcup \Omega_j$. The formal germ Φ from NA-2 will be called *asymptotic for the normalizing atlas* τ. A normalizing atlas is called *normed*, if each of its charts is normed. Narrowing and refinement of normalizing atlases are defined in a natural way.

Thus the normalizing atlas for F is an atlas on a cut neighborhood of the origin such that in each of its charts the germ $\mathscr{F}$ coincides with the formal normal form F_0; the charts agree in the sense of condition NA-2.

4.2. Systems of transition functions. Using the information on formal normalizing transformations provided by the Corollary from 1.4, we deduce from NA-2 that for any normalizing atlas τ it is possible to find its narrowing $(\widetilde{\omega}, H)$ such that the system of domains $H(\widetilde{\omega})$ will also be a regular covering. Moreover, we can assume that the latter system has a narrowing $\omega = \{\Omega_j\}_{j \in J}$ consisting of sectorial domains such that for all $(i, j) \in J'(\omega)$ on the intersection $\Omega_{ij} = \Omega_i \cap \Omega_j \in \omega'$ the superposition $H_{ij} = H_i \circ H_j^{-1}$ is well defined. From the NA conditions we get the set of conditions satisfied by the system of transition functions $\{H_{ij}\}$:

T-1. $H_{ij} \circ F_0 = F_0 \circ H_{ij}$.

T-2. The identical germ is strictly asymptotic for all H_{ij} on Ω_{ij}.

T-3. $H_{ij} \circ H_{ji} = \text{id}$ (in particular, all such superpositions are well defined).

DEFINITION 1. A pair $(\omega, \delta H)$, $\omega = \{\Omega_j\}$, $\delta H = \{H_{ij}\}$, satisfying T-1 through T-3, will be called a *system of transition functions*. The system of transition functions constructed above for the normalizing atlas τ will be called the *system of transition functions associated with the normalizing atlas τ*. Narrowings and refinements of such systems are defined in a natural way.

DEFINITION 2. Two systems of transition functions are called *equivalent* if there exist refinements $\{\omega, \delta H\}$, $\{\tilde{\omega}, \delta \tilde{H}\}$ of their narrowings, $\omega = \{\Omega_j\}$, $\tilde{\omega} = \{\tilde{\Omega}_j\}$, $\delta H = \{H_{ij}\}$, $\delta \tilde{H} = \{\tilde{H}_{ij}\}$, biholomorphic maps $G_j: \Omega_j \to \tilde{\Omega}_j$, and a formal invertible affine germ $\Psi \in \mathscr{F}_0^2$ such that the following set of conditions holds:

E-1. $G_j \circ F_0 = F_0 \circ G_j$ on $\Omega_j \cap F_0^{-1}(\Omega_j)$.
E-2. The germ Ψ is strictly asymptotic for G_j.
E-3. $G_i \circ H_{ij} = \tilde{H}_{ij} \circ G_j$ on $\Omega_{ij} = \Omega_i \cap \Omega_j$. In particular, all the superpositions are well defined.

LEMMA 10. *Let τ and $\tilde{\tau}$ be any two normalizing atlases for the same germ with coinciding coverings. Then their systems of transition functions are equivalent.*

PROOF. Let H_j, $\tilde{H}_j$ be the charts of the normalizing atlases $\tau, \tilde{\tau}$ on Ω_j^0, and $\Phi, \tilde{\Phi}$ their strictly asymptotic formal germs. Let $\Omega_j = H_j(\Omega_j^0)$, $\tilde{\Omega}_j = \tilde{H}_j(\Omega_j^0)$, $G_j = \tilde{H}_j \circ H_j^{-1}: \Omega_j \to \tilde{\Omega}_j$, $\Psi = \tilde{\Phi} \circ \Phi^{-1}$. All the properties now follow from the definition of the normalizing atlas.

REMARK. It is easy to check that the inverse statement also holds: if two systems of transition functions are equivalent, and one of them corresponds to a certain normalizing atlas, then the other also corresponds to some (other) normalizing atlas.

4.3. Functional invariants. The main $\mathscr{B}_1$ Theorem.

DEFINITION 1. Let T_1 be the space of all system of transition functions. The equivalence classes of systems of transition functions will be called *functional invariants*. The equivalence class of a system of transition functions η will be denoted by $[\eta]$, the total quotient space, by $[T_1]$.

Take any germ $F \in \mathscr{B}_1$. Its functional invariant m_F is the equivalence class of the system of transition functions associated with any normalizing atlas for the germ F.

REMARK. Lemma 10 implies that the functional invariant of a germ of class $\mathscr{B}_1$ depends neither on the choice of the representative nor on the choice of the normalizing atlas and of the corresponding system of transition functions.

DEFINITION 2. A family of equivalence classes $\{m_t\}|_{t \in U}$ (U is an open domain in $\mathbb{C}^n$) is called *analytic*, if for each point $t_0 \in U$ one can find an analytic family of representatives μ_t, $[\mu_t] = m_t$ (analyticity of the family

μ_t means that the transition functions depend analytically on the parameter in the domains which are subsets of the Cartesian product of the phase and the parameter spaces).

$\mathscr{B}_1$ THEOREM. *The space of functional invariants $[T_1]$ is the moduli space of analytic classification for germs of class $\mathscr{B}_1$. Namely, the following assertions hold*:

1. Equimodality and equivalence. *Any two germs $F, G \in \mathscr{B}_1$ are equivalent if and only if $m_F = m_G$.*
2. Realization. *For any $m \in [T_1]$ there exists an $F \in \mathscr{B}_1$ such that $m = m_F$.*
3. Analytical dependence on parameters. *The functional invariants of an analytic family of germs of class $\mathscr{B}_1$ constitute an analytic family in the sense of Definition 2. Any analytic family $[m_t] \subset [T_1]$ can be realized as the invariant of a certain analytic family of germs belonging to $\mathscr{B}_1$.*

The proof will be given in the next three subsections.

4.4. Equimodality and equivalence. Note that if H_j is a normalizing transformation for a germ F and H is any germ of a biholomorphism, then $H_j \circ H^{-1}$ is the normalizing transformation for the germ $H \circ F \circ H^{-1}$. This immediately implies that equivalent germs are equimodal.

Conversely, let the systems of transition functions associated with the normalizing atlases H and $\widetilde{H}$ of two germs F, $\widetilde{F}$ be equivalent. Let them be conjugated by $\{G_{ij}\}$ in the sense of Definition 2 in 4.2. Then the map G, which in the charts of the normalizing atlases ($\{H_j\}$ in the target and $\{\widetilde{H}_j\}$ in the source space) coincides with the corresponding maps from $\{G_j\}$, is in fact a certain well-defined (by condition E-3) map between cut neighborhoods. Extending G to the entire neighborhood by using the removable singularity theorem, we obtain the transformation conjugating F and $\widetilde{F}$.

4.5. Realization of functional moduli. Let us prove that each element m of the "moduli space" $[T_1]$ is the functional invariant for a certain germ from $\mathscr{B}_1$. The proof follows the usual plan and relies on the machinery of almost complex structures (for a detailed description of the plan together with all the necessary information concerning almost complex structures see Paper I, 2.4 and 3.3; in particular, we do not recall here the definitions of an almost complex structure, and of its integrability, presented there, as well as the formulation of the Newlander-Nirenberg theorem).

Let $m \in [T_1]$ be an arbitrary element of the moduli space, $\Delta = (\omega, \delta H)$ be a representative of the class m. Here ω is a collection of sectorial domains Ω_j covering a cut neighborhood of the origin in $\mathbb{C}^2$, as before. Next, δH is a collection of maps $H_{j,j+1}$ defined in the intersection $\Omega_j \cap \Omega_{j+1}$ (the numbering is assumed cyclic). Besides, the map id is strictly asymptotic for the maps $H_{j,j+1}$ and $H_{j,j-1} = H_{j-1,j}^{-1}$:

$$H_{j,j\pm 1} - \mathrm{id} = o(x^N) \quad \text{for } x \to 0, \ (x,y) \in \Omega_j \cap \Omega_{j\pm 1}, \qquad (4.1)$$

uniformly in y for any $N \in \mathbb{N}$.

Contract the domains Ω_j in such a way that the collection $\{\Omega_j \times \{j\}\}$ becomes a topological manifold, the points of the domains $\Omega_j \times \{j\}$ and $\Omega_{j+1} \times \{j+1\}$ being glued together by means of the map

$$\widetilde{H}_{j,j+1}: z \times \{j\} \mapsto H_{j,j+1}(z) \times \{j+1\}.$$

Denote this manifold by S. The covering ω is transformed into a covering $\widetilde{\omega}$ of the manifold S by the domains $\widetilde{\Omega}_j$ obtained from Ω_j by the gluing process. On each of the $\widetilde{\Omega}_j$ a natural chart $z_j: \widetilde{\Omega}_j \to \Omega_j \subset \mathbb{C}^2$ is defined. The atlas obtained defines a complex structure on S, since the transition functions $H_{j,j+1}$ are biholomorphic.

Let us prove that the manifold S (possibly, after an appropriate contraction of the sectorial domains Ω_j achieved by reducing their radii) is biholomorphically equivalent to a cut neighborhood of the origin in $\mathbb{C}^2$. To do this, we carry out a construction similar to that described in 2.6 of Paper I. Let us start by finding a diffeomorphism of S to a cut neighborhood V of the origin in $\mathbb{C}^2$.

Let $\{\theta_j\}$ be an infinitely smooth partition of unity on

$$\Omega = \bigcup \omega_j \subset \mathbb{C}^2 = \{(x,y)\}$$

subordinate to the covering $\{\Omega_j\}$: $\mathrm{supp}\,\theta_j \subset \Omega_j$, $\sum \theta_j = 1$. Besides, suppose that $\theta_j: \Omega_j \to \mathbb{R}$ is a function of $\arg x$ and that all of its derivatives increase no faster than a certain (depending on the order of the derivative) negative power of $|x|$ as $x \to 0$ in Ω_j. Put

$$G = \sum \theta_j \circ z_j, \qquad z_j: S \to \mathbb{C}^2, \qquad G(S) = \mathring{V}.$$

The map G belongs to the class C^∞. Let $G_j = G|_{\widetilde{\Omega}_j}$. Observe that the difference $H_{j,j\pm 1} - \mathrm{id}$ tends to zero together with all of its derivatives as $x \to 0$ faster than any power of x (this fact follows from (4.1) and the Cauchy estimates). Hence similar estimates hold for $G_j \circ z_j^{-1} - \mathrm{id}$ (since the derivatives of the functions θ_j increase as some negative powers of x for $x \to 0$). Therefore, one can assume that the maps G_j are injective (possibly, after an appropriate contraction of the domains Ω_j), similar estimates hold also for the differences $z_j \circ G_j^{-1} - \mathrm{id}$, and G is injective on S.

The diffeomorphism $G: S \to \mathring{V}$ endows the domain $\mathring{V}$ with an almost complex structure induced by the map G from the complex structure on S. Namely, let

$$\alpha_1^j = (z_j \circ G_j^{-1})^* dx, \qquad \alpha_2^j = (z_j \circ G_j^{-1})^* dy$$

and define 1-forms α_1, α_2 on $\mathring{V}$:

$$\alpha_k = \sum_j \theta_j \circ z_j \circ G_j^{-1} \cdot \alpha_k, \qquad k = 1, 2.$$

Then the 1-forms α_1, α_2 are the generators of the subspace of "1-forms holomorphic with respect to the almost complex structure". Extend these forms to the deleted line $x = 0$ by putting $\alpha_1 = dx$, $\alpha_2 = dy$. Since the differences $z_j \circ G_j^{-1} - \mathrm{id}$, $\sum \theta_j \circ z_j \circ G_j^{-1} - 1$ are flat, the above extension is smooth. Hence one obtains an almost complex structure on the entire neighborhood V of the origin in $\mathbb{C}^2$ such that $V_0 = V \setminus \{x = 0\}$. This almost complex structure is integrable. Indeed, the almost complex structure on $\mathring{V}$ is generated by the complex structure existing on S via the diffeomorphism G, hence it is integrable; therefore, by continuity, it is integrable at any point of the domain V as well. Now the Newlander-Nirenberg theorem implies the existence of a smooth local chart (φ_1, φ_2) in a neighborhood of the origin that is holomorphic in the sense of the above almost complex structure. Without loss of generality, one can assume the smooth map $\Phi = (\varphi_1, \varphi_2)$ to be defined (and injective) on the entire domain $\mathring{V}$. The compositions $\Phi \circ G_j$ are holomorphic, hence $\Phi \circ G$ is a biholomorphism of the manifold S to the domain $\Phi(\mathring{V})$.

Observe that the maps G_j are holomorphic at some point, provided that $\theta_j \circ z_j = 1$, $\theta_{j-1} \circ z_{j-1} = \theta_{j+1} \circ z_{j+1} = 0$ in a neighborhood of this point. Let the domain $\widetilde{U}_j \subset \widetilde{\Omega}_j$ consist of all such points, $U_j = z_j(\widetilde{U}_j)$. If the domains Ω_k are narrow enough, then the domain U_j is nonempty for any j and contains some sectorial domain U. Since $G_j = z_j$ on $\widetilde{U}_j$, Φ is holomorphic in the domain U_j.

LEMMA 11. *Let $\Phi\colon V \to \mathbb{C}^2$ be a smooth map of a neighborhood of the origin that is holomorphic in a sectorial domain $U \subset V$. Then its Taylor series in the variable x belongs to the class of semiformal maps $\mathscr{F}^2$. This series $\widehat{\Phi}$ is strictly asymptotic for Φ in the domain U.*

PROOF. The first assertion of the lemma follows from the uniform continuity (in the variable y) of the partial derivatives of any order in x and $\bar{x}$ at the points of $\{x = 0\}$. The second assertion is obtained easily by applying the Taylor formula with remainder in Lagrange form to the real and the imaginary parts of the components of Φ.

COROLLARY. *Without loss of generality, one can assume that the map Φ defined above has the following form:*

$$\Phi(x, y) = (x + o(x^2), y + o(x)) \quad \text{as } x \to 0. \tag{4.2}$$

Indeed, it suffices to replace Φ by the composition $\Phi_2^{-1} \circ \Phi$, where Φ_2 is the corresponding partial sum of the series $\widehat{\Phi}$ (it is holomorphic, since $\widehat{\Phi} \in \mathscr{F}^2$).

Relation (4.2) implies, in particular, that the domain $\mathring{W} = \Phi(\mathring{V})$ is a cut neighborhood of the origin in $\mathbb{C}^2$. Thus, the required domain $\mathring{W}$ and the biholomorphism $H = \Phi \circ G$ of the manifold S and the domain $\mathring{W}$ are constructed.

On S define a holomorphic map F_S that coincides with F_0 in the natural charts z_j (the map F_S is well defined, since the transition functions $z_j \circ z_{j+1}^{-1}$ commute with F_0). Let $\mathring{F} = H \circ F_0 \circ H^{-1}$; without loss of generality, we can assume $\mathring{F}$ to be defined in the entire domain $\mathring{W}$. The asymptotics obtained above for the maps $G_j \circ z_j^{-1}$ and $z_j \circ G_j^{-1}$ together with (4.2) enable us to extend $\mathring{F}$ to a map F that is holomorphic in the entire neighborhood $W = \Phi(V)$. The points of the line $W \setminus \mathring{W}$ are fixed points of F.

Since the identity map is asymptotic for both $G \circ z_j^{-1}$ and $z_j \circ G^{-1}$, the semiformal map $\hat{\Phi}$ defined in Lemma 11 conjugates the semiformal series $\hat{F}_0$ and $\hat{F}$ asymptotic to the maps F_0 and F, i.e., we have $F = \hat{\Phi} \circ F_0 \circ \hat{\Phi}^{-1}$. Now (4.2) implies $F(x, y) = (x + o(x^2), y + x + o(x))$, hence $F \in \mathscr{B}_1$.

It remains to observe that the holomorphic maps $H_j = z_j \circ (\Phi \circ G_j)^{-1}$ conjugate F with F_0 and possess a common asymptotic series $\tilde{\Phi}^{-1} \in \mathscr{F}^2$. Replacing the domains $\Phi \circ G_j(\widetilde{\Omega}_j)$ of the maps H_j by appropriate sectorial domains, and restricting H_j to these new domains, one obtains a normalizing atlas for F_0. The collection of transition functions for this atlas is equivalent to the initial collection $(\omega, \delta H)$ in the sense of Definition 2 in 4.3. This means that $m = m_F$. Therefore, the realization is proved.

4.6. Analytic dependence on parameters. The analyticity of functional invariants can be established by using the analyticity of the normalizing atlases constructed in 4.1 (the latter in turn follows from the uniform character of all the estimates of §§2 and 3 with respect to parameters). The inverse statement is easily obtainable from 4.5: one must add the t coordinate, $t \in \mathbb{C}^n$, assuming that all the maps preserve it.

In the following three subsections we shall consider another description of functional invariants of germs from $\mathscr{B}_1$.

4.7. Local cohomology of the quotient space. As before let

$$\pi_0 \colon (x, y) \mapsto (x, \exp(2\pi i y/x))$$

be the projection of the cut plane $\mathbb{C}_*^2 = \{(x, y) : x \neq 0\}$ onto the quotient space $\mathbb{C}_*^2/F_0$. The image of a cut neighborhood of the origin will be called the *quotient cut neighborhood*. The quotient analogs of sectorial domains, regular coverings, etc. are defined in a natural way and are denoted by a hat. This notation should not be confused with the one for the semiformal series; the difference will be always clear from the context.

Due to condition T-1, for each system of transition functions $\eta = (\omega, \delta H)$ we can define the π_0-*projection* $\pi_0(\eta) = (\pi_0(\omega), \pi_0(\delta H))$, where $\pi_0(\omega)$ is the covering of the quotient cut neighborhood by the images of the domains from ω, and the transition maps $\pi_{0*}(\delta H) = \{\widehat{H}_{ij}\}$ between the charts of the quotient covering are defined to satisfy the condition

$$\pi_0 \circ H_{ij} = \widehat{H}_{ij} \circ \pi_0$$

for all i, j. Other terms for this object which we shall sometimes use are the *quotient system of transition functions*, or *1-cocycle map*, or *gluings*. Refinements and narrowings for gluings are defined in the same manner as in 4.1, 4.2.

Let $\widehat{T}_1$ be the space of all the quotient systems of transition functions. From the definitions we obtain immediately the quotient analogs of conditions T-2 and T-3 of 4.2 (condition T-1 becomes trivial after factorization). We have the following lemma.

LEMMA 12. *The space $\widehat{T}_1$ consists of all pairs $(\widehat{\omega}, \delta\widehat{H})$ such that $\widehat{\omega} = \{\widehat{\Omega}_j\}_{j \in J}$ is a quotient regular covering and $\delta\widehat{H} = \{\widehat{H}_{ij}\}_{(i,j) \in J'(\widehat{\omega})}$ is a collection of maps satisfying the conditions*

$\widehat{T}$-2. $\widehat{H}_{ij}(x, t) = (x + o(x^n), t(1 + o(x^n)))$ as $x \to 0$, $(x, t) \in \widehat{\Omega}_{ij}$, *uniformly in t for all $n \in \mathbb{N}$.*

$\widehat{T}$-3. $\widehat{H}_{ij} \circ \widehat{H}_{ji} = \text{id}$ *on* $\widehat{\Omega}_{ij} = \widehat{\Omega}_i \cap \widehat{\Omega}_j$.

DEFINITION 1. The gluings are called *cohomologous* if for some refinements of their narrowings with transition maps $\{H_{ij}\}$, $\{\widetilde{H}_{ij}\}$ there exist a quotient sectorial covering $\{U_j\}$, a finite family of biholomorphic mappings $\widehat{G}_j: U_j \to \widehat{G}_j(U_j) \subset \mathbb{C}^2$ and formal power series $\varphi(x)$, $\lambda(x)$, $\varphi(0) = 0$, $\varphi'(0) \neq 0$, $\lambda(0) \neq 0$, such that

Ê-2. The formal map $\widehat{\Psi}: (x, t) \mapsto (\varphi(x), t\lambda(x))$ is strictly asymptotic for $\widehat{G}_j$ in $\widehat{\Omega}_j$: if the polynomials φ_n, λ_n are such that $\text{ord}(\varphi - \varphi_n) > n$, $\text{ord}(\lambda - \lambda_n) > n$, then $G_j(x, t) = (\varphi_n(x) + o(x^n), t(\lambda_n(x) + o(x^n)))$ uniformly in t as $x \to 0$, $(x, t) \in U_j$, for any $n \in \mathbb{N}$.

Ê-3. $\widehat{G}_i \circ \widehat{H}_{ij} = \widetilde{\widehat{H}}_{ij} \circ \widehat{G}_j$.

If in this definition $\arg \varphi'(0) = 0$, then the gluings are called *strictly cohomologous*.

If in this definition any map $\widehat{G}_j$ is t-linear:

$$G_j: (x, t) \mapsto (\varphi_j(x), t\lambda_j(x)),$$

then the gluings are called *linearly cohomologous*.

REMARK 1. It can be easily checked that the equivalence of two systems of transition functions holds if and only if their projections are cohomologous.

DEFINITION 2. The class of gluings cohomologous in the above sense is called the modulus. $\mathscr{H}^1$ is the set of moduli; $\widehat{m}_F$ is the modulus containing the projection of the system of transition functions m_F of the normalizing atlas for F.

Due to Remark 1 and Lemma 12, $\mathscr{B}_1$ Theorem (see 4.3) remains true if in its formulation the class $[T_1]$ is replaced by $\mathscr{H}^1$.

REMARK 2. The meaning of the modulus $\widehat{m}_F \in \mathscr{H}^1$ is the following: for any representation $(\widehat{\omega}, \delta\widehat{H})$ of $\widehat{m}_F$, there exists a cut neighborhood U of the origin such that the quotient space U/F is obtained by gluing together the domains of the quotient regular covering $\widehat{\omega}$ using the maps of the family $\delta\widehat{H}$ for identification.

4.8. Coverings of class $\mathscr{W}_4$. The subsequent construction will be used in §5 to produce nontrivial examples of functional moduli. Let $\mathscr{W}_4$ be the class of all sectorial coverings ω consisting exactly of four domains Ω_j, $j = 1, 2, 3, 4$, whose projections on the x-plane contain the sectors
$$V_j = \{x \in \mathbb{C} : |x| < \varepsilon, \ \pi(j-1)/2 - \delta < \arg x < \pi j/2 + \delta\},$$
the numbers $\varepsilon > 0$, $0 < \delta < \pi/4$ being so far undetermined. For coverings $\omega \in \mathscr{W}_4$, the set $J'(\omega)$ is independent of ω: $J'(\omega) = J'(\mathscr{W}_4)$ consists of all pairs (i, j), $i, j = 1, \ldots, 4$, of different parity. Let $\widehat{T}_1(\mathscr{W}_4)$ be the set of all quotient systems of transition functions $(\widehat{\omega}, \widehat{\Delta\lambda}) \in \widehat{T}_1$ such that $\widehat{\omega}$ is the projection $\pi_0(\omega)$ on the quotient space of a certain covering $\omega \in \mathscr{W}_4$. Denote by $\mathscr{H}^1(\mathscr{W}_4)$ the space of cohomology classes of systems of transition functions from $\widehat{T}_1(\mathscr{W}_4)$.

The sectorial normalization theorem (§3) permits us to construct the normalizing atlas for any sectorial covering consisting of sectorial domains with angles less than π and sufficiently small diameters. Moreover, one may assume that for all the charts H_j of this normalizing atlas the condition $H_j(x, 0) = (x, 0)$ holds.

Therefore the main theorem from 4.3 immediately implies the following assertion.

PROPOSITION. $\mathscr{H}^1 = \mathscr{H}^1(\mathscr{W}_4)$, *or, in other words,* each system of *transition functions is equivalent to a certain* system of *transition functions from the class* $\mathscr{H}^1(\mathscr{W}_4)$.

REMARK. It is clear that an analogous assertion is valid for any class of coverings $\mathscr{W}_3$ consisting of three domains with angles less than π. But for us it is more convenient to work with coverings consisting of an even number of domains.

§5. Nontriviality of the moduli space

As it can be seen from the definitions of §4, the space of functional invariants $[T_1]$ seems rather complicated; in particular, it is not evident whether there exist different (in the sense of Definition 1 in 4.3) functional invariants.

In this section we show that actually the homological equivalence of *systems of* transition functions is a rigid condition, so the class of functional invariants is rather rich. First we prove some auxiliary assertions.

5.1. Subharmonicity lemma. The following statement is a simple corollary of the Phragmen-Lindelöf principle.

LEMMA 13. *A function φ subharmonic in a domain*
$$U \supset U_0 = \{t \in \mathbb{C} : \operatorname{Re} t > 0\}$$
and satisfying $\varphi(t) \leq -|t|$ for $t \in U_0$, is equal to $-\infty$ at all points $t \in U$.

PROOF. Let $v(t) = \varphi(t) + c \operatorname{Re} t^\alpha$, where
$$c = c(\alpha) = [\alpha^\alpha (1-\alpha)^{1-\alpha} \cos(\pi\alpha/2)]^{-1}, \qquad 0 < \alpha < 1.$$
We can easily check that the function v is subharmonic and bounded from above on U_0, and on the boundary ∂U_0 its values do not exceed 1. Therefore by the Phragmen-Lindelöf principle [22] we have $v(t) \leq 1$ for all $t \in U_0$, hence $\varphi(t) \leq 1 - c(\alpha) \operatorname{Re} t^\alpha$ for $t \in U_0$. Passing to the limit as $\alpha \to 1 - 0$, we obtain $c(\alpha) \to +\infty$, and then $\varphi(t) \leq -\infty$ on U_0. But a subharmonic function φ such that $\varphi = -\infty$ on a set of positive measure is equal to $-\infty$ on its entire domain of definition [13]. □

5.2. Lemma on analytic continuation.

LEMMA 14. *Suppose that the domain $V \subset \mathbb{C} \setminus \{0\}$ contains the half-disk $\{\operatorname{Re} x > 0, |x| < \varepsilon\}$, and the domain $S \subset \mathbb{C}^2$ is of the form*
$$S = \{(x, t) \in \mathbb{C}^2 : x \in V, \ \exp(-c/|x|) < |t| < \exp(c/|x|)\}$$
with a certain $c > 0$. Then any function φ holomorphic in S can be analytically extended to the domain $V \times (\mathbb{C} \setminus \{0\})$. If in addition φ is bounded on S, then $\varphi(x, t) \equiv \varphi_0(x)$ for some $\varphi_0(x)$ holomorphic in V.

PROOF. The set S is a Hartogs domain [24], and its projection is an analyticity domain, since $V \subset \mathbb{C}^1$. Therefore by the Hartogs theorem [24], the holomorphic envelope of S has the form
$$\widehat{S} = \{(x, t) \in \mathbb{C}^2 : x \in V, \ \exp u(x) < |t| < \exp U(x)\},$$
where $u(x)$ is the best subharmonic minorant for $\rho(x) = -c/|x|$ and $U(x)$ is the best superharmonic majorant for $P(x) = c/|x|$ on V. After making an appropriate substitution of the form $z = A + Bx^{-1}$, we conclude from Lemma 13 that $u(x) = -\infty$, $U(x) = +\infty$. Therefore the function φ can be analytically extended to the domain $V \times (\mathbb{C} \setminus \{0\})$. Note that the set of values of such a continuation $\widehat{\varphi}$ on $\widehat{S}$ coincides with that for φ on S (see [24, Russian page 153]), therefore the boundedness of φ implies that of $\widehat{\varphi}$. Hence the function $\alpha_x(t) = \widehat{\varphi}(x, t)$ is constant by the Liouville theorem. □

COROLLARY 1. *Any function f analytic and bounded on the quotient sectorial domain of aperture greater than π is independent of t: $f(x, t) \equiv c(x)$.*

From this and Remark 2 in 2.4 we get immediately

COROLLARY 2. *Lemma* 7' *from* 2.4 *is true.*

COROLLARY 3. *If a map* $F: U \to \mathbb{C}^2$ *of the sectorial domain* U *of the quotient space has the form* $F(x, t) = (O(1), tO(1))$ *as* $x \to 0$, *uniformly in* t, $(x, t) \in U$, *then* $F(x, t) = (\varphi(x), t\lambda(x))$ *on* U *for certain analytic functions* $\varphi(x)$, $\lambda(x)$.

5.3. Simple and special gluings. The equivalence classes of cohomologous gluings (systems of transition functions) from $\widehat{T}_1$ and even from $\widehat{T}_1(\mathscr{W}_4)$ are too large to be simply described. Here we consider subsets of $\widehat{T}_1$ inside which these equivalence classes admit a simple description.

DEFINITION 1. We shall call a quotient system of transition functions $(\widehat{\omega}, \delta\widehat{H}) \in \widehat{T}_1(\mathscr{W}_4)$ *simple*, if all the maps $H_{ij} \in \delta\widehat{H}$ except $H_{23} = H_{32}^{-1}$ and $H_{14} = H_{41}^{-1}$ are the identity. The notation $s = s(h_+, h_-)$ will mean that s is a simple gluing with $H_{14} = h_+$, $H_{23} = h_-$.

The gluing $s = (h_+, h_-)$ is called *special*, if the maps $h_\pm$ are extendable onto any domain of the form

$$\{(x, t) \in \mathbb{C}^2 : |x| < \varepsilon, |\arg(\pm x)| < \pi/2 - \delta, \exp(-c/|x|) < |t| < \exp(c/|x|)\}$$

for all $\delta > 0$ and certain ε, c (depending on δ and sufficiently small) while preserving the asymptotics $\widehat{T}$-2, see 4.7.

Let $\widehat{T}_1^*(\mathscr{W}_4)$ be the space of all special gluings.

REMARK 1. If $(\widehat{\omega}, \delta\widehat{H})$ is a refinement of the narrowing of a simple gluing and the domains of the quotient covering $\widehat{\omega}$ have a cyclic numbering, then there are at most two maps among the $H_{j, j+1} \in \delta\widehat{H}$ that differ from the identity map.

LEMMA 15. *Strictly cohomologous simple gluings are linearly cohomologous. Cohomologous special gluings are linearly cohomologous.*

PROOF. Suppose that two simple gluings are cohomologous, and $\{H_{ij}\}$, $\{\widetilde{H}_{ij}\}$, $\{G_j\}$ are the corresponding collections of maps, as in Definition 1 in 4.7. In the setting of the lemma, each of the maps H_{ij}, $\widetilde{H}_{ij}$ is either the identity (see Remark 1), or admits an analytic continuation to a sectorial domain of a larger aperture. Therefore equalities $\widehat{E}$-3 of Definition 2 in 4.7, enable us to extend each of the "conjugating" maps G_j to a sectorial domain of aperture $> \pi$. The asymptotic property $\widehat{E}$-2 (see Definition 1 in 4.7) remains true for the extended maps G_j as well. By Corollary 3 in 5.2, all such maps are t-linear: $G_j : (x, t) \mapsto (\varphi_j(x), t\lambda_j(x))$. Hence the initial gluings are linearly cohomologous in the sense of Definition 1 in 4.7. □

REMARK 2. It can be easily checked that the two simple gluings $s(h_+, h_-)$ and $\widetilde{s}(\widetilde{h}_+, \widetilde{h}_-)$ are linearly cohomologous if and only if there exist some functions φ_j, λ_j analytic in the sectors

$$V_j = \{x \in \mathbb{C} : |x| < \varepsilon,\ \psi - \varepsilon + \pi(j-1) < \arg x < \psi + \varepsilon + \pi j\},$$
$$j = 1, 2,\ |\psi| < \pi/2,\ 0 < \varepsilon < \pi/2 - |\psi|,$$

and satisfying the following set of conditions (we denote by $V_\pm$ the intersections $V_1 \cap V_2 \cap \{\pm \operatorname{Re} x > 0\}$):

1. There exist formal series $\varphi(x)$, $\lambda(x)$ asymptotic respectively for φ_j and λ_j in V_j, $j = 1, 2$, subject to the conditions $\varphi(0) = 0$, $\varphi'(0) \neq 0$, $\lambda(0) \neq 0$;
2. The maps $g_j : (x, t) \mapsto (\varphi_j(x), t\lambda_j(x))$ are injective on $V_j \times \mathbb{C}$ and either

$$g_j(V_\pm) \subset \{\pm \operatorname{Re} x > 0\},\quad \tilde{h}_\pm \circ g_2 = g_1 \circ h_\pm \quad \text{on } V_\pm, \tag{5.1+}$$

or

$$g_j(V_\pm) \subset \{\mp \operatorname{Re} x > 0\},\quad \tilde{h}_\mp^{-1} \circ g_2 = g_1 \circ h_\pm \quad \text{on } V_\pm. \tag{5.1$-$}$$

5.4. Examples of nontrivial gluings. For any even function $p(x)$ holomorphic at the origin with $p(0) \neq 0$, we set

$$c_p(x) = \exp(-x^{-1})p(x),\qquad h^p_\pm(x, t) = (x,\ 1 + (t-1)(1 + c_p(\pm x)(t-1))^{-1}).$$

Finally let s_p be the gluing (h^p_+, h^p_-).

Then s_p is a special gluing.

LEMMA 16. *Two gluings s_p and $s_{\tilde{p}}$ are cohomologous if and only if $p \equiv \tilde{p}$.*

PROOF. By virtue of Lemma 15, the cohomological equivalence of two special gluings is the same as their linear cohomological equivalence. So one of the two conditions $(5.1\pm)$ must hold. Comparing the first components in $(5.1\pm)$, we conclude that $\varphi_1 = \varphi_2$ on $V_\pm$, therefore $\varphi_j = \varphi|_{V_j}$ for a certain φ holomorphic at the origin. Comparing the second components in $(5.1\pm)$, we conclude that either

$$c_p(\pm x) = c_{\tilde{p}}(\pm \varphi(x)),\quad x \in V_\pm \qquad (\text{and then } \lambda_1 = \lambda_2 \equiv 1),$$

or

$$c_p(\pm x) = -c_{\tilde{p}}(\pm \varphi(x)),\quad x \in V_\pm \qquad \left(\text{and then } \lambda_1 = \lambda_2^{-1} = \frac{-1 + c_p(\pm x)}{1 + c_p(\pm x)}\right).$$

But the second alternative is impossible. Indeed, from the second relation we conclude that the functions λ_1 and λ_2 can be extended analytically to a punctured neighborhood of the origin. But asymptotic series for these functions equal -1 identically, since the function c_p is flat. Hence $\lambda_1 \equiv \lambda_2 \equiv -1$, and $c_p \equiv 0$. This contradicts the assumption.

Since both p and $\tilde{p}$ are even, the relations $c_p(\pm x) = c_{\tilde{p}}(\pm\varphi(x))$ imply
$$\tilde{p}\circ\varphi(x)/p(x) = \exp(1/\varphi(x) - 1/x) = \exp(1/x - 1/\varphi(x)).$$
Therefore $\varphi(x) \equiv x$, and $p(x) \equiv \tilde{p}(x)$. □

REMARK 1. Let us stress that the gluings $s_p(h_+^p, h_-^p) = (h_+, h_-)$ satisfy the relations
$$h_\pm^p(x, 1) = (x, 1), \qquad h_+^{-1} = \widehat{I}_0 \circ h_- \circ \widehat{I}_0,$$
where $\widehat{I}_0\colon (x, t) \mapsto (-x, t^{-1})$. Moreover, if the function p is real analytic at the origin, then $\widehat{\sigma}_0 \circ h_\pm = h_\pm^{-1} \circ \widehat{\sigma}_0$, where $\widehat{\sigma}_0\colon (x, t) \mapsto (\overline{x}, \overline{t}^{-1})$.

REMARK 2. An even wider class of pairwise noncohomologous special gluings may be obtained by considering the gluings of the form
$$s(\varphi_c, \mathrm{id}), \qquad \varphi_c(x, t) = (x, c(x, t)),$$
where
$$c(x, t) = \sum_{k\in\mathbb{Z}} c_k(x) t^k$$
is the series with coefficients $c_k(x) = \exp(-k/x) p_k(x)$, p_k being holomorphic and uniformly bounded in a certain common neighborhood of the origin:
$$\exists U \subset \mathbb{C}\colon \quad |p_k(x)| < M \quad \forall x \subset U, \qquad p_0 = p_1 = 1.$$
Indeed, this representation implies that (5.1−) is impossible, while (5.1+) implies $\lambda_j = 1$, $\varphi_j \equiv x$, so homologous gluings must coincide.

§6. Germs of maps preserving additional structures

In this section we give the analytic classification for subsets of the germ class $\mathscr{B}_1$ that consist of germs preserving some additional structures.

6.1. The structures. We will consider the following four structures in $(\mathbb{C}^2, 0)$.

1. The distinguished curve: the germ of a holomorphic map
$$\gamma\colon (\mathbb{C}^1, 0) \to (\mathbb{C}^2, 0).$$

2. The antisymmetry: the germ of an antiholomorphic map
$$\sigma\colon (\mathbb{C}^2, 0) \to (\mathbb{C}^2, 0) \quad \text{such that} \quad \sigma\circ\sigma = \mathrm{id}.$$

3. The symmetry: the germ of a holomorphic map
$$I\colon (\mathbb{C}^2, 0) \to (\mathbb{C}^2, 0) \quad \text{such that} \quad I\circ I = \mathrm{id}.$$

4. The almost symplectic structure: the germ of a holomorphic closed 2-form on $(\mathbb{C}^2, 0)$.

A *structure* is any of the above four structures.

EXAMPLES OF STRUCTURES.
$$\gamma_0\colon x \mapsto (x, 0), \qquad \sigma_0\colon (x, y) \mapsto (\overline{x}, \overline{y}), \qquad I_0\colon (x, y) \mapsto (-x, y),$$
$$\alpha_0 = x\, dx \wedge dy.$$
These structures are called *standard*.

A structure will be denoted by s, a standard structure by s_0, so that $s_0 \in \{\gamma_0, \sigma_0, I_0, \alpha_0\}$.

DEFINITION 1. We say that the germ of a map $H: (\mathbb{C}^2, 0) \to (\mathbb{C}^2, 0)$ takes a structure $s \in \{\gamma, \sigma, I, \alpha\}$ into another structure $\tilde{s}$, if:

$$H \circ \gamma = \tilde{\gamma}, \quad H \circ \sigma = \tilde{\sigma} \circ H, \quad H \circ I = \tilde{I} \circ H, \quad \alpha = H^* \tilde{\alpha}.$$

DEFINITION 2. A structure s is a *deformation* of the standard structure s_0 if they differ by higher than third order terms, i.e.,

$$\gamma(x) - \gamma_0(x) = o(x^3),$$
$$\sigma(x, y) - \sigma_0(x, y) = o(x^3),$$
$$I(x, y) - I_0(x, y) = o(x^3),$$
$$\alpha = a(x, y)\,dx \wedge dy, \qquad a(x, y) = x + o(x^3),$$

as $x \to 0$.

The following assertion is evident.

LEMMA 17. *For any deformation s of the standard structure s_0, there exists a germ of a transformation $H(x, y) = \mathrm{id} + o(x^2)$ holomorphic at the origin and taking s to the standard structure s_0.*

REMARK. If the germ H from the lemma conjugates a germ $F \in \mathcal{B}_1$ with another germ $\tilde{F}$, then $\tilde{F} \in \mathcal{B}_1$.

6.2. Definitions and notation. For each of the four structures we give definitions analogous to those of §4. These definitions take into account the additional requirement of preserving the structures.

A germ $H: (\mathbb{C}^2, 0) \to (\mathbb{C}^2, 0)$ *preserves* a structure s, if it takes the structure into itself. The germ *agrees* with the structure "form, symmetry or antisymmetry" if it preserves the form, or swaps the orbits of the symmetry or antisymmetry according to the type of the structure. In particular, $H \in \mathcal{B}_1$ agrees with σ or I if it commutes with the antisymmetry or anticommutes with the symmetry:

$$H \circ \sigma = \sigma \circ H, \quad \text{or} \quad H \circ I = I \circ H^{-1}.$$

We will assume that any germ *agrees* with the structure "distinguished curve".

Let $\mathcal{B}_1^s$ be the subclass of germs of $\mathcal{B}_1$ that agree with the structure s (for example, $F_0 \in \mathcal{B}_1^{s_0}$ for any standard structure s_0). Two germs $F, G \in \mathcal{B}_1^s$ are *formally s-equivalent*, if there exists a germ of a formal transformation conjugating these germs and preserving the structure s.

A regular covering ω is said to be an *s-regular covering*, if it is *s-symmetric*: this means that for the symmetry/antisymmetry structure the image

$s(\Omega_j)$ of any domain $\Omega_j \in \omega$ must again be an element of ω (in the case of the curve/form structure no additional conditions are implied).

For an s-regular covering ω in the case when s is σ or I, the permutation $j \mapsto j_s$ arises if $s(\Omega_j) = \Omega_{j_s}$. Clearly $(j_s)_s = j$.

A system of maps $h = \{H_j\}_{j \in J(\omega)}$ or $\delta H = \{H_{ij}\}_{(i,j) \in J'(\omega)}$ associated with a regular covering ω *preserves the structure* s if

a) in the case of the curve/form structure, all the maps of the system preserve s, or

b) in the case of the symmetry/antisymmetry structure, ω is an s-regular covering, and

$$\forall j \in J(\omega), \quad s \circ H_j = H_{j_s} \circ s$$

or $\quad \forall (i,j) \in J'(\omega), \quad s \circ H_{ij} = H_{i_s j_s} \circ s.$

A normalizing atlas (a system of transition functions) is said to be an *s-normalizing atlas* (a *system of s-transition functions*), if it preserves s. Two systems of s-transition functions are *s-equivalent*, if they are equivalent in the sense of Definition 1 in 4.3, and the system $\{G_j\}$ from this definition preserves s.

Next, let $\pi_0\colon (x,y) \mapsto (x, \exp(2\pi i y/x))$ be the projection of $\mathbb{C}^2 \setminus \{x = 0\}$ onto the quotient space $S = (\mathbb{C}^2, 0)/F_0$. The projection of a structure s on S is the direct image $\hat{s}$ of s by π_0: this projection is well defined, provided that F_0 agrees with s.

EXAMPLES OF PROJECTIONS OF STANDARD STRUCTURES.

$$\hat{\gamma}_0\colon x \mapsto (x,1), \quad \hat{\sigma}_0\colon (x,t) \mapsto (\overline{x}, \overline{t}^{-1}), \quad \hat{I}_0\colon (x,t) \mapsto (-x, t^{-1}),$$
$$\hat{\alpha}_0 = x^2(2\pi i t)^{-1} dx \wedge dt.$$

In the same way as before, on the projection $\hat{s}$ of the structures we define s-*gluings* and s-*cohomologous* gluings. The spaces of all systems of s-transition functions and s-gluings will be denoted by T_1^s and $\hat{T}_1^s$ respectively, while the set of s-equivalence (s-homological equivalence) classes by $[T_1^s]$ (respectively, by $[\hat{T}_1^s]$). The latter notation will sometimes be replaced by $\mathscr{H}_1^s$. Note that the map taking a system of s-transition functions into its projection on S is a bijection between T_1^s and $\hat{T}_1^s$ mapping s-equivalence classes of systems of s-transition functions into s-homological equivalence classes of s-gluings.

6.3. s-normalizing atlases.

LEMMA 18. *For any $F \in \mathscr{B}_1^{s_0}$ there exists an s_0-normalizing atlas. The system of transition functions corresponding to any s_0-normalizing atlas, is the system of s_0-transition functions. Two systems of transition functions corresponding to two s_0-normalizing atlases for the germ F, are s_0-equivalent.*

PROOF. The second and the third assertions are evident. Let us prove the first. Note that the γ_0-normalizing atlas was in fact constructed in 4.1. For the case $s_0 = \sigma_0$ (or $s_0 = \gamma_0$) construct the normalizing atlas starting from a certain regular covering $\omega \in \mathscr{W}_4$, see 4.8 (this covering will actually be an s_0-regular covering). Then we construct normed normalizing maps H_j, $j = 1, 2$, and define $H_{j_{s_0}}$ by the relations $H_{j_{s_0}} = s_0 \circ H_j \circ s_0$. Since $F \in \mathscr{B}_1^{s_0}$, the maps $H_{j_{s_0}}$ will be the normalizing map for F. Uniqueness of the semiformal normed normalizing transformation implies that we have thus obtained an s_0-normalizing atlas. Finally let $F \in \mathscr{B}_1^{\alpha_0}$, and (ω, H) be a normalizing atlas for F. Let $H = \{H_j\}_{j \in J}$ and assume that H_j takes the structure α_0 into $\beta_j = b_j(x, y) \, dx \wedge dy = ((H_j)^{-1})^* \alpha_0$. Since H_j conjugates F and F_0, while F preserves α_0, we see that F_0 preserves β_j so that $b_j(x, y + x) = b_j(x, y)$. This in turn implies that the formal asymptotic series $\widehat{b}$ for b_j has the form $\widehat{b}(x, y) = \widehat{b}(x) = kx + \cdots$, $k \neq 0$. Without loss of generality we can assume that $k = 1$.

Let

$$\psi_j(x, y) = \int_0^y b_j(x, z) \, dz, \qquad d_j(x) = \int_0^x 3\psi_j(t, t) \, dt,$$

$$k_j(x) = (d_j(x))^{1/3},$$

where the branch of the cubic root is chosen to satisfy the condition $k_j(x) = x + o(x)$ as $x \to 0$. Let us introduce the maps

$$G_j(x, y) = (k_j(x), \psi_j(x, y)k_j(x)/\psi_j(x, x)).$$

It can be easily seen that G_j commutes with F_0 and $G_j^* \alpha_0 = \beta_j$, and all G_j have the same asymptotic germ. Therefore for a certain narrowing $\widetilde{\omega}$ of the regular covering ω the system $(\widetilde{\omega}, \widetilde{H})$ with $\widetilde{H} = \{G_j \circ H_j\}_{j \in J}$ is the required α_0-normalizing atlas. □

REMARK. Actually we proved the existence of a formal s_0-normalizing transformation for any $F \in \mathscr{B}_1^{s_0}$.

6.4. The main $\mathscr{B}_1^s$ Theorem. Note that Lemma 18 implies that for each standard structure s_0 there exists a well-defined map $\Pi^{s_0} \colon \mathscr{B}_1^{s_0} \to [T_1^{s_0}]$ taking the germ $F \in \mathscr{B}_1^{s_0}$ into the s_0-equivalence class of systems of s_0-transition functions corresponding to the s_0-normalizing atlas for the germ F.

$\mathscr{B}_1^{s_0}$ THEOREM. *For each of the standard structures the corresponding space $[T_1^{s_0}] = \mathscr{H}_{s_0}^1$ is the moduli space of analytic classification: all the assertions of $\mathscr{B}_1$ Theorem from §4 hold also for s_0-symmetric objects.*

PROOF. We prove only the realization part, since all the rest is demonstrated exactly as in §4. All the transition functions constitute the system of

s_0-transition functions, while the germ F_0 preserves the structure s_0. Therefore the map $F \in \mathscr{B}_1$ constructed in 4.5 preserves a certain structure s, while the maps G_j of the normalizing atlas for F take s into s_0. All the maps G_j may be put close to the identity due to (4.2), so s is the deformation of s_0. The assertion of the theorem comes from Lemma 17 and the subsequent Remark. □

6.5. Concerted structures and their combinations. The structures introduced in 6.1 are in a sense "elementary". They may be considered in pairs. We will call *concerted* the following pairs of elementary structures:

(1) any pair (γ, α);
(2) (γ, σ) if $\sigma \circ \gamma(x) = \gamma(\bar{x})$;
(3) (γ, I) if $I \circ \gamma(x) = \gamma(-x)$;
(4) (σ, I) if $I \circ \sigma = \sigma \circ I$;
(5) (σ, α) if the forms α and $\sigma^* \alpha$ are complex conjugate: $\bar{\alpha} = \sigma^* \alpha$;
(6) (I, α) if $I^* \alpha = \alpha$.

In particular, any two standard elementary structures are concerted.

From now on we shall call *structure* any collection of pairwise concerted elementary structures. The *standard structure* is the tuple of standard elementary structures. In the same way as in 6.1, 6.4, we define *deformations of* (nonelementary) *structures*, the notions of *preservation, agreement, equivalence*, etc. In short, all the constructions described earlier may be implemented with elementary structures replaced by nonelementary structures; we preserve in the notation the symbol s for the structure.

Both Lemmas 17, 18 and the $\mathscr{B}_1^s$ Theorem hold true in this generalized framework. The proof requires some minor modifications. Namely, when proving Lemma 17, one needs to put into standard form (to normalize) the elementary structures according to the following order: $\sigma, \gamma, I, \alpha$. The concerting conditions allow us to choose the new transformation in such a manner that the preceding elementary structures will be preserved. All the rest goes more or less *verbatim*.

6.6. Examples of nontrivial s-gluings. In this section we consider some subspaces of the space T_1^s. In particular, we show the nontriviality of the s-classification problems: the examples below provide vast classes of gluings that are pairwise not homologous. For instance, Example 1 shows nontriviality of the γ_0-classification, Example 2 that of the (γ_0, I_0)- and $(\gamma_0, \sigma_0, I_0)$-classifications, Example 3 that of the α_0-classification. Finally, Example 4 provides nontriviality for all the s-classifications such that $\alpha_0 \in s$. The detailed study of the examples requires calculations, which are omitted because of space limitations.

EXAMPLE 1. γ_0-cohomologous simple gluings coincide. This fact is implied by the first assertion of Lemma 15 and the following two simple remarks. If two gluings are γ_0-homologous, then they are strictly homologous.

Any t-linear map $G: (x, t) \mapsto (\varphi(x), t\lambda(x))$ that preserves $\widehat{\gamma}_0$ (that is, such that $\widehat{\gamma}_0 = \widehat{\gamma}_0 \circ G$) is the identity: $G \equiv \text{id}$.

Lemma 16 and Remark 1 in 5.4 yield

EXAMPLE 2. The gluings s_p from 5.4 are pairwise nonequivalent (γ_0, I_0)-gluings. If in addition the function p is real analytic then s_p are $(\gamma_0, \sigma_0, I_0)$-gluings, also pairwise nonequivalent.

LEMMA 19. *Let the functions $v = b(x, t)$ and $t = \varphi(x, v)$ be inverse to each other for any x, i.e., $b(x, \varphi(x, v)) \equiv v$. Assume that the function g satisfies the condition*

$$[g^3(x, v)]'_x = 3x^2 v \varphi'_v(x, v)/\varphi(x, v), \qquad (6.1)$$

and $a(x, t) = g(x, b(x, t))$, $\Phi(x, t) = (a(x, t), b(x, t))$. Then the map Φ preserves the projection $\widehat{\alpha}_0$ of the structure α_0.

PROOF. A simple computation. □

EXAMPLE 3. Let $p(x) = \exp(-1/x)d(x)$, where $d(x)$ is a function holomorphic in $(\mathbb{C}^1, 0)$ with $d(0) = d_0 \neq 0$. Denote

$$b(x, t) = 1 + \frac{t - 1}{1 + p(x)(t - 1)}, \qquad \varphi(x, v) = 1 + \frac{v - 1}{1 - p(x)(v - 1)},$$

and define g, a and Φ according to Lemma 19:

$$g(x, v) = \left[\int_0^x 3\tau^2(1 - p(\tau)(v - 1))(1 + p(\tau)(v^{-1} - 1)) \, d\tau\right]^{1/3},$$

where the integration is over the line segment, and the branch of the cubic root is chosen so that $g(x, v) = x + o(x)$ as $x \to 0$. Put

$$a(x, t) = g(x, b(x, t)), \qquad \Phi_p = (a, b).$$

Then

$$\Phi_p = \Phi_p(x, t)$$
$$= \left(\left[\int_0^x \frac{3\tau^2(t + p(x)(t - 1))(1 + p(x)(t - 1))}{(1 + (p(x) - p(\tau))(t - 1))(t + (p(x) - p(\tau))(t - 1))} d\tau\right]^{1/3},\right.$$
$$\left.\frac{t + p(x)(t - 1)}{1 + p(x)(t - 1)}\right)$$

and the gluing $\tau_p = s(\Phi_p, \text{id})$ is the special α_0-gluing. As before, the α_0-homological equivalence of a pair τ_p, $\tau_{\widetilde{p}}$ means that $p = \widetilde{p}$. This may be proved in the same way as in Lemma 16.

EXAMPLE 4. Let Φ_p be the map from the previous example, and

$$\psi_p^+ = \widehat{\sigma}_0 \circ \Phi_p^{-1} \circ \widehat{\sigma}_0 \circ \Phi_p, \qquad \psi_p^- = \widehat{I}_0 \circ (\psi_p^+)^{-1} \circ \widehat{I}_0, \qquad \tau_p^p = s(\psi_p^+, \psi_p^-).$$

Then τ_p^* is the special $(\gamma_0, \sigma_0, I_0, \alpha_0)$-gluing, and $\widehat{\alpha}_0$-homological equivalence of τ_p^* and $\tau_{\widetilde{p}}^*$ implies $p \equiv \widetilde{p}$.

6.7. Remarks and agreements.

REMARK 1. Since any deformation γ (in the sense of the definition from 6.1) of the structure γ_0 may be transformed into the standard structure (Lemma 17), the γ_0-classification of germs of class $\mathscr{B}_1$ is actually the classification of pairs (F, γ) with $F \in \mathscr{B}_1$ and γ being a deformation of γ_0 with respect to the following equivalence relation: $(F, \gamma) \sim (\widetilde{F}, \widetilde{\gamma})$ if the transformation conjugating F with $\widetilde{F}$ takes γ into $\widetilde{\gamma}$. Such a classification problem looks very natural, at least for the reason of the uniqueness of the conjugating transformation.

REMARK 2. As it was shown in Chapter I, 1.1, the classification problem for pairs of involutions is reduced to that of the analytic classification of I_0-symmetric germs of maps (ibid.) Similarly, the I_0-classification of germs from $\mathscr{B}_1^{I_0}$ is in fact the classification of pairs of involutions (I, J) formally equivalent to the pair (I_0, J_0):

$$I_0: (x, y) \mapsto (-x, y), \qquad J_0: (x, y) \mapsto (-x, y + x)$$

(the pairs are equivalent if the corresponding involutions are conjugated by the same holomorphism).

REMARK 3. The σ_0-classification of germs from $\mathscr{B}_1^{\sigma_0}$ is simply the classification of real analytic germs from $\mathscr{B}_1$ with respect to the group of real analytic local transformations. In the same way as above, we see that the (σ_0, I_0)-classification is in fact the classification of pairs of real analytic involutions.

REMARK 4. Speaking more generally, any s_0-classification of germs of class $\mathscr{B}_1^{s_0}$ is the classification of pairs

(germ of a map, structure)

formally equivalent to the pair (F_0, s_0). In particular, the $(\sigma_0, I_0, \alpha_0)$-classification of germs from $\mathscr{B}_1^{(\sigma_0, I_0, \alpha_0)}$ is the classification of all triples (I, J, α_0) formally equivalent to (I_0, J_0, α_0), where I, J are real analytic involutions preserving the real analytic 2-form α_0; equivalence of the triples means that there exists a real analytic coordinate change $(\mathbb{R}^2, 0) \to (\mathbb{R}^2, 0)$ taking one form into the other and conjugating the corresponding involutions.

REMARK 5. Note that the class $\mathscr{B}_1^{\sigma_0}$ consists exactly of the (complexifications of the) real analytic germs. Let $\mathscr{F} \in \mathscr{B}_1^{\sigma_0}$. We mention another way of finding a σ_0-normalizing atlas for $\mathscr{F}$. To do this, consider a σ_0-covering of a cut neighborhood of the origin by sectorial domains of small diameter and of aperture less than π such that it has two σ_0-invariant domains Ω_1, Ω_2: $\sigma_0(\Omega_j) = \Omega_j$, $j = 1, 2$. According to the third assertion of Theorem 3 in 3.1, find normalizing sectorial maps for $\mathscr{F}$ in these domains. Normalizing maps in all the other domains of the covering are defined in the same way as in Lemma 18 in 6.3. The σ_0-normalizing atlas obtained possesses the following property: the restriction of its charts to the real plane (extended by continuity to $x = 0$) is a smooth (real) diffeomorphism that conjugates $\mathscr{F}$ with its formal normal form F_0 (see Corollary in 3.1).

REMARK 6. A similar reasoning for germs of class $\mathscr{B}_1^{I_0, \sigma_0, \alpha_0}$ together with Remark 4 yields the following

COROLLARY. *Any triple* (I, J, α) *from Remark 4 is smoothly equivalent to the standard triple* (I_0, J_0, α_0).

This corollary is an important step in the proof of the smooth version of the Darboux-Whitney theorem.

§7. The spaces $\mathscr{B}_q$ and $\mathscr{B}_q^s$ (the general case)

7.1. The class $\mathscr{B}_q$. The $\mathscr{B}_q$ Theorem. Let $\mathscr{B}_q$ be the class of germs of holomorphic maps $F: (\mathbb{C}^2, 0) \to (\mathbb{C}^2, 0)$ such that

$$F(x, y) = (x + \varphi_1(x, y), y + \varphi_2(x, y)),$$
$$\varphi_1(x, y) = x^{q+1}\psi_1(x, y), \quad \varphi_2(x, y) = x^q \psi_2(x, y), \quad \psi_2(0, 0) \neq 0$$

for some ψ_i, $i = 1, 2$, holomorphic in $(\mathbb{C}^2, 0)$. Denote by F_q the map

$$F_q: (x, y) \mapsto (x, y + x^q)$$

so that F_1 now coincides with the formal normal form previously denoted by F_0.

In the construction of domains, atlases, etc. described in §§1–5 we must make the following modifications. The formal normal form F_0 must be replaced by F_q. The sectorial domains of aperture less (greater) than π turn into sectorial domain of aperture less (greater) than π/q. The maps which were affine, now take the form

$$(x, y) \mapsto (\varphi(x), y(\varphi(x)/x)^q + k(x)).$$

Instead of the projection $\pi_0: U \to U/F_0$, $\pi_0: (x, y) \mapsto (x, \exp(2\pi i y/x))$ we shall use the projection

$$\pi_q: (x, y) \mapsto (x, \exp(2\pi i y/x^q)).$$

The notations $\mathscr{B}_q$, T_q, $\widehat{T}_q$, $\mathscr{H}_q^1$, Π^q are the counterparts of $\mathscr{B}_1$, T_1, $\widehat{T}_1$, $\mathscr{H}_1^1 \stackrel{\text{def}}{=} \mathscr{H}^1$, Π^1.

PROPOSITION. *For the spaces $\mathscr{B}_q$, $q \geq 2$, the following theorems hold true*:
- *the formal classification theorem*;
- *the sectorial normalization theorem*;
- *the analytic classification theorem ($\mathscr{B}_q$ Theorem)*.

These theorems are proved exactly as for the case $q = 1$.

7.2. The class $\mathscr{B}_q$: examples. In §§4–6 we make the following modifications. Instead of the class $\mathscr{W}_4$, we introduce the class $\mathscr{W}_{4q}$ of regular coverings consisting of $4q$ sectorial domains

$$\Omega_j = \{|x| < \varepsilon, \ |y| < \varepsilon, \ (\pi/2q)(j-1) - \delta < \arg x < (\pi/2q)j + \delta\},$$
$$j = 1, \ldots, 4q, \ 0 < \delta < \pi/4q.$$

Thus the classes $\widehat{T}(\mathscr{W}_{4q})$ and $\mathscr{H}_q^1(\mathscr{W}_{4q})$ naturally appear. In the definition of simple and special gluings in 5.3, the maps H_{14}, H_{41}, H_{23}, H_{32} must be replaced by $H_{1,4q}$, $H_{4q,1}$, $H_{2q,2q+1}$, $H_{2q+1,2q}$ respectively, while the quotient sectorial domain must have the form

$$\{|x| < \varepsilon, \ |\arg(\pm x)| < \pi/2q - \delta, \ \exp(-c|x|^{-q}) < |t| < \exp(c|x|^{-q})\}.$$

The class of special gluings will be denoted by $\widehat{T}_q^*(\mathscr{W}_{4q})$. Replace the function $c_p(x)$ from 5.4 by the function $\exp(-x^{-q})p(x)$ and make the corresponding corrections in all the formulas from 5.4 containing the term $\exp(-1/x)$, substituting for it the expression $\exp(-x^{-q})$. Then all the corollaries of Lemmas 14–16, the Proposition from 4.8 and the remarks from 5.5 and 4.9 remain true. In particular, we obtain the nontriviality of the cohomology space $\mathscr{H}_q^1(\mathscr{W}_{4q})$.

7.3. The class $\mathscr{B}_q^s$. Instead of the standard almost symplectic structure $\alpha_0 = x\, dx \wedge dy$ from §6, we introduce the more general $\alpha_k = x^{k+1}\, dx \wedge dy$, $k \geqslant -1$. In Definition 2 of deformations of the standard structure in 6.1, we replace the condition $s - s_0 = o(x^3)$ by the following:

$$\begin{cases} s - s_0 = o(x^{q+2}) & \text{for } \gamma_0, \sigma_0, I_0, \\ \alpha - \alpha_k = b(x,y)\, dx \wedge dy, \ b(x,y) = o(x^{k+q+2}) & \text{for } \alpha_k. \end{cases}$$

The conditions imposed on the conjugating map H in Lemma 17 take the form $H(x,y) = \mathrm{id} + o(x^{q+1})$. This modifications made, the conclusion of the lemma will be valid also for the class $\mathscr{B}_q$.

The definition of concerted structures goes without changes. In particular, $F \in \mathscr{B}_q$ agrees with the structure I_0 if $F \circ I_0 = I_0 \circ F^{-1}$ for k odd and $F \circ I_0 = I_0 \circ F$ for k even; compare this with the Corollary in 6.2. The structures I_0 and α_k agree for even k.

Since we have replaced the projection π_0 by π_q, the expressions for the projections of the standard structures will also change:

$$\widehat{\alpha}_k = \frac{x^{k|q|1}}{2\pi i t} dx \wedge dt, \qquad \widehat{I}_0 : (x,t) \mapsto \begin{cases} (-x, t^{-1}) & \text{for } q \text{ even}, \\ (-x, t) & \text{for } q \text{ odd}. \end{cases}$$

PROPOSITION. *For the $\mathscr{B}_q^s$ space (the class $\mathscr{B}_q$ endowed with the structure s) the analogs of Lemmas 17, 18 and $\mathscr{B}_1^s$ Theorem remain true for both elementary and nonelementary structures, with any $k \geqslant -1$, $q \geqslant 1$.*

These assertions are proved exactly as in the case $q = 1$, $k = 0$.

Finally, let us replace the condition (6.1) from Lemma 19 by the condition

$$(\partial/\partial x)[\gamma(x, v)]^{k+q+2} = (k + q + 2)x^{k+q+1}v\varphi'_v(x, v)(\varphi(x, v))^{-1},$$

and replace the function $p(x) = \exp(-x^{-1})d(x)$ by the new function $p(x) = \exp(-x^{-q})d(x)$. Then all the assertions from 6.5, 6.6 also remain valid. Moreover, all the remarks from 6.7 hold for q odd.

7.4. Pairs of real analytic involutions preserving the almost symplectic structure. Here we formulate some particular cases of the assertions obtained earlier.

DEFINITION. Suppose that $\mathscr{J}$ is the space of all triples (I, J, α), where $I, J: (\mathbb{R}^2, 0) \to (\mathbb{R}^2, 0)$ are germs of real analytic involutions, and α is a 2-form with real analytic coefficients, preserved by both I and J: $I^*\alpha = J^*\alpha = \alpha$.

Two triples from the class $\mathscr{J}$ are analytically (formally, smoothly) equivalent if their constituents are componentwise conjugated by a certain analytic (respectively, formal, smooth) germ.

Let $\mathscr{J}_5$ be the subspace of triples formally equivalent to the standard triple

$$\tau_0 = (I_0, J_0, \alpha_-),$$
$$I_0: (x, y) \mapsto (-x, y), \quad J_0: (x, y) \mapsto (-x, y + x^5), \quad \alpha_- = x\,dx \wedge dy,$$

and let $\widetilde{\mathscr{J}_5} \subset \mathscr{J}_5$ stand for the subset of triples of the form (I_0, J, α).

$\mathscr{J}_5$ THEOREM. *The space* $[T_5^{\sigma_0, I_0, \alpha_-}] \cong \mathscr{H}^1_{5, \sigma_0, I_0, \alpha_-}$ *is the moduli space for the analytic classification problem on the classes* $\mathscr{J}_5$ *and* $\widetilde{\mathscr{J}_5}$. *All the triples from* $\mathscr{J}_5$ *are smoothly equivalent to the standard one.*

PROOF. The first assertion is implied by Proposition in 7.3, $\mathscr{B}_q^{\sigma_0, I_0, \alpha_-}$ Theorem and Remark 4 in 6.6. The second statement is implied by Corollary in 6.7 with $q = 5$. □

This theorem is used in 1.6 of Paper III.

REMARK. The moduli space appearing in the classification problem for the classes $\mathscr{B}_q^5$ and similar problems is rather rich, as can be seen from the examples given earlier: these examples are easily modified for the case $q > 1$, see 7.3.

CHAPTER III. REDUCTION OF THE DARBOUX-WHITNEY PROBLEM
TO THE SECOND CLASSIFICATION PROBLEM FOR INDICATORS

In this chapter we obtain different versions of the Darboux-Whitney theorem (see 2.9, Chapter I). All the computations are implemented in the real analytic case. But, since the only technical tools used here are the Weierstrass preparation theorem, the division theorem, and the implicit function theorem, then replacing the real analytic case by the complex or formal counterpart (in the complex or formal framework), or using the Malgrange preparation theorem (in the smooth category), we obtain simultaneously all the other versions of the Darboux-Whitney theorem.

The detailed scheme of the reduction of the Darboux-Whitney theorem to the second classification problem is presented in §1. The proofs of all the propositions of §1 can be found in §§2 and 3.

§1. The scheme of the reduction

1.1. Statement of the problem. Recall that the swallowtail is the hypersurface with singularities Γ consisting of all points $(A, B, C) \in \mathbb{R}^3$ such that the equation $x^4 + Ax^2 + Bx + C = 0$ has multiple roots (see Figure 4); the Cartesian product $\Lambda = \Gamma \times \mathbb{R} \subset \mathbb{R}^4$ is called the *extended swallowtail* in $\mathbb{R}^4$.

The germ of a diffeomorphism $H: (\mathbb{R}^4, 0) \to (\mathbb{R}^4, 0)$ is a Λ-*diffeomorphism*, if H takes the germ of Λ at the origin into itself.

The symplectic structure in $(\mathbb{R}^4, 0)$ is the germ of a nondegenerate closed 2-form with real analytic coefficients. Two symplectic structures are Λ-equivalent (smoothly, formally or analytically), if one of them can be taken into the other by a (smooth, analytic or formal) Λ-diffeomorphism.

Recall the formal Darboux-Whitney theorem.

FORMAL DARBOUX-WHITNEY THEOREM (Arnold [2, 3]). *A generic (=nondegenerate on the plane $B = C = 0$) symplectic structure is formally Λ-equivalent to the standard structure*

$$\omega_0 = dA \wedge dD + dC \wedge dB.$$

Our goal is to obtain smooth and analytic variants of this theorem.

1.2. The parametrized swallowtail $\tilde{\Lambda}$ and $\tilde{\Lambda}$-diffeomorphisms. Recall the definition of the *standard parametrization* of Λ from 2.10, Chapter I: this is the map $p: \mathbb{R}^3 \to \mathbb{R}^4$, $(t, x, v) \mapsto (A, B, C, D)$, given by

$$A = t, \quad B = -2tx - 4x^3, \quad C = 3x^4 - x^2 t, \quad D = v. \tag{1.1}$$

We always use coordinates (t, x, v) in the parameter space $\mathbb{R}^3$ and (A, B, C, D) in $\mathbb{R}^4 \supset \Lambda$. The Jacobian matrix p_* is singular at points of the hypersurface

$$\gamma_1 = \{(t, x, v) \in \mathbb{R}^3 : t^2 + 6x^2 = 0\},$$

and the null space of p_* at the points of γ_1 consists of the vectors collinear to the vector $\partial/\partial x$. The surface γ_1 and its p-image $p(\gamma_1) \subset \mathbb{R}^4$ for the embedding $i_1: \mathbb{R}^2 \to \gamma_1 \subset \mathbb{R}^3$, $i_1(x, v) = (-6x^2, x, v)$ will be called the *extended return edge* for Λ. The subbundle of the tangent bundle with the base γ_1 and the fiber $\ker p_*$ is the *isotropic bundle* (denoted also by $\ker p_*$).

The map p is not injective: the images of the points (t, x, v) and $(t, -x, v)$ on the surface

$$\gamma = \{(t, x, v) : t + 2x^2 = 0\}$$

coincide. The surface γ and its image $p(\gamma)$ together with the embedding $i: \mathbb{R}^2 \to \mathbb{R}^3$, $i(x, v) = (-2x^2, x, v)$ will be called the *extended curve of self-intersection* of Λ. The *symmetry of the self-intersection curve* γ is the map

$$I: \gamma \to \gamma, \qquad I(t, x, v) = (t, -x, v).$$

This map swaps preimages of points on the self-intersection curve $p(\gamma)$. The same does the map

$$I_0: \mathbb{R}^2 \to \mathbb{R}^2, \qquad I_0 = i^{-1} \circ I \circ i, \qquad I_0(x, v) = (-x, v).$$

The four objects constructed above (the isotropic bundle, the extended return edge, the self-intersection curve and the symmetry of the latter) will be called the *swallowtail-structures*. The space $\mathbb{R}^3$ endowed with these structures is the *parametrized extended swallowtail* $\widetilde{\Lambda}$.

DEFINITION 1. The $\widetilde{\Lambda}$-diffeomorphism is the germ of a transformation of $(\mathbb{R}^3, 0)$ preserving the germs of the swallowtail-structures. That is, the $\widetilde{\Lambda}$-diffeomorphism H must take the germs of γ and γ_1 into themselves, the restriction of H to γ must commute with the germ of the symmetry I and the tangent map H_* must take each fiber of the isotropic bundle $\ker p_*$ over a point $M \in \gamma_1$ into the fiber of $\ker p_*$ over the point $H(M) \in \gamma_1$.

DEFINITION 2. The *p-restriction* of a Λ-diffeomorphism $H: (\mathbb{R}^4, 0) \to (\mathbb{R}^4, 0)$ to $\widetilde{\Lambda}$ is the diffeomorphism $\widetilde{H}: (\mathbb{R}^3, 0) \to (\mathbb{R}^3, 0)$ satisfying

$$H \circ p = p \circ \widetilde{H}. \tag{1.2}$$

The initial diffeomorphism H is called the *p-extension* of $\widetilde{H}$ onto $(\mathbb{R}^4, 0)$.

REMARK. The letter p in the definition of *p*-restrictions and *p*-extensions points out a connection between (1.2) and the standard parametrization p. The existence of a *p*-restriction for a Λ-diffeomorphism is not evident, since the standard parametrization p is not injective.

THEOREM 1. *The p-restriction of any Λ-diffeomorphism always exists and is the $\widetilde{\Lambda}$-diffeomorphism. Any $\widetilde{\Lambda}$-diffeomorphism is the p-restriction of a certain Λ-diffeomorphism.*

The proof is postponed till 2.3.

1.3. $\widetilde{\Lambda}$-forms. Let δ be a differential k-form on $\mathbb{R}^4$ with coefficients analytic in $(\mathbb{R}^4, 0)$. Its *p-restriction* to $\widetilde{\Lambda}$ is the k-form $\widetilde{\delta} = p^*\delta$. The form δ is the *p-extension* of $\widetilde{\delta}$.

It can be easily seen that any p-restriction $\tilde{\delta}$ satisfies the following conditions:

(1) $\tilde{\delta} = 0$ on $\ker p_*$, that is, for any point $M \in \gamma_1$ and any vectors $\xi_1, \ldots, \xi_k \in T_M \mathbb{R}^3$ such that $\xi_1 \in \ker p_*(M)$, one has the following relation: $\tilde{\delta}(M)(\xi_1, \ldots, \xi_k) = 0$.
(2) $d\tilde{\delta} = 0$ on $\ker p_*$.
(3) The symmetry I_0 of the extended self-intersection curve preserves the restriction of the form $\tilde{\delta}$ to this curve.

DEFINITION. A differential k-form $\tilde{\delta}$ with coefficients analytic in $(\mathbb{R}^3, 0)$ that satisfies the above conditions is called a $\tilde{\Lambda}$-*form*. The first two conditions are called the *agreement conditions with the structure* $\ker p_*$, and the third the *agreement condition with the structure of the self-intersection curve*.

THEOREM 2 (on p-extension of $\tilde{\Lambda}$-forms). *For any $\tilde{\Lambda}$-form its p-extension exists.*

COROLLARY. *The differential of a $\tilde{\Lambda}$-form is again a $\tilde{\Lambda}$-form.*

However, this assertion could be derived directly from the definitions.

DEFINITION. A differential form is said to be $\tilde{\Lambda}$-*exact* if it is a differential of a $\tilde{\Lambda}$-form.

THEOREM 3 (the Poincaré lemma for $\tilde{\Lambda}$-forms). *Any closed $\tilde{\Lambda}$-form is $\tilde{\Lambda}$-exact.*

COROLLARY 1 (relative Poincaré lemma for $\tilde{\Lambda}$-forms). *Let a differential form δ be closed and its p-restriction to $\tilde{\Lambda}$ be equal to zero. Then δ is the differential of a form whose p-restriction to $\tilde{\Lambda}$ is zero:*

$$d\delta = 0, \ p^*\delta = 0 \implies \exists \alpha\colon \delta = d\alpha, \ p^*\alpha = 0.$$

REMARK. Corollary 1 is weaker that the analogous statement for smooth submanifolds [7], which claims that the form α vanishes on the entire tangent bundle and not only on the subbundle tangent to the submanifold. A similar statement for the swallowtail is not true.

COROLLARY 2 (on p-extensions of closed $\tilde{\Lambda}$-forms). *A closed $\tilde{\Lambda}$-form is the p-restriction to $\tilde{\Lambda}$ of a certain closed form.*

Theorems 2 and 3, as well as their corollaries, are proved in 2.4.

1.4. $\tilde{\Lambda}$-symplectic structures and their equivalence.

DEFINITION. Two $\tilde{\Lambda}$-forms are $\tilde{\Lambda}$-*equivalent* if one of them can be taken into the other by a $\tilde{\Lambda}$-diffeomorphism. The p-restrictions of generic (=nondegenerate on the plane $B = C = 0$) symplectic structures to $\tilde{\Lambda}$ are called $\tilde{\Lambda}$-*symplectic structures*.

AGREEMENT. In this and in the two following subsections $\tilde{\Lambda}$-equivalent means smoothly, real-analytically, or holomorphically equivalent.

THEOREM 4 (relative Darboux theorem for $\widetilde{\Lambda}$). *Generic symplectic structures are Λ-equivalent if and only if their p-restrictions to $\widetilde{\Lambda}$ are $\widetilde{\Lambda}$-equivalent.*

THEOREM 5. *A 2-form on $(\mathbb{R}^3, 0)$ is a $\widetilde{\Lambda}$-symplectic structure if and only if it is formally $\widetilde{\Lambda}$-equivalent to the p-restriction $p^*\omega_0$ of the standard symplectic structure $\omega_0 = dA \wedge dD + dC \wedge dB$ to $\widetilde{\Lambda}$.*

Theorems 4 and 5 are proved in 2.5.

1.5. The trace of a $\widetilde{\Lambda}$-symplectic structure on the extended self-intersection curve. The direction determined by a vector $v \in T_M\mathbb{R}^3$ is called *characteristic* (at the point M) for a representative $\widetilde{\omega}$ of the $\widetilde{\Lambda}$-symplectic structure, if $\widetilde{\omega}(M)(\xi, v) = 0$ for any $\xi \in T_M\mathbb{R}^3$. For example, the direction $\ker p_*$ is characteristic at any point of the extended self-intersection curve γ_1. The germ of this direction at the origin does not depend on the choice of the representative of the $\widetilde{\Lambda}$-symplectic structure. We call it the *characteristic direction field*. In the standard coordinates $(t, x, v) \in \mathbb{R}^3$ the germ of the characteristic direction field for the $\widetilde{\Lambda}$-symplectic structure

$$\widetilde{\omega} = P\, dt \wedge dx + Q\, dt \wedge dv + R\, dx \wedge dv$$

is generated by the germ of the analytic vector field

$$R\,\partial/\partial t - Q\,\partial/\partial x + P\,\partial/\partial v.$$

LEMMA 1. *For any representative ξ of the germ of the characteristic direction field for the $\widetilde{\Lambda}$-symplectic structure $\widetilde{\omega}$ one can find a neighborhood of the origin $U \subset \mathbb{R}^3$ with the following property.*

Any connected component of intersection of the integral curve of the characteristic direction field with this neighborhood, passing through a point M on the extended self-intersection curve γ, meets γ again at exactly one point (if $M \notin \gamma \cap \gamma_1$) or does not meet γ any more (if $M \in \gamma \cap \gamma_1$).

Let $\widetilde{J}_{\widetilde{\omega}}$ be the germ of the map swapping the intersection points of characteristic curves with γ as this is described in Lemma 1 (we put $\widetilde{J}_{\widetilde{\omega}} = \mathrm{id}$ on $\gamma \cap \gamma_1$). Clearly, this germ does not depend on the choice of the representative ξ.

DEFINITION. The germ $\widetilde{J}_{\widetilde{\omega}}$ is the *characteristic map* for the $\widetilde{\Lambda}$-symplectic structure $\widetilde{\omega}$. Let

$$i: \mathbb{R}^2 \to \gamma \subset \mathbb{R}^3, \qquad i(x, v) = (-2x^2, x, v)$$

be the parametrization of γ, $I_0: (x, v) \mapsto (-x, v)$ be the symmetry of γ, and $J_{\widetilde{\omega}} = i^{-1} \circ \widetilde{J}_{\widetilde{\omega}} \circ i$, $\widehat{\omega} = i^*\widetilde{\omega}$. The triple $(I_0, J_{\widetilde{\omega}}, \widehat{\omega})$ is called the *trace* of the $\widetilde{\Lambda}$-symplectic structure $\widetilde{\Lambda}$ on the extended self-intersection curve γ.

As before (see 7.4, Chapter II), let $\mathscr{J}$ be the space of all triples (I, J, α) such that α is the germ at the origin of a 2-form with real analytic coefficients, and I, $J: (\mathbb{R}^3, 0) \to (\mathbb{R}^3, 0)$ is a pair of real analytic involutions preserving α. Recall that two triples from $\mathscr{I}$ are *equivalent*, if one of them can be taken into the other by a local transformation (7.4, Chapter II). Finally, let $\widetilde{\mathscr{I}_5}$ be the space of triples from $\mathscr{I}$ having the form (I_0, J, α) and formally equivalent to (I_0, J_0, α_-), where $J_0: (x, v) \mapsto (-x, v+x^5)$, $\alpha_- = x\, dx \wedge dv$, see 7.4, Chapter II.

THEOREM 6. *The trace of a $\widetilde{\Lambda}$-symplectic structure belongs to the class $\widetilde{\mathscr{I}_5}$. Any triple from $\widetilde{\mathscr{I}_5}$ is the trace of a certain $\widetilde{\Lambda}$-symplectic structure.*

THEOREM 7. *Two $\widetilde{\Lambda}$-symplectic structures are equivalent if and only if their traces are equivalent.*

Theorems 6, 7 and Lemma 1 are proved in 3.2, 3.3 and 3.2, respectively.

Thus Theorems 4 through 7 reduce the Darboux-Whitney problem to the problem of classification of triples from the class $\widetilde{\mathscr{I}_5}$ already solved in §7, Chapter II.

Indeed, Theorem 4 reduces the Darboux-Whitney problem to the problem of $\widetilde{\Lambda}$-classification of $\widetilde{\Lambda}$-symplectic structures. Theorems 6 and 7 reduce the latter problem to that of classification of triples in $\widetilde{\mathscr{I}_5}$, which, in turn, is solved in 7.4, Chapter II, $\mathscr{I}_5$ Theorem.

The results obtained ($\Lambda_\mathbb{R}$ and $\Lambda_\mathbb{C}$ Theorems) are formulated in the two next subsections.

1.6. Real versions of the Darboux-Whitney theorem. Theorems 4 through 7 (respectively, the smooth or analytic variant, see the remark in the beginning of Chapter III) together with $\mathscr{I}_5$ Theorem imply the following statement.

$\Lambda_\mathbb{R}$ THEOREM. 1 (Smooth version). *All generic real analytic symplectic structures are smoothly Λ-equivalent.*

2 (Analytic version). *The space $\mathscr{H}_{5, \sigma_0, I_0, \alpha_-}$ is the moduli space for the problem of analytic classification of generic real analytic symplectic structures with respect to real analytic Λ-equivalence.*

REMARK 1. It can be easily verified (§§2, 3 and 7, Chapter II) that in the analytic case the modulus of a symplectic structure depends analytically on the parameters in the same sense as in §4, Chapter II.

REMARK 2. The remark after $\mathscr{I}_5$ Theorem, Chapter II, implies that the moduli space for symplectic structures is rather rich. In particular, we obtain the following consequence.

COROLLARY. *As a rule, the normalizing series in the formal version of the Darboux-Whitney theorem diverge.*

1.7. The complex case. From the holomorphic versions of Theorems 4 through 7 (with the class $\widetilde{\mathscr{I}_5}$ in Theorem 6 replaced by its complex counterpart), and Theorem $\mathscr{B}_5^s$ from 7.3, Chapter II, we derive the following.

$\Lambda_{\mathbb{C}}$ THEOREM (the holomorphic version of the Darboux-Whitney theorem). *The space $\mathscr{H}^1_{5, I_0, \alpha_-}$ is the moduli space for the problem of analytic classification of generic holomorphic symplectic structures with respect to holomorphisms preserving the extended swallowtail $\Lambda \subset \mathbb{C}^4$.*

REMARK. In the complex case all the remarks and corollaries from 1.6 also hold true.

§2. Inner geometry of $\widetilde{\Lambda}$ and theorems on continuation from $\widetilde{\Lambda}$

2.1. Rings and modules associated with $\widetilde{\Lambda}$. Let K_m be the ring of germs of functions real analytic in $(\mathbb{R}^m, 0)$, $m = 3$ or $m = 4$. The standard parametrization of the swallowtail $p: (t, x, v) \mapsto (A, B, C, D)$ given by (1.1) induces the ring homomorphism

$$j_1: K_4 \to K_3, \qquad j_1: f \mapsto \widetilde{f} = f \circ p.$$

Thus K_3 turns out to be a module and an algebra over the ring K_4. Define the homomorphisms $j_s: (K_4)^s \to K_3$ of the free modules $(K_4)^s$, $s = 2, 3$, to K_3 by the formulas

$$j_2(\beta, \gamma) = x\widetilde{\beta} + \widetilde{\gamma}, \qquad j_3(\alpha, \beta, \gamma) = x^2\widetilde{\alpha} + x\widetilde{\beta} + \widetilde{\gamma}.$$

Denote $\mathscr{R}_s = \ker j_s$ the submodules in $(K_4)^s$, $s = 1, 2, 3$.

EXAMPLES. 1. Let

$$e_1 = (2A, 3B, 4C), \qquad e_2 = (6B, 8C - 4A^2, -AB),$$
$$e_3 = (16C, -2AB, 8AC - 3B^2).$$

Then $e_1, e_2, e_3 \in \mathscr{R}_3$.

2. Let

$$Q_1 = 9B^2 - 8AC + 2A^3, \qquad Q_2 = 12BC + A^2B,$$
$$Q_3 = 16C^2 + 3AB^2/2 - 4A^2C, \qquad Q_4 = 9B^3/4 - 8ABC$$

and let

$$e_4 = (Q_1, Q_2), \qquad e_5 = (Q_2, Q_3), \qquad e_6 = (Q_3, Q_4),$$
$$e_7 = 3Be_4 + 2Ae_5 = (27B^3 + 8A^3B, 32C^2A - 8A^3C + 36B^2C + 6A^2B^2).$$

Then $e_4, \ldots, e_7 \in \mathscr{R}_2$.

3. Let e_0 be the discriminant of the polynomial $x^4 + Ax^2 + Bx + C$:

$$e_0 = 256C^3 - 128A^2C^2 + 144AB^2C + 16A^4C - 4A^2B^2 - 27B^4.$$

Then $e_0 \in \mathscr{R}_1$.

Moreover, in 2.2 we show that the following holds.

2.2. Main lemma and its corollaries. Note that the local algebra $K_4/\mathscr{R}_1$ of the map p at the point 0 is the algebra of truncated polynomials of degree ≤ 2 with the generators 1, x, x^2.

Applying the Weierstrass preparation theorem [5], we conclude that any element $\varkappa \in K_3$ can be represented as $\varkappa = ax^2 + bx + c$ with coefficients from the ring $j_1(K_4)$. In other words, $j_3(K_4^3) = K_3$, and K_3 is finitely generated over $j_1(K_4)$. However, this statement may be significantly strengthened.

LEMMA 2. *Any element $\varkappa \in K_3$ can be represented in the form*

$$\varkappa(t, x, v) = \beta_0(D)x^2 + x[C\beta_1(D) + B^2\beta_2(A, D) + B\beta_3(A, D) + \beta_4(A, D)] \\ + \beta_5(A, B, C, D), \tag{2.1}$$

where $(A, B, C, D) = p(t, x, v)$, $\beta_s \in K_4$, $s = 0, \ldots, 5$.

PROOF. Let us simplify successively the coefficients a and b in the representation $\varkappa = ax^2 + bx + c$, $a, b, c \in j_1(K_4)$. Let $a(A, B, C, D) = 2A\tilde{a}(A, B, C, D) + a_1(B, C, D)$. Since $2Ax^2 + 3Bx + 4C = 0$ (this relation is equivalent to $e_1 \in R_3$, see Example 1 in 2.1), one can obtain another representation: $\varkappa = a_1 x^2 + b_1 x + c_1$, where $b_1 = b - 3B\tilde{a}$, $c_1 = c - 4C\tilde{a}$. Now using the special form of the elements $e_2, e_3 \in R_3$ (ibid.) and acting in the same way, we obtain a new representation of $\varkappa$, and the leading coefficient of this representation has the prescribed form. To normalize the coefficient of the linear term, we must act in the same way, using the special form of the elements $e_4, \ldots, e_7 \in R_2$ (see Example 2 in 2.1) and the Weierstrass division theorem [12]. □

COROLLARY 1. *The equation $x^2\tilde{\alpha} + x\tilde{\beta} + \tilde{\delta} = \varkappa$ has the solution $(\alpha, \beta, \delta) \in K_4^3$ for any $\varkappa \in K_3$.*

COROLLARY 2. *The equation $x\tilde{\beta} + \tilde{\delta} = \varkappa$ admits the solution $(\beta, \delta) \in K_4^2$ for any $\varkappa \in K_3$ provided that $\varkappa''_{xx}(0, 0, v) \equiv 0$.*

Indeed, differentiating (2.1) w.r.t. x twice, we obtain $\beta_0 = 0$.

COROLLARY 3 (on continuation of functions from $\tilde{\Lambda}$). *The equation $\tilde{\delta} = \varkappa$ has the solution $\delta \in K_4$ for any $\varkappa \in K_3$, provided that*

(1) $\varkappa'_x = 0$ on γ_1, and
(2) $\varkappa$ is symmetric on γ in the following sense: $\varkappa \circ I = \varkappa$ on γ.

Indeed, the second condition implies that the coefficient β_4 in the representation (2.1) vanishes. Differentiating relation (2.1) w.r.t. x, we conclude that $\beta_0 = \cdots = \beta_3 = 0$.

COROLLARY 4. *The system of equations*

$$\varphi = x\tilde{\alpha} - \tilde{\beta}, \qquad \psi = \tilde{\delta} - 2x\tilde{\beta} + x^2\tilde{\alpha}$$

admits a solution $(\alpha, \beta, \gamma) \in K_4^3$ for any $\varphi, \psi \in K_3$ provided that
(1) $\psi'_x = 2\varphi$ on γ_1, and
(2) $\Delta = 2x\varphi - \psi$ is symmetric on γ: $\Delta \circ I = \Delta$ on γ.

PROOF. Comparing lower order terms, we conclude from the above two conditions that $\varphi''_{xx}(0, 0, v) = 0$ and find the solution (α, β) to the first equation by applying Corollary 2. But then the assumptions of Corollary 3 hold for the germ $\psi_1 = \psi + 2x\widetilde{\beta} - x^2\widetilde{\alpha}$, therefore $\psi_1 = \widetilde{\delta}$ for a certain $\delta \in K_4$. □

COROLLARY 5. *The system of the equations*

$$\varphi = (2t + 12x^2)(x\widetilde{\delta} - \widetilde{\beta}), \qquad \psi = \widetilde{\alpha} - 2x\widetilde{\beta} + x^2\widetilde{\delta}$$

admits a solution $(\alpha, \beta, \gamma) \in K_4^3$ for any $\varphi, \psi \in K_3$ provided that
(1) $\varphi = 0$ on γ_1,
(2) $\psi'_x = \varphi'_t$ on γ_1, and
(3) $\Delta = \varphi - 4x\psi$ is symmetric on γ in the same sense as above.

PROOF. This easily comes from the previous corollary after applying the preparation theorem to φ. □

2.3. Proof of Theorem 1. 1. Let H be a Λ-diffeomorphism. Then the homomorphism $H^*: f \mapsto f \circ H$ is an automorphism of the ring K_4 whose restriction to the ideal $\mathscr{R}_1 = \ker j_1$ (to the maximal ideal $m_4 \subset K_4$) is the automorphism of $\mathscr{R}_1$ (respectively, of m_4). Therefore a well-defined homomorphism h_0 of the local ring $K_0 = j_1(K_4) = K_4/\mathscr{R}_1$ appears; it satisfies

$$j_1 \circ H^* = h_0 \circ j_1. \qquad (2.2)$$

Note that the ring K_3 can be embedded into the field of fractions k_0 of the ring K_0, see Example 2 in 2.1. Therefore the homomorphism h_0 can be extended to a homomorphism $h: K_3 \to k_0$, and since K_0 is a subring of the ring K_3, one may regard the image $h(K_3)$ as belonging to the field of fractions k of the ring K_3.

On the other hand, all elements of K_3 are integers over K_0, hence so are the elements of $h(K_3) \subset k$. But the ring K_3 is factorial, hence integer closed [16]. Therefore $h(K_3) \subset K_3$. Applying the same arguments to H^{-1} we conclude that h is an automorphism of K_3. Evidently $h(m_3) \subseteq m_3$, where m_3 is the maximal ideal of the ring K_3.

Thus the automorphism h of the ring K_3 is determined by a certain local diffeomorphism $\widetilde{H}$ in $\mathbb{R}^3$: $h(\varphi) = \varphi \circ \widetilde{H}$ for all $\varphi \in K_3$. Then (2.2) implies (1.2), which in turn immediately implies the preservation of all the structures associated with the swallowtail by the map $\widetilde{H}$. Hence the p-restriction $\widetilde{H}$ of the diffeomorphism H exists and is a Λ-diffeomorphism.

2. Let the map $\widetilde{H}$ be a $\widetilde{\Lambda}$-diffeomorphism. Let $p \circ \widetilde{H} = (G_1, G_2, G_3, G_4)$. Preservation of all the structures associated with the swallowtail implies that Corollary 3 in 2.2 is applicable to G_j, hence $G_j = H_j \circ p$ for certain germs $H_j \in K_4$, $j = 1, \ldots, 4$. Therefore the condition (1.2) holds for the map $H = (H_1, \ldots, H_4)$. Comparing linear terms of maps $\widetilde{H}$ and H, we obtain invertibility of H. Finally, (1.2) means that H preserves Λ, and we conclude that H is the p-extension of $\widetilde{H}$ from $\widetilde{\Lambda}$ to $(\mathbb{R}^4, 0)$. □

2.4. Theorems on p-extension of $\widetilde{\Lambda}$-forms.

PROOF OF THEOREM 2. Let δ be a k-form on $(\mathbb{R}^3, 0)$, satisfying the conditions 1 through 3 in 1.3 (in other words, δ is a $\widetilde{\Lambda}$-form). Then the existence of the p-extension immediately follows from the following assertions stated in 2.2:

for $k = 0$, from Corollary 3;
for $k = 1$, from Corollaries 3 and 5;
for $k = 2$, from the preparation theorem and Corollaries 1 and 5;
for $k = 3$, from the preparation theorem and Corollary 1.

PROOF OF THEOREM 3. For a closed differential k-form δ which is a $\widetilde{\Lambda}$-form, we explicitly write the $\widetilde{\Lambda}$-form α such that $d\alpha = \delta$. Here and below we denote by $\int f \, dv$ the germ $F \in K_3$ such that $F(0) = 0$, $F'_v = f$. The answer can be expressed as

$k = 1:$ $\delta = a \, dt + b \, dx + c \, dv$, $\alpha = \int c \, dv$;

$k = 2:$ $\delta = q \, dt \wedge dx + r \, dt \wedge dv + s \, dx \wedge dv$,

$\alpha = a \, dt + b \, dx$, $a = -\int r \, dv$, $b = -\int s \, dv$;

$k = 3:$ $\delta = M \, dt \wedge dx \wedge dv$, $\alpha = q \, dt \wedge dx$, $q = \int M \, dv$.

The simple computation shows that if the conditions 1 through 3 from 1.3 hold for the form δ, then they hold also for α; therefore, α is a $\widetilde{\Lambda}$-form and δ is $\widetilde{\Lambda}$-exact. □

REMARK. Corollary 1 (the relative Poincaré lemma) immediately follows from the p-extension theorem (Theorem 2) and the Poincaré lemma for $\widetilde{\Lambda}$ (Theorem 3), while Corollary 2 (on the p-extension of closed $\widetilde{\Lambda}$-forms) follows from the p-extension theorem and Corollary 1. The proofs are standard (see [7]) and we omit them.

2.5. Theorems on $\widetilde{\Lambda}$-symplectic structures.

PROOF OF THEOREM 4. 1. The equivalence of p-restrictions of the general symplectic structures follows from the definition of p-restrictions and Theorem 1.

2. The equivalence of generic symplectic structures ω and ω_1 having the $\widetilde{\Lambda}$-equivalent p-restrictions $\widetilde{\omega} = p^* \omega$ and $\widetilde{\omega}_1 = p^* \widetilde{\omega}_1$ can be proved by using the homotopy method as in [7].

Let there be a $\tilde{\Lambda}$-diffeomorphism $\tilde{H}_1$, satisfying $\tilde{H}_1^*\tilde{\omega} = \tilde{\omega}_1$. By virtue of Theorem 1, $\tilde{H}_1$ is the p-restriction to $\tilde{\Lambda}$ of a certain Λ-diffeomorphism H_1. Denote $\omega_0 = H_1^*\omega$. Then ω_0 is the generic symplectic structure (equivalent to ω), whose p-restriction $p^*\omega_0$ coincides with $p^*\omega_1$. Moreover, one can assume that the 1-jets of the forms ω and ω_1 at the origin coincide as well (one can achieve this coincidence by a linear change of coordinates). Let Δ be the difference $\omega_1 - \omega_0$, and denote $\omega_t = \omega_0 + t\Delta$. The 2-form Δ is closed and its p-restriction to $\tilde{\Lambda}$ is equal to zero: $p^*\Delta = 0$. By virtue of the relative Poincaré lemma (Corollary 1), there exists a 1-form α such that $\Delta = d\alpha$, $p^*\alpha = 0$. The forms ω_t are nondegenerate at the origin for all $t \in [0, 1]$, and hence for any $t \in [0, 1]$ in a certain neighborhood of the origin there exist vector fields ξ_t such that $\omega_t(\xi_t, \cdot) + \alpha(\cdot) = 0$. Since $p^*\alpha = 0$, ξ_t are tangent to the swallowtail Λ at smooth (regular) points of Λ.

Denote g_0^t the flow map for the (nonautonomous) vector field ξ_t: the map g_0^t takes the point M into $M(t)$, where $M(\cdot)$ is the solution to the initial value problem $\dot{M} = \xi_t(M)$.

This flow map is the identity on $p(\gamma) \cup p(\gamma_1)$ by the uniqueness theorem, and preserves Λ, while conjugating ω_0 with ω_t, see [7]. Therefore all ω_t including ω_1 are Λ-equivalent to ω_0. □

PROOF OF THEOREM 5. This theorem immediately follows from the formal counterpart of Theorem 4 (see the Remark at the beginning of Chapter III) together with the formal Darboux-Whitney theorem and the following argument (already used while proving the previous theorem): any symplectic structure formally equivalent to the standard one, is nondegenerate on the plane $B = C = 0$. □

§3. Theorems on traces of $\tilde{\Lambda}$-symplectic structures

3.1. Seminormalizing transformations.

DEFINITION. A local transformation $H: (\mathbb{R}^3, 0) \to (\mathbb{R}^3, 0)$ preserving the extended self-intersection curve and the cuspidal edge, is called *seminormalizing* for a $\tilde{\Lambda}$-symplectic structure $\tilde{\omega}$, if $H^*\tilde{\omega} = \delta_0$, where $\delta_0 = dt \wedge dv$.

Clearly the seminormalizing transformation preserves the isotropic subbundle for $\tilde{\Lambda}$ and, in general, does not preserve the symmetry of the extended self-intersection curve, thus not being a $\tilde{\Lambda}$-diffeomorphism.

For a seminormalizing transformation H its restriction $h_H = i^{-1} \circ H \circ i$ to the extended self-intersection curve γ is well defined; here $i: (x, v) \mapsto (-2x^2, x, v)$ is the parametrization of γ. The involution $I_H = h_H^{-1} \circ I_0 \circ h_H$ is called the *deformation* of the standard involution $I_0: (x, v) \mapsto (-x, v)$.

EXAMPLE. For the restriction to $\tilde{\Lambda}$ of the standard symplectic structure $\omega_0 = dA \wedge dD + dC \wedge dB$, we have

$$p^*\omega_0 = \tilde{\omega}_0 = x^2(2t + 12x^2)\,dt \wedge dx + dt \wedge dv,$$

and the transformation

$$H: (t, x, v) \mapsto (t, x, v - (2/3)tx^3 - (12/5)x^5)$$

will be seminormalizing. The restriction h_H takes the form

$$h_H: (x, v) \mapsto (x, v - (32/15)x^5),$$

and the associated deformation of the standard symmetry and the characteristic involution are

$$I_H: (x, v) \mapsto (-x, v - (32/15)x^5), \qquad J_{\tilde{\omega}_0}: (x, v) \mapsto (-x, v + (32/15)x^5),$$

and finally the trace on $\tilde{\Lambda}$ is the triple $(I_0, J_{\tilde{\omega}_0}, \alpha_0)$, $\alpha_0 = -4x\, dx \wedge dv$.

REMARK. From this example and Theorem 5, it follows that a formal seminormalizing transformation exists for each $\tilde{\Lambda}$-symplectic structure.

LEMMA 3. *For any $\tilde{\Lambda}$-symplectic structure there exists an analytic seminormalizing transformation.*

PROOF. Denote by $\tilde{\omega}$ the $\tilde{\Lambda}$-symplectic structure.

1. *Normalization of the form $\tilde{\omega}$.* Let H_1 be the polynomial transformation obtained from the formal seminormalizing transformation (existing by virtue of the above Remark) by truncation of all terms of sufficiently large order. Then the characteristic direction field for the form $H_1^*\tilde{\omega}$ is close to the standard one defined by the vector field $v_0 = \partial/\partial x$. Hence the direction field can be "straightened" (put into standard form) by a diffeomorphism H_2 close to the identity. Therefore the characteristic direction field for the form $\omega_1 = (H_1 \circ H_2)^{-1}\tilde{\omega}$ is standard, and this form must be proportional to $dt \wedge dv$: $\omega_1 = k(t, x, v)\, dt \wedge dv$. Since ω_1 is closed, $k'_x \equiv 0$. Let $K = K(t, v)$ be the function satisfying $K'_v = k$, $K|_{v=0} = 0$, and denote by H_3 the map $(t, x, v) \mapsto (t, x, K(t, v))$. The map $\mathcal{H}_3 = H_1 \circ H_2^{-1} \circ H_3^{-1}$ normalizes $\tilde{\omega}$: $\mathcal{H}_3^*\tilde{\omega} = \delta_0$ and is close to the formal seminormalizing transformation H_0 (in particular, $\mathcal{H}_3$ is invertible).

2. *Normalization of the extended self-intersection curve.* By the shifts

$$(t, x, v) \mapsto (t, x - a(v), v), \qquad (t, x, v) \mapsto (t + b(v), x, v)$$

and the expansions

$$(t, x, v) \mapsto (t, \lambda(t, x, v)x, v)$$

preserving δ_0, we subsequently normalize

(1) the projection l of the set L of contacts of the characteristic direction field v_0 with the deformed self-intersection curve $\tilde{\gamma} = \mathcal{H}_3^{-1}(\gamma)$ onto the plane (x, v),
(2) the set of contacts L itself, and
(3) the entire curve $\tilde{\gamma}$.

We find the functions a, b, λ using the implicit function theorem applicable because $\mathscr{H}_3$ is close to a formal seminormalizing transformation.

3. *Normalization of the extended cuspidal edge γ_1.* Note that γ_1 is tangent with the characteristic direction field for the form $\widetilde{\omega}$ at the same points as γ is. Therefore the transformations constructed earlier take γ_1 into the curve $\widetilde{\gamma}_1$ which can be defined by the equation $t = -6x^2\psi_1(x, v)$, where the function ψ_1 is real analytic near the origin and equal to 1 for $x = 0$. Define the functions ψ_2, ψ_3, g analytic near the origin by the relations

$$\psi_2^2 = \psi_1, \quad \psi_2(0, v) = 1, \quad \psi_2 = 1 + x\psi_3(x, v), \quad g(x, v) = \frac{\psi_2(x, v)}{2 - 6\psi_1(x, v)}$$

and let H_4 be the map

$$H_4: (t, x, v) \mapsto (t, x + (t + 2x^2)g(x, v), v).$$

The map H_4 preserves δ_0 and γ, is invertible and takes $\widetilde{\gamma}_1$ into γ_1. The superposition of all the transformations yields the required seminormalizing transformation. □

3.2. Proofs of Lemma 1 and Theorem 6.

PROOF OF LEMMA 1. The seminormalizing transformation H of the $\widetilde{\Lambda}$-symplectic structure $\widetilde{\omega}$ takes the standard direction field $v_0 = \partial/\partial x$ into the characteristic direction field for the form $\widetilde{\omega}$. Next, H preserves the self-intersection curve and the line on which both the standard and the characteristic direction fields touch the self-intersection curve. Together with Lemma 3 this implies the required assertion. □

REMARK. In particular, we see that the restriction $h = H|_\gamma$ of the seminormalizing transformation for $\widetilde{\omega}$ conjugates the characteristic map $J_{\widetilde{\omega}}$ and the symmetry I_0 of the self-intersection curve:

$$h \circ I_0 = J_{\widetilde{\omega}} \circ h.$$

Moreover, the characteristic map I_0 of the standard form $\delta_0 = dt \wedge dv$ preserves the restriction $\alpha_0 = i^*\delta_0 = -4x\,dx \wedge dv$. Therefore the characteristic map $J_{\widetilde{\omega}}$ also preserves the restriction $\widehat{\omega} = i^*\widetilde{\omega}$: $J_{\widetilde{\omega}}^*\widehat{\omega} = \widehat{\omega}$.

PROOF OF THEOREM 6. 1. Note that the trace on γ of the restriction to $\widetilde{\Lambda}$ of the standard symplectic structure ω_0 belongs to the class $\widetilde{\mathscr{F}}_5$, see Example in 3.1 and the preceding Remark. Hence the trace of any $\widetilde{\Lambda}$-symplectic structure also belongs to $\widetilde{\mathscr{F}}_5$ (this follows from the formal analogs of Theorem 4 and Theorem 7 proved below).

2. Let the triple $\tau = (I_0, J, \alpha)$ be of class $\widetilde{\mathscr{F}}_5$. For the involution J find a local analytic transformation H_1 such that $H_1 \circ J = I_0 \circ H_1$. Denote $\alpha_1 = (H_1^{-1})^*\alpha = k(x, v)\,dx \wedge dv$. Since the involution J preserves the form α, the map I_0 preserves α_1, therefore the function $k(x, v)$ is odd in x. Define an analytic in $(\mathbb{R}^2, 0)$ germ $K(x, v)$ by the conditions

$$K(x, 0) = 0, \qquad K'_v = -\widetilde{k}/4,$$

where $\tilde{k}(x, v) = k(x, v)/x$. The function K is even in x, and from the definition of the class $\widetilde{\mathscr{I}_5}$ we get the invertibility of the germ

$$H_2 \colon (x, v) \mapsto (x, K(x, v))$$

analytic in $(\mathbb{R}^2, 0)$. Hence the map H_2^{-1} commuting with I_0 is well defined and

$$(H_2^{-1})^* \alpha_1 = \alpha_0 = -4x\, dx \wedge dv.$$

Let us show that the map $h = H_2 \circ H_1 = (h_1, h_2)$ is the restriction to the self-intersection curve of a certain seminormalizing transformation H. Indeed, from the definition of the class $\widetilde{\mathscr{I}_5}$ it follows that $J(x, v) = I_0(x, v) + o(x^4)$ as $x \to 0$. Comparing lower order terms in the relation $h \circ J = I_0 \circ h$, we conclude that

$$h_1(x, v) = xk_1(v) + x^3 k_3(v) + o(x^4) \quad \text{as } x \to 0,$$
$$h_2(x, v) = k_0(v) + x^2 k_2(v) + o(x^3) \quad \text{as } x \to 0$$

for certain functions k_j analytic at the origin, $j = 0, \ldots, 3$.

Using the preparation theorem [5, 12], one can find germs a, b, g, φ analytic in $(\mathbb{R}^2, 0)$ such that

$$h'_{2,x}(x, v) = -4xa(-6x^2, v) + 8x^4 b(-6x^2, v),$$
$$[h_1(x, v)]^2 = x^2 g(x, v), \qquad g'_x(x, v) = -8x\varphi(x, v).$$

Denote $\psi(t, x, v) = a(t, v) + x^3 b(t, v)$ and let

$$H(t, x, v) = \big(tg(x, v) + (t + 2x^2)(t + 6x^2)\varphi(x, v), h_1(x, v),$$
$$h_2(x, v) + (t + 2x^2)\psi(t, x, v)\big).$$

Invertibility of h implies invertibility of H in $(\mathbb{R}^3, 0)$. Next, H preserves both γ and γ_1, as well as the isotropic subbundle of the extended swallowtail. The restriction of H to γ is well defined and coincides with h. Therefore H is the seminormalizing transformation for the 2-form $\tilde{\omega} = (H^{-1})^* \delta_0$, and the initial triple τ is the trace of $\tilde{\omega}$ on γ. In particular, the symmetry of the extended self-intersection curve preserves $\tilde{\omega}$ (the third condition from 1.4).

Since H preserves the isotropic subbundle $\ker p_*$, $\tilde{\omega} = 0$ on $\ker p_*$. Moreover, $\tilde{\omega}$ is closed. Therefore $\tilde{\omega}$ is a closed 2-form on $\tilde{\Lambda}$, that is, a $\tilde{\Lambda}$-form, see 1.4. Finally, from the formal version of Theorem 7 below, we obtain the formal equivalence between $\tilde{\omega}$ and the restriction of the standard structure ω_0 to $\tilde{\Lambda}$. By Theorem 5, this means that $\tilde{\omega}$ is the $\tilde{\Lambda}$-symplectic structure. $\square$

3.3. Proof of Theorem 7.

1. Equivalence of the traces of two $\widetilde{\Lambda}$-equivalent $\widetilde{\Lambda}$-symplectic structures is evident.

2. Let us show that the equivalence of the traces τ and $\widetilde{\tau}$ of two $\widetilde{\Lambda}$-symplectic structures ω and $\widetilde{\omega}$ implies the $\widetilde{\Lambda}$-equivalence of these structures. Indeed, let H and $\widetilde{H}$ be the seminormalizing transformations for these two structures (they exist by Lemma 3), and I_H and $I_{\widetilde{H}}$ the corresponding deformations of symmetries of self-intersection curves. Denote as usual

$$\alpha_0 = -4x\, dx \wedge dv, \qquad \delta_0 = dt \wedge dv, \qquad I_0 \colon (x, v) \mapsto (-x, v).$$

Then the triples $\tau_1 = (I_H, I_0, \alpha_0)$ and $\widetilde{\tau}_1 = (I_{\widetilde{H}}, I_0, \alpha_0)$ are equivalent in $\mathscr{J}$ to the triples τ and $\widetilde{\tau}$ respectively, so all of them are equivalent to each other.

Let $h = (h_1, h_2)$ be the local transformation taking τ_1 into $\widetilde{\tau}_1$, in other words,

$$h \circ I_H = I_{\widetilde{H}}, \qquad h \circ I_0 = I_0 \circ h, \qquad h^* \alpha_0 = \alpha_0. \tag{3.1}$$

From the second of these relations, using the Preparation theorem, we get

$$h_1(x, v) = x a(-2x^2, v), \qquad h_2(x, v) = b(-2x^2, v)$$

for certain germs a, b of analytic functions. Define the map

$$H_0(t, x, v) = (t a^2(t, v), x a(t, v), b(t, v)).$$

This map preserves δ_0 by the third relation from (3.1). Finally let G be the superposition $H \circ H_0 \circ \widetilde{H}^{-1}$. From the definitions of H_0 and the seminormalizing transform, we conclude that $G^* \omega = \widetilde{\omega}$, while G preserves the extended self-intersection curve and the cuspidal edge. Moreover, G also preserves the symmetry of the extended self-intersection curve. G is invertible, since h is. All these properties mean that G is the required diffeomorphism realizing $\widetilde{\Lambda}$-equivalence of $\widetilde{\Lambda}$-symplectic structures ω and $\widetilde{\omega}$. □

References

1. V. I. Arnold, *Supplementary chapters in the theory of ordinary differential equations*, "Nauka", Moscow, 1978; English transl., *Geometrical methods in the theory of ordinary differential equations*, Springer-Verlag, Berlin and New York, 1988.

2. _____, *Singularities in variational calculus*, Itogi Nauki i Tekhniki: Sovremennye Problemy Matematiki: Noveishie Dostizheniya, vol. 22, VINITI, Moscow, 1983, pp. 3–55; English transl. in J. Soviet Math. **27** (1985), no. 3.

3. _____, *Singular Lagrangian varieties, asymptotic rays, and an open swallowtail*, Funktsional. Anal. i Prilozhen. **15** (1981), no. 4, 1–14; English transl. in Functional Anal. Appl. **15** (1981).

4. _____, *On the theory of envelopes*, Uspekhi Mat. Nauk **31** (1976), no. 3, 249. (Russian)

5. V. I. Arnold, A. N. Varchenko, and S. M. Gusein-Zade, *Singularities of differential maps*. I, "Nauka", Moscow, 1982; English transl., Birkhäuser, Basel, 1985.

6. V. I. Arnold, V. A. Vassiliev, V. V. Goryunov, and O. V. Lyashko, *Singularities. II. Classification and applications*, Itogi Nauki i Tekhniki: Sovremennye Problemy Mat.: Fundamental′nye Napravleniya, vol. 39, VINITI, Moscow, 1989, pp. 5–254; English transl. in Encyclopaedia of Math. Sci., vol. 39, Springer-Verlag, Berlin and New York (to appear).
7. V. I. Arnold and A. B. Givental′, *Symplectic geometry*, Itogi Nauki i Tekhniki: Sovremennye Problemy Mat.: Fundamental′nye Napravleniya, vol. 4, VINITI, Moscow, 1986, pp. 7–139; English transl. in Encyclopaedia of Math. Sci., vol. 4, Springer-Verlag, Berlin and New York, 1989.
8. V. I. Arnold and Yu. S. Il′yashenko, *Ordinary differential equations*, Itogi Nauki i Tekhniki: Sovremennye Problemy Mat.: Fundamental′nye Napravleniya, vol. 1, VINITI, Moscow, 1985, pp. 7–150; English transl. in Encyclopaedia of Math. Sci., vol. 1, Springer-Verlag, Berlin and New York, 1988.
9. G. D. Birkhoff, *On the periodic motions of dynamical systems*, Acta Math. **50** (1927), 359–379.
10. D. Cerveau and R. Moussu, *Groupes d'automorphismes de* $(\mathbb{C}, 0)$ *et équations différentielles* $ydy + \cdots = 0$, Bull. Soc. Math. France **116** (1988), 459–488.
11. J. Ecalle, *Sur les fonctions résurgentes*, I, II, Publ. Math. d'Orsay, Université de Paris-Sud, Orsay, 1981.
12. R. Gunning and H. Rossi, *Analytic functions of many complex variables*, Prentice Hall, Englewood Cliffs, N.J., 1965.
13. W. Heyman and P. Kennedy, *Subharmonic functions*, Academic Press, London and New York, 1976.
14. M. Kuczma, *Functional equations in a single variable*, PWN, Warsawa, 1968.
15. A. Kufner and S. Fucik, *Nonlinear differential equations*, Elsewier, Amsterdam and New York, 1980.
16. S. Lang, *Algebra*, Addison-Wesley, Reading, Mass., 1965.
17. J. Martinet, *Sur les singularités des formes différentielles*, Ann. Inst. Fourier (Grenoble) **20** (1970), no. 1, 95–178.
18. J. Martinet and J.-P. Ramis, *Classification analytique des équations différentielles non linéaires résonnantes du premier ordre*, Ann. Sci. École Norm. Sup. (4) **16** (1983), no. 4, 571–621.
19. _____, *Problème de modules pour des équations différentielles non linéaires du premier ordre*, Inst. Hautes Études Sci. Publ. Math. **55** (1982), 63–164.
20. R. Melrose, *Equivalence of glancing hypersurfaces*, Invent. Math. **37** (1976), 165–191.
21. _____, *Equivalence of glancing hypersurfaces. II*, Math. Ann. **255** (1981), no. 2, 159–198.
22. I. I. Privalov, *Subharmonic functions*, ONTI, Moscow and Leningrad, 1937. (Russian)
23. V. H. Reznick, *Darboux-Whitney theorem*, Dissertation, Chelyabinsk State University, 1988. (Russian)
24. B. V. Shabat, *Introduction to complex analysis. II. Functions in several complex variables*, "Nauka", Moscow, 1985; English transl., Amer. Math. Soc., Providence, R.I., 1992.
25. S. M. Voronin, *Analytic classification of pairs of involutions and its applications*, Funktsional. Anal. i Prilozhen. **16** (1982), no. 2, 21–29; English transl. in Functional Anal. Appl. **16** (1982).
26. M. Zhitomirskii, *Typical singularities of differential 1-forms and Pfaff functions*, Amer. Math. Soc., Providence, R.I., 1992.

Translated by A. VAĬNSHTEĬN

Nonlinear Stokes Phenomena in Smooth Classification Problems

YU. S. IL'YASHENKO AND S. YU. YAKOVENKO

§0. Introduction

The Stokes phenomenon is a global term describing effects arising mostly in classification problems in the complex domain. It consists in the fact that a complex dynamical system (say, a vector field) cannot be put in its formal normal form by an analytic transformation in an entire neighborhood of a singularity, though it is possible to find such a transformation in domains of a certain special shape. This transformation turns out to be almost unique, hence on the intersections of these domains the transition functions between the normalizing charts constitute a system of invariants.

Analogous effects may also occur in the case of real dynamical systems. By the latter term we mean either vector fields or diffeomorphisms on a real phase space. When speaking of singularities of dynamical systems, we mean either zeros of vector fields or fixed points of diffeomorphisms respectively. If the system is considered in an arbitrarily small neighborhood of a singularity, we use the term "local dynamical system".

Singularities of local dynamical systems are naturally ordered by their codimensions. The classification of singularities of codimensions 0 and 1 with respect to C^∞-smooth transformations is described in [AI], where the list of polynomial normal forms is given.

Another kind of problem arises when passing from individual singularities to their *deformations*, i.e., families of dynamical systems depending on a finite number of parameters which contain the given singularity as corresponding to a certain critical value (usually zero) of the parameters. To understand effects occurring in such deformations, one usually adopts the topological level of description of phase portraits and their structural changes. This topic is covered by the term "bifurcation theory" [AAIS].

Nevertheless it turns out that the analysis of topological properties of deformations of more complicated systems can often be simplified provided

---

1991 *Mathematics Subject Classification.* Primary 32G34, 34A20; Secondary 43C05.

that the smooth normal forms for some simple singularities and their deformations are at hand. The most natural way is to start by analyzing the case of the smallest codimension.

Let us proceed with the description of the hierarchy of codimensions. Suppose a local dynamical system is given. The linear terms of its Taylor expansion form the linearization matrix, whose set of eigenvalues will be referred to as the *spectrum of the singularity*. In the case of diffeomorphisms, the eigenvalues are also called *multipliers*.

The spectrum of a singularity of a vector field is said to be *hyperbolic*, if all the eigenvalues have nonzero real parts; respectively, the hyperbolicity of a local diffeomorphism means that there are no modulus 1 multipliers. The spectrum is said to be *nonresonant*, if there are no vanishing integer combinations between the eigenvalues (respectively, no equal to 1 monomial expressions composed of multipliers); the coefficients of these combinations and exponents of monomials are subject to well-known restrictions [A], which we do not recall here.

A generic (i.e., codimension 0) vector field at its singular point, is hyperbolic nonresonant, the same is true for diffeomorphisms. It was shown by Sternberg and Chen that such generic systems can be linearized by C^∞-smooth transformations, see [AI] and references there. Moreover, every deformation of a hyperbolic nonresonant singularity can be linearized for all sufficiently small values of the parameter in the C^k-category with k as large as we wish although always finite.

The following list includes all possible types of codimension one singularities (when defining these types we impose certain conditions on their eigenvalues, implicitly assuming that the remaining part of the spectrum is hyperbolic nonresonant and the nonlinear terms are generic):

(1) Hyperbolic singularities of vector fields and diffeomorphisms with a single resonance (this means that all the arithmetical identities between the eigenvalues or multipliers are consequences of a certain single identity).
(2) Vector fields having exactly one zero eigenvalue (the saddle-node case).
(3) Diffeomorphisms with exactly one modulus 1 multiplier (being real, it must equal either 1 or -1).
(4) Vector fields with a single pair of pure imaginary eigenvalues $\pm i\omega$, $\omega \neq 0$.
(5) Diffeomorphisms with a single pair of modulus 1 complex multipliers $e^{\pm i\varphi}$, where the angle φ is nontrivial: $\varphi \neq 0, \pi$.

It follows from the general theory that codimension 1 singularities are to be investigated together with their generic one-parameter families, otherwise some phenomena may be missed [A]. Therefore we proceed with a description of the state of the art in the classification theory for *generic one-parameter families* of vector fields and diffeomorphisms.

The first two cases were studied in [IY], where it was shown that although the C^∞-smooth classification of typical families is not possible, nevertheless the transformation to certain polynomial integrable normal forms can be achieved by C^k-substitutions, where k is finite but as large as we wish. This seems to be sufficient for most applications in bifurcation theory.

The fifth case apparently does not admit even a reasonable topological classification of deformations.

The main body of the present paper is devoted to the investigation and classification of generic one-parameter deformations of dynamical systems in the remaining two cases. Using the fundamental result by F. Takens [T1] on smooth saddle suspensions over a central manifold, one may without loss of generality restrict oneself to the case of lowest possible dimension, i.e., that of real line diffeomorphisms and planar vector fields. The general multidimensional case can be considered as a semidirect product of a low-dimension system and a system linear in the remaining variables. However, the transformation taking the initial system to such a form is only finitely differentiable. The exact order of differentiability depends on the arithmetical properties of the eigenvalues of the initial system and the size of the domain of the normalizing transformation.

Let us begin with some definitions. Denote by x the coordinate in the phase space $(\mathbb{R}^1, 0)$ and by $\varepsilon \in (\mathbb{R}^1, 0)$ the parameter.

DEFINITION 0. A *smooth local family of line diffeomorphisms* is the germ of a map $(\mathbb{R}^1, 0) \times (\mathbb{R}^1, 0) \to (\mathbb{R}^1, 0) \times (\mathbb{R}^1, 0)$ preserving the parameter: in the coordinates x, ε the local family is represented by the map

$$\mathbf{F}: (x, \varepsilon) \mapsto (F(x, \varepsilon), \varepsilon), \tag{0.1}$$

where $F: (\mathbb{R}^1, 0) \times (\mathbb{R}^1, 0) \to (\mathbb{R}^1, 0)$ is the germ of a smooth function.

Two families $\mathbf{F}, \widetilde{\mathbf{F}}$ of the form (0.1) will be called *equivalent*, or *conjugate*, if there exists the smooth germ of a transformation

$$\mathbf{H}: (x, \varepsilon) \mapsto (H(x, \varepsilon), \mu(\varepsilon))$$

fibered over the parameter axis and such that $\mathbf{F} \circ \mathbf{H} = \mathbf{H} \circ \widetilde{\mathbf{F}}$.

This definition differs from the one given in [AAIS] only by the smoothness requirement. Here and further on we shall use the adjective "smooth" as a synonym of "C^∞-differentiable". Note that all $H(\cdot, \varepsilon)$ must be defined in some common neighborhood of the origin in the phase space, but in general $H(0, \varepsilon) \neq 0$ for $\varepsilon \neq 0$. The morphism $\mathbf{H}$ is called the *conjugacy* between the two families.

Analogous notions in the case of vector fields are defined *mutatis mutandis*: the local family $\mathbf{V}$ of planar vector fields is a germ of a vector field in the Cartesian product of the phase space $(\mathbb{R}^2, 0)$ by the parameter space $(\mathbb{R}^1, 0)$, which is parallel to the phase 2-plane. The equivalence of two such families of fields means that one can be transformed into the other by a change of coordinates and parameter fibered over the parameter axis and subsequent multiplication by a smooth nonvanishing function. Using the complex variable z as a coordinate on the plane $\mathbb{R}^2 \simeq^{\mathbb{R}} \mathbb{C}$, one may write

$$\mathbf{V} = V(z, \varepsilon) \partial/\partial z + 0 \cdot \partial/\partial \varepsilon,$$

where $V(z, \varepsilon)$ is also complex-valued.

REMARK. We adopt the following agreement in our notation. When speaking about local families of diffeomorphisms of the line, we denote by boldface letters the corresponding maps of the (x, ε)-plane which are the identity in their second component (see (0.1)). Therefore only the first components are to be specified, which we do using ordinary italics.

Ambiguity in the notation may arise when we speak of families of vector fields on the real line. In that case we denote a family of vector fields on the line by the same symbol as a vector field on the (x, ε)-plane parallel to the ε-axis. Under these circumstances, the symbol g_v^t denoting the time t map for the flow of a vector field v may be interpreted either as a family of line diffeomorphisms, or as a two-dimensional map. Nevertheless, each time it will be clear from the context, what possibility we had in mind.

Denote by $\mathscr{E}$ the space of all smooth germs $(\mathbb{R}^1, 0) \to \mathbb{R}^1$. According to the list of singularities of codimension 1, we introduce the following subspaces of the space $\mathscr{E}$:

$$\mathscr{SN} = \{f \in \mathscr{E} : f(0) = 0, \ f'(0) = 1\},$$
$$\mathscr{F} = \{f \in \mathscr{E} : f(0) = 0, \ f'(0) = -1\}.$$

These are subspaces of germs tangent at the origin to the identity map and to the standard involution $x \mapsto -x$ respectively. Clearly, both are invariant under smooth transformations.

DEFINITION 1. The local family $\mathbf{F} = (F, \mathrm{id})$ of line diffeomorphisms is said to be *the saddle-node family* (in short, *SN-family*), if:

(1) the germ $f = F(\cdot, 0)$ belongs to the subspace $\mathscr{SN} \subset \mathscr{E}$ of germs tangent to the identity (the identity map $x \mapsto x$) at the origin;
(2) f has a fixed point of multiplicity 2 at the origin;
(3) the family $F(\cdot, \varepsilon)$ is transversal to $\mathscr{SN}$.

Choosing the appropriate coordinates, one may describe SN-families by the following set of conditions imposed on the family of maps F:

$$F(x, 0) = x + ax^2 + O(x^3), \qquad a > 0, \ \partial F/\partial \varepsilon(0, 0) < 0 \qquad (0.2)$$

(the given combination of signs may be obtained by direction reversal for x or ε).

The iteration square of any map $f \in \mathscr{E}$ having a fixed point at the origin with the multiplier equal to -1 is a map tangent to the identity. But such a square must have a fixed point at the origin with multiplicity no less than 3

(for generic maps the equality holds). An explicit computation shows that a germ of the form
$$x \mapsto -x + c_2 x^2 + c_3 x^3 + \cdots$$
after a quadratic substitution of the form $x \mapsto x + kx^2$ with an appropriate k is transformed into the map
$$x \mapsto f(x) = -x + ax^3 + \cdots. \tag{0.3}$$

DEFINITION 2. The local family $\mathbf{F} = (f, \text{id})$ of line diffeomorphisms is said to be a *flip family*, or F-family, if:
   (1) the germ $F(\,\cdot\,, 0)$ belongs to the subspace $\mathscr{F} \subset \mathscr{E}$ of germs tangent to the standard involution $x \mapsto -x$ at the origin;
   (2) the origin is a fixed point for $f = F(\,\cdot\,, 0)$, and $f \circ f$ has a triple fixed point there;
   (3) the family $F(\,\cdot\,, \varepsilon)$ is transversal to $\mathscr{F}$.

The second item in the list means that in (0.3) $a \neq 0$.

Together with the functional subsets $\mathscr{SN}, \mathscr{F} \subset \mathscr{E}$ we introduce yet another one, namely

$\mathscr{AH} = \{\text{planar vector fields with a pair of imaginary eigenvalues } \pm i\omega\}$.

This is a subspace of the space of germs of planar vector fields.

DEFINITION 3. The local family $\mathbf{V} = V(z, \varepsilon)\partial/\partial z + 0 \cdot \partial/\partial \varepsilon$ of planar vector fields is said to be an *Andronov-Hopf family* (in short, *AH-family*), if:
   (1) the germ $v = V(\,\cdot\,, 0) \cdot \partial/\partial z$ belongs to the subspace $\mathscr{AH}$;
   (2) the second focal value of the field v at the singularity is nonzero;
   (3) the family $v(\,\cdot\,, \varepsilon)$ is transversal to $\mathscr{AH}$.

In an appropriate complex coordinate z, normalizing the 2-jet, the AH-family $\mathbf{V}$ takes the form

$$V(z, \varepsilon) = z(i\omega(\varepsilon) + a(\varepsilon)z\bar{z} + O(|z|^4))\partial/\partial z, \qquad z \in \mathbb{C}, \tag{0.4}$$
$$\operatorname{Re} a(0) < 0, \ \operatorname{Im} \omega(0) = 0, \ \omega \neq 0, \ \partial(\operatorname{Im}\omega)/\partial\varepsilon(0, 0) < 0.$$

The transversality theorem implies that saddle-node, flip and Andronov-Hopf families are generic among 1-parameter ones.

Now we can state the main result in the smooth classification of these three types of families. Unlike the preceding hyperbolic case, it is impossible to obtain any smooth classification with a finite number of parameters in normal form: each time functional invariants appear. The nature of this phenomenon is exactly the same as in the analytic classification of (individual) saddle-node type complex line holomorphisms (see Paper I). Let us explain this in general terms.

The topological description of the above three types of local families is widely known and transparent. In each deformation, the nonhyperbolic singularity at the origin splits into at least two distinct hyperbolic invariant subsets. In the SN-case these are two fixed points, one of them being an attractor, the other—a repeller. In the flip case there is a period 2 hyperbolic cycle which is born at the fixed point, causing the latter to change its stability. Finally, in the AH-case a hyperbolic limit cycle is born from the steady state.

As the established theory of hyperbolic singularities claims, the system can be linearized in some small neighborhoods of such invariant subsets. Moreover, it is easy to show (we do it below) that the linearizing chart is uniquely defined (up to a "small" one-parameter group of linear transformations). These normalizing charts are uniquely extended to the whole basin of attraction (respectively, repulsion). On the other hand, in all three types of families, heteroclinic orbits with distinct α- and ω-limit sets occur. Hence the linearizing charts are defined on certain domains with nonempty intersections. Thus the normalizing atlas arises, since the above domains of attraction form a covering of the phase space. As explained in detail in Paper I, the transition functions constitute a natural system of functional invariants.

The above reasoning explains the nature of real Stokes phenomena. Nevertheless some difficulties arise. The most important of them is the following. Since we are interested in the classification of families rather than that of individual systems, it is necessary to find a normal form for a family of hyperbolic singularities that lose their hyperbolicity for the limiting value of the parameter. This is technically the most difficult part of the whole paper. The normal form is polynomial (in a natural sense) and integrable; it can be useful in itself for different applications (an example is given in the paper). The corresponding result is analogous to different sectorial normalization theorems scattered over Paper I. We postpone its proof until the last sections of the paper.

We conclude this introduction by the following remark on terminology. Let Ω be a (closed) subset of the real plane, which coincides with the closure of its interior.

DEFINITION 4. A real function $f\colon \Omega \to \mathbb{R}$ is called *smooth on* Ω, if it is smooth on $\operatorname{int}\Omega$ and all its derivatives admit continuous extensions to Ω.

The Whitney continuation theorem implies that if the boundary of Ω consists of analytic curves, then any function smooth in the above sense can be represented as the restriction to Ω of a certain function defined in some open neighborhood of Ω and smooth on it in the usual sense.

PART I. CLASSIFICATION THEOREMS FOR LOCAL FAMILIES

§1. Preliminary normal forms of local families

In the above definitions of equivalence we assumed that the reparameterization of families is possible. But the subgroup of reparameterizations is small and trivial in the group of all the fibered conjugacies. Our goal in the first stage is to provide the so-called *preliminary normal form* such that for any two families already in this form any conjugation between them (if it exists) must preserve the parameter.

1.1. Preliminary normal form for saddle-node families. First consider the most important example of an SN-family. Let

$$v(x, \tilde{\varepsilon}) = (x^2 - \tilde{\varepsilon})(1 + a(\tilde{\varepsilon})x)^{-1} \partial/\partial x \qquad (1.1)$$

be the *standard family* of polynomial vector fields on the line. From the results of [IY] it follows that any smooth deformation of the germ $v(x) = (x^2 + \cdots)\partial/\partial x$ may be put into such a form by a smooth coordinate transformation (0.2) (this is also true in the analytical category, see [K]). One can easily check that the time 1 map F_0 for the family (1.1) is indeed a saddle-node family of diffeomorphisms. The fixed points of F_0 belong to the parabola $\{x^2 - \tilde{\varepsilon} = 0\}$.

Denote the eigenvalues of the family (1.1) by

$$\lambda_{\pm}(\tilde{\varepsilon}) = \pm 2\sqrt{\tilde{\varepsilon}}(1 \pm a(\tilde{\varepsilon})\sqrt{\tilde{\varepsilon}})^{-1},$$

so that the multipliers $\mu_{\pm}(\tilde{\varepsilon})$ of the standard family F_0 are equal to $\mu_{\pm} = \exp \lambda_{\pm}$. This equality allows us to express $\tilde{\varepsilon}$ and $a(\tilde{\varepsilon})$ in terms of $\lambda_{\pm} = \ln \mu_{\pm}$ explicitly:

$$\tilde{\varepsilon} = (\lambda_+^{-1} - \lambda_-^{-1})^{-2}, \qquad (1.2)$$

$$a(\tilde{\varepsilon}) = \lambda_+^{-1} + \lambda_-^{-1}. \qquad (1.2')$$

These formulas provide a sort of *normalizing condition* relating the parameter of an SN-family to its multipliers. We can always reparametrize any such a family so that the condition (1.2) is satisfied.

DEFINITION 1A. An SN-family F is said to be in *preliminary normal form*, if

$$F(x, \varepsilon) = x + (x^2 - \varepsilon) f(x, \varepsilon), \qquad (1.3)$$

where f is a smooth nonvanishing function, and the fixed points $x_{\pm}(\varepsilon) = \pm\sqrt{\varepsilon}$ for $\varepsilon > 0$ have the multipliers $\mu_{\pm}(\varepsilon) = F_x'(\pm\sqrt{\varepsilon}, \varepsilon)$ related to the parameter ε of the family by formula (1.2), where $\lambda_{\pm} = \ln \mu_{\pm}$, $\tilde{\varepsilon} = \varepsilon$.

LEMMA 1A. (1) *Each local SN-family is conjugated with a certain family in preliminary normal form.*

(2) *Any conjugacy between two families in the preliminary normal form must preserve the positive values of the parameter ε.*

PROOF. The second assertion is a direct consequence of the invariance of the multipliers; so we need to prove only the first one.

1. For a given SN-family we first normalize its fixed points. Since the function $F(x, 0) - x$ has a double zero at the origin while $F_\varepsilon' \neq 0$, we conclude that the locus

$$\Gamma = \{F(x, \varepsilon) - x = 0\}$$

on the (x, ε)-plane is the graph of a function $\varepsilon = \phi(x)$ with $\phi(0) = \phi'(0) = 0$, $\phi''(0) \neq 0$. Applying the Morse lemma, we find a transformation of the x-axis which takes the function ϕ to the standard quadratic form $\phi(x) = x^2$. So in the new coordinates the fixed points of the family form the standard parabola $\{x^2 - \varepsilon = 0\}$.

2. Now we prove that the expression (1.2) can be used to introduce the new parameter of the family. We *set* $\tilde{\varepsilon}$ equal to the right-hand side of (1.2) where $\lambda_{\pm}$ are the logarithms of the multipliers of the fixed points $\pm\sqrt{\varepsilon}$. The problem is to show that the function $\varepsilon \mapsto \tilde{\varepsilon}$ is a smooth (and nondegenerate) reparametrization.

To prove this fact we write $F(x, \varepsilon)$ as $x + (x^2 - \varepsilon)f(x, \varepsilon)$ with a smooth and nonvanishing f; the last property follows from the definition of an SN-family. Split the smooth function $\ln(1 + 2xf(x, \varepsilon))$ into the sum of even and odd terms:

$$\ln(1 + 2xf(x, \varepsilon)) = \lambda_e(x^2, \varepsilon) + x\lambda_o(x^2, \varepsilon),$$

so that

$$\lambda_{\pm} = \lambda_e(\varepsilon, \varepsilon) \pm \sqrt{\varepsilon}\lambda_o(\varepsilon, \varepsilon).$$

Let

$$\tilde{\lambda}_e = \lambda_e(\varepsilon, \varepsilon), \qquad \tilde{\lambda}_o = \lambda_o(\varepsilon, \varepsilon).$$

The definition of the functions λ_e, λ_o and the inequality $f(0, 0) \neq 0$ imply that

$$\lambda_e(0, 0) = 0, \qquad \lambda_o(0, 0) \neq 0,$$

therefore $\tilde{\lambda}_e$ is divisible by ε. Simplifying expression (1.2) for $\tilde{\varepsilon}$, we obtain

$$\tilde{\varepsilon} = (\tilde{\lambda}_e^2 - \varepsilon\tilde{\lambda}_o^2)^2/(4\varepsilon\tilde{\lambda}_o^2),$$

which is a smooth function vanishing at the origin with nonzero derivative. Therefore it can be taken as the new parameter of the family, with (1.2) automatically satisfied.

3. The reparametrization procedure had ruined the previous normalization of the fixed points, but it can be regained by repeating the first step. Since the transformation normalizing it involves only the x-variable, the multipliers remain exactly the same, so that condition (1.2) is preserved.

COROLLARY. *For any SN-family in preliminary normal form, the expression* (1.3) *constructed from the multipliers of the fixed points of the family defines a smooth germ* (*in the sense of Definition* 4).

PROOF. It is sufficient to compute the right-hand side of (1.3) using the above splitting:

$$a(\varepsilon) = 2\tilde{\lambda}_e(\varepsilon)(\tilde{\lambda}_e^2(\varepsilon) - \varepsilon\tilde{\lambda}_o^2(\varepsilon))^{-1}.$$

The divisibility of $\tilde{\lambda}_e$ by ε and the condition $\tilde{\lambda}_o(0, 0) \neq 0$ guarantee the smoothness of the expression for $a(\varepsilon)$. $\square$

The above lemma restricts our investigation to the case of preliminary normal forms with a smaller group of transformations.

1.2. Preliminary normal form for flip families. As in the case of SN-families, we start with the most important example. Consider an odd family of vector fields

$$w(x, \varepsilon) = x(x^2 - \varepsilon)(1 + b(\varepsilon)x^2)\partial/\partial x, \qquad x \in (\mathbb{R}^1, 0), \ \varepsilon \in (\mathbb{R}^1, 0). \quad (1.4)$$

The standard involution $\sigma : x \mapsto -x$ preserves w, therefore commutes with the corresponding flow maps. Define a family

$$\mathbf{F} = (F, \text{id}), \qquad F = \sigma \circ g_w^{1/2}.$$

One can easily verify that $\mathbf{F}$ is indeed a flip family in the sense of Definition 2. By analogy with the SN-case, we shall call it the *standard flip family*, or the *formal normal form*. Its 2-periodic points $x = \pm\sqrt{\varepsilon}$ belong to the standard parabola Γ, while the origin is a fixed point for all ε. Explicit computation of the multipliers μ_0, μ_* of the 1- and 2-periodic points yields:

$$\mu_0 = -\exp(-\varepsilon), \qquad \mu_* = \exp(4\varepsilon(1 + b(\varepsilon)\varepsilon^2)). \quad (1.5)$$

(Recall that a multiplier of a T-periodic cycle for a map is by definition the product of the derivatives of the map over all T points constituting the cycle.)

These relations permit us to express both the local parameter ε of the family and the germ $b(\varepsilon)$ in invariant terms as functions of the multipliers. As in the case of SN-families, we introduce the notion of a preliminary normal form for flip.

DEFINITION 1B. A smooth flip family $\mathbf{F} = (F, \text{id})$ of line diffeomorphisms is said to be in *preliminary normal form*, if:

(1) the origin is a fixed point for $F(\cdot, \varepsilon)$ for all $\varepsilon \in (\mathbb{R}^1, 0)$;
(2) the points $x = \pm\sqrt{\varepsilon}$ constitute a 2-periodic cycle for $F(\cdot, \varepsilon)$ for all positive ε;
(3) the local parameter ε is connected to the multiplier μ_0 of the fixed point by the first relation from (1.5).

REMARK. A flip family in preliminary normal form can be written as (F, id), where F is a smooth family of functions of the form

$$F(x, \varepsilon) = -x + x(x^2 - \varepsilon)f(x, \varepsilon) \quad (1.6)$$

with $f(0, \varepsilon) \underset{\varepsilon}{\equiv} 1$.

LEMMA 1B. (1) *Any flip family can be put into preliminary normal form*;
(2) *Any smooth conjugacy between two families in the preliminary normal form must preserve the local parameter.*

PROOF. The second assertion is trivial since the multiplier μ_0 is invariant under smooth transformations, and the parameter ε is expressed via μ_0.

To prove the first assertion, one needs to normalize all fixed and periodic points of the family. It follows from the implicit function theorem that the fixed points of $F(\cdot, \varepsilon)$ lie on a smooth curve Λ passing through the origin and transversal to the x-axis. Using the same arguments as in the proof of Lemma 1A above, one can conclude that the 2-periodic points constitute a smooth curve Γ tangent to the x-axis with second order. Our goal is to find a smooth transformation fibered over ε which takes Γ into the standard parabola $\{x^2 - \varepsilon = 0\}$ and Λ into the ε-axis at the same time.

The curve Λ can be taken as the new parameter axis by virtue of the implicit function theorem. Afterwards we must normalize Γ while preserving Λ, and this is done by using the Morse lemma again: we find the transformation $x \mapsto \tilde{x}$ with $\tilde{x}(0) = 0$ taking Γ into the standard quadratic parabola. The latter condition guarantees that the $\tilde{\varepsilon}$-axis will be preserved. Thus the "trident" $\Lambda \cup \Gamma$ is taken to the standard one $\{x(x^2 - \varepsilon) = 0\}$.

The rest of the proof reproduces essentially that of Lemma 1A. We omit the details. □

COROLLARY. *Let* **F** *be a smooth flip family in preliminary normal form. Then the function* $b(\varepsilon)$ *defined by the system* (1.5), *where* μ_0, $\mu_\pm$ *stand for the multipliers of this family, is smooth for* $\varepsilon > 0$ *and admits a smooth continuation for negative values of the parameter* ε.

PROOF. The multipliers of the family **F** can be explicitly computed in terms of the smooth function f, see (1.6). The system (1.5) allows us to express b via these multipliers. The smoothness of the result can be seen from this expression if we decompose f into sum of its even and odd parts and take into account the fact that $f(0, 0) = 1$ for the family F in the corollary. □

The above results mean that there is a regular way to associate to every flip family **F** a certain standard flip family $\widetilde{\mathbf{F}} = \sigma \circ g_w^{1/2}$, where w is given by (1.4), the parameter ε and the function $b(\varepsilon)$ can be found from (1.5). Both families have the same 1- and 2-periodic points with coinciding multipliers. The smooth classification of flip families is based on the possibility of conjugation between **F** and $\widetilde{\mathbf{F}}$ in certain sector-like domains. This program is implemented in §5 below.

1.3. Remark on a homotopy method and smooth classification of families of vector fields on the line.
In the above sections we have associated to every SN- or flip family of diffeomorphisms a certain family of polynomial vector fields on the line. This family is determined by a single germ denoted by $a(\cdot)$ in the SN-case and by $b(\cdot)$ in the flip one. Recall that in both cases this germ is defined only for $\varepsilon > 0$. Let us show that the choice of a smooth continuation of the germ for negative values of ε makes no difference with respect to

the smooth classification. In order to prove this fact we use the so-called homotopy method, which seems to be useful in many different situations arising in smooth classification theory.

Consider a smooth local family $v = A(x, \varepsilon)\partial/\partial x$, $x \in (\mathbb{R}^1, 0)$, $\varepsilon \in (\mathbb{R}^k, 0)$, defined on a closed domain $S \subseteq (\mathbb{R}^1, 0) \times (\mathbb{R}^k, 0)$ (recall that this implies existence of derivatives inside S and their continuity on the boundary). We assume that S contains the origin, and $A(0, 0) = 0$.

THEOREM. *Let $w = B(x, \varepsilon)\partial/\partial x$ be another smooth family of vector fields on the line. Suppose that the function B is divisible by A^2 in S (so that the ratio is smooth on S). Then the family v is smoothly conjugated with $v + w$.*

REMARK. The divisibility condition means that w is in a sense small with respect to v: in particular, it has the same zeros as v. Moreover, for all hyperbolic singularities of v this condition means that $v + w$ has the same linear terms as v.

PROOF. Consider the Cartesian product of the domain S and the closed interval $I = [0, 1]$, t being the coordinate on the latter. On the product define a vector field parallel to the ε and t-axis as the linear homotopy:

$$W(x, t, \varepsilon) = v(x, \varepsilon) + tw(x, \varepsilon) = (A + tB)\partial/\partial x. \tag{1.7}$$

The flow maps of the field W preserve both foliations $\varepsilon = $ const and $t = $ const. Let T be another field on $S \times I$ which is parallel to the ε-axis and has the t-component identically equal to 1:

$$T(x, t, \varepsilon) = P(x, t, \varepsilon)\partial/\partial x + 1 \cdot \partial/\partial t.$$

Suppose that the field T commutes with W. Then the flow maps of T take the hyperplane $t = 0$ to hyperplanes of the form $t = $ const and conjugate the restrictions of W on these hyperplanes. In particular, the time 1 map transforms $v = W|_{t=0}$ into $v + w = W|_{t=1}$.

To find the field T with the desired commutation property, one must solve the equation

$$[T, W] = 0 \tag{1.8}$$

with respect to T in the class of smooth vector fields. Denote $A + tB$ by C. Then equation (1.8) yields

$$\frac{\partial P}{\partial x} C - P \frac{\partial C}{\partial x} = B. \tag{1.9}$$

This is a linear equation, and we shall seek the solution in the form $P = QC$. Equation (1.9) implies

$$\partial Q/\partial x = BC^{-2}.$$

From our assumptions it follows that $B = A^2 D$, so $C = A(1 + tAD)$ and finally one obtains the equation

$$\frac{\partial Q}{\partial x} = \frac{A^2 D}{A^2(1 + tAD)^2} = \frac{D}{(1 + tAD)^2}. \tag{1.10}$$

Since we had assumed that $A(0, 0) = 0$, the right-hand side of (1.10) is smooth on $S \times I$. Integrating it, one obtains Q and finally the field T which is clearly smooth on its domain. □

COROLLARY 1. *Let S be an entire neighborhood of the origin and suppose that v is a smooth family of vector fields on the line, $v|_{\varepsilon=0} = (x^p + \cdots)\partial/\partial x$. Then v is equivalent to a polynomial family of degree $\leq 2p - 1$.*

PROOF. Using the Weierstrass preparation theorem, one may write
$$v = A_p(x, \varepsilon)\widetilde{A}(x, \varepsilon)\partial/\partial x,$$
where A_p is a polynomial of degree p in x and $\widetilde{A}$ a smooth nonvanishing function. Applying the division theorem, one obtains $\widetilde{A} = A_{p-1} + A_p A^*$ with a polynomial A_{p-1} of degree $p - 1$ and a certain smooth A^*.

The above theorem now implies that v is conjugated to $A_p A_{p-1} \partial/\partial x$. □

COROLLARY 2. *Let v_1, v_2 be two families of the form*
$$v_i = (x^2 - \varepsilon)(1 + a_i(\varepsilon)x)^{-1}\partial/\partial x$$
with smooth functions a_i coinciding identically for $\varepsilon > 0$. Then these families are smoothly conjugate for all ε.

PROOF. The difference $v_1 - v_2$ is smooth and flat on the ε-axis, being identically zero for $\varepsilon \geq 0$. Since the function $(x^2 - \varepsilon)^{-1}$ is finite for $\varepsilon < 0$ and grows polynomially as $\varepsilon \to 0^-$, the ratio mentioned in the theorem is smooth and flat on $\varepsilon = 0$, hence the assertion. □

§2. Constructing functional invariants for saddle-node families

2.1. Embeddable families. The preceding section started with an example of an SN-family F_0 which is the time 1 map for the standard family of vector fields. Families of line diffeomorphisms which can be represented as time 1 maps will be called *embeddable*. Note that the real germ $a(\cdot)$ in (1.1) can be expressed as a function of the two multipliers of the family F_0, see (1.3); we replace $\widetilde{\varepsilon}$ by ε:

$$a(\varepsilon) = (\ln \mu_+(\varepsilon))^{-1} + (\ln \mu_-(\varepsilon))^{-1}, \qquad \varepsilon > 0. \tag{2.1}$$

By the corollary from 1.1, the same formula defines a smooth germ, which can be extended to the point $\varepsilon = 0$ if the $\mu_\pm$ are the multipliers of some SN-family in preliminary normal form.

So it is possible to associate to every SN-family **F** in preliminary normal form the family $\widetilde{\mathbf{F}}$ which is the time 1 map for the field v:

$$\widetilde{\mathbf{F}} = g_v^1(x, \varepsilon) = (F, \mathrm{id}), \qquad v = \frac{x^2 - \varepsilon}{1 + a(\varepsilon)x} \frac{\partial}{\partial x},$$
$$a(\varepsilon) = \frac{1}{\ln \mu_+} + \frac{1}{\ln \mu_-}; \qquad \mu_\pm(\varepsilon) = \left.\frac{\partial F}{\partial x}\right|_{x=\pm\sqrt{\varepsilon}}, \qquad \varepsilon > 0.$$
(2.2)

The field v will be called the *associated family*, while the family $\widetilde{\mathbf{F}}$ will be referred to as the *formal normal form* of the family $\mathbf{F}$ for reasons to be clarified later. Note that both the associated family and the formal normal form are defined by the germ $a(\cdot)$ given by (2.1) for $\varepsilon \geq 0$ and admitting a smooth extension to the entire neighborhood $\varepsilon \in (\mathbb{R}^1, 0)$. Sometimes by the formal normal form we shall mean this very germ. Recall that Corollary 2 in 1.3 implies that the choice of smooth continuation does not give rise to any difference between two formal normal forms with respect to the smooth equivalence.

The local families $\mathbf{F}$ and $\widetilde{\mathbf{F}}$ have the same fixed points for all values of parameter (for $\varepsilon < 0$ neither family has such points at all); the multipliers of these points coincide identically. One might hope that any SN-family is conjugated with its associated family, in which case the classification problem for SN-families would be reduced to that of vector fields (see above). In part this conjecture is true: indeed, these two families are conjugated *but only in sectors of special form in the (x, ε)-plane*. The precise formulation will be given below. Now we explain why the conjugation is in general impossible in the entire plane.

2.2. Hyperbolic local families of line diffeomorphisms. Note that both fixed points $x_\pm = \pm\sqrt{\varepsilon}$ of the preliminarily normalized SN-family for $\varepsilon > 0$ are hyperbolic: $|\mu_\pm(\varepsilon)| \neq 1$.

In the following lemma f should not to be confused with the same letter in (1.3).

LEMMA 2. *Let $f : (\mathbb{R}^1, 0) \to (\mathbb{R}^1, 0)$ be a smooth orientation-preserving hyperbolic germ: $\lambda = f'(0) > 0$, $\lambda \neq 1$. Then:*

(1) *In some neighborhood of the origin the representative of the germ is linearizable: there exists a smooth transformation $h : (\mathbb{R}^1, 0) \to (\mathbb{R}^1, 0)$ such that $h \circ f = \lambda h$.*

(2) *The above linearizing chart is unique up to linear transformations: if $\widetilde{h}$ is another linearizing chart, then $\widetilde{h} = ch$, $c \in \mathbb{R}$, $c \neq 0$.*

(3) *If the germ f depends smoothly on some parameters while remaining hyperbolic, then h may be chosen to depend smoothly on them.*

(4) *There exists a germ of a vector field $v = X(x)\partial/\partial x$, $X'(0) = \ln\lambda$, such that f is the time 1 map for v: $f = g_v^1$. This field will be called* the local generator *of the germ f.*

(5) *The local generator is invariantly associated with the germ with respect to C^1-smooth transformations: if there are two germs $f, \widetilde{f}$ conjugated by $h \in C^1$, that is, $f \circ h = h \circ \widetilde{f}$, and $v, \widetilde{v}$ denote the corresponding local generators, then $h_*\widetilde{v} = v \circ h$.*

(6) *Any C^1-smooth conjugacy between two C^∞-smooth hyperbolic germs is necessarily C^∞-smooth.*

(7) *The centralizer of the germ* f *(i.e., the set of all germs commuting with* f*) is one-dimensional and consists only of flow maps* g_v^t, $t \in \mathbb{R}$, *for the local generator* v.

PROOF. The basic fact is the existence of the smooth linearization. For the class C^k, $k < \infty$, the proof is presented in [IY]. The general smooth case gives rise to no additional difficulties.

All the remaining assertions can be deduced by simple explicit computations from the basic linearization principle. For example, the local generator for the linear germ $f\colon x \mapsto \lambda x$ is the linear vector field $v = \ln \lambda \cdot x\, \partial/\partial x$. Let us prove the uniqueness of the local generator. If $f\colon x \mapsto \lambda x$ is a linear map and $v = X(x)\partial/\partial x$ is its local generator, then v is preserved by f: $f_* v = v$, that is,

$$X(\lambda x) = \lambda X(x). \qquad (*)$$

Since $X \in C^1$, and $X(0) = 0$, by the Hadamard lemma, $X(x) = xb(x)$, $b(\cdot) \in C^0$. Therefore $b(\lambda x) = b(x)$. Since $\lambda \neq 1$, we have $\lambda^n \to 0$ as $n \to \infty$ or $n \to -\infty$, and from the continuity of b at the origin it follows that $b = \text{const}$, so the generator must be a linear field. But there is only one *linear* generator, hence the assertion follows. The same computation proves the statements concerning the centralizer and the uniqueness of the linearizing chart. The remaining statements are proved in a similar way, because all of them are reduced to certain statements concerning the equation $(*)$, which has only linear solutions in the class C^1. □

COROLLARY. *Let* f *be a smooth map of an interval* $I \subset \mathbb{R}$ *into itself having a unique hyperbolic fixed point on it. Then there exists a smooth chart* $h\colon I \to \mathbb{R}$ *linearizing* f *globally (i.e., on the entire interval):* $h \circ f = \lambda h$, $\lambda \in \mathbb{R} \setminus \{0, \pm 1\}$. *This chart is unique up to linear transformations of the form* $h \mapsto ch$, $c \neq 0$.

Moreover, there exists a smooth vector field v *on* I *such that* f *is the time* 1 *map for* v. *This field is linearized by* h *and any other linearizing chart* $\tilde{h}$ *differs from* h *by the time* t *map of the field* v *for a certain* $t \in \mathbb{R}$: $\tilde{h} = h \circ g_v^t$.

PROOF. Assume for simplicity that $\lambda \in (0, 1)$ (the remaining cases are treated in a similar way). The existence of the linearizing chart h in a small neighborhood of the fixed point $a \in I$ is the claim of Lemma 2. Next, by iterating the equality $h \circ f = \lambda \cdot h$ one obtains on the domain of h the relation

$$\forall n \in \mathbb{Z}, \quad h \circ f^{[n]} = \lambda^n h \qquad (\dagger)$$

(we denote the iteration power by square brackets). Now note that the uniqueness of the singularity implies that the entire interval is the basin of attraction. So after a sufficient number of iterations each point enters into the neighborhood, hence the left-hand side of $(\dagger)$ is well defined. Using $(\dagger)$, we can also define its right-hand side, thus proving the existence of the global linearizing chart. The same reasoning proves uniqueness. Indeed, the germ of h at the fixed point uniquely determines the map on the entire interval: since all local linearizing charts differ only by scalar factors, the same is true for the global ones, because of (†). The second part of the corollary becomes evident if we choose v as the inverse image by h of the linear field $\ln \lambda \cdot x \, \partial/\partial x$ on $\mathbb{R}$, since the set of time t maps for a linear field is precisely the set of linear orientation-preserving maps of the line.

2.3. Normalizing maps and transition functions.
Now we apply these statements to SN-families in preliminary normal form. Let $\mathbf{F}$ be such a family. Denote by v the associated family of vector fields defined by (2.2). From the hyperbolicity of fixed points of both $\mathbf{F}$ and its formal normal form $\widetilde{\mathbf{F}} = g_v^1$ and the coincidence of the corresponding multipliers in the domain $\varepsilon > 0$, it follows that $\mathbf{F}$ and $\widetilde{\mathbf{F}}$ are smoothly conjugated in a certain neighborhood S_+ of the positive branch $\Gamma_+ = \{x = +\sqrt{\varepsilon}, \, \varepsilon > 0\}$ of the parabola Γ: there exists a transformation $\mathbf{H}_+$ defined on S_+ of the form $(x, \varepsilon) \mapsto (H_+(x, \varepsilon), \varepsilon)$ such that $\mathbf{H}_+ \circ \mathbf{F} = \widetilde{\mathbf{F}} \circ \mathbf{H}_+$ in $S_+ \cap \mathbf{F}^{-1}(S_+)$. Since the fixed point $x = +\sqrt{\varepsilon}$ is unstable, its basin of repulsion includes the interval $(-\sqrt{\varepsilon}, \sqrt{\varepsilon})$ between the fixed points. By the corollary to Lemma 2, the map $H(\cdot, \varepsilon)$ can be uniquely extended to this interval. In a similar way the map $\mathbf{H}_- = (H_-, \mathrm{id})$ can be defined in a certain neighborhood S_- of the set $\Gamma_- = \{x = -\sqrt{\varepsilon}, \, \varepsilon > 0\}$ and then extended to the domain $D = \{\varepsilon > 0, \, |x^2| < \varepsilon\}$.

Thus we obtain two maps $\mathbf{H}_\pm$, both defined on D and conjugating the given family $\mathbf{F}$ with its formal normal form $\widetilde{\mathbf{F}}$. Their ratio

$$\mathbf{\Phi} = \mathbf{H}_+ \circ \mathbf{H}_-^{-1} : D \to D, \qquad \mathbf{\Phi} = (\Phi(x, \varepsilon), \varepsilon) \tag{2.3}$$

is a smooth map defined in the (open) domain D which commutes with $\widetilde{\mathbf{F}}$.

If $\widetilde{\mathbf{H}}_\pm$ denote any other pair of smooth maps conjugating F with its formal normal form in the domains $S_\pm$, then by Lemma 2 they must differ only by the flow map of the standard field: there exist two functions $\tau_\pm(\varepsilon)$

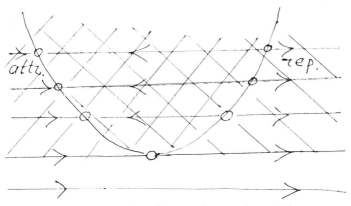

FIGURE 1. Basins of attraction and repulsion.

smooth on a certain interval $0 < \varepsilon < \varepsilon_0$ such that
$$\widetilde{h}_\pm(x, \varepsilon) = H_\pm(g_{v(\cdot, \varepsilon)}^{\tau_\pm(\varepsilon)} x, \varepsilon).$$
The ratio $\widetilde{\Phi} = \widetilde{H}_+ \circ \widetilde{H}_-^{-1}$ is related to Φ as
$$\widetilde{\Phi} = g_v^{\tau_+} \circ \Phi \circ g_v^{-\tau_-}, \tag{2.4}$$
where $g_v^{\tau_\pm}$ stand for the flow maps $(x, \varepsilon) \mapsto (g_{v(\cdot, \varepsilon)}^{\tau_\pm(\varepsilon)} x, \varepsilon)$.

Since the normalizing charts are invariantly associated with SN-families, the transition function Φ is also invariant under at least C^1-smooth transformations up to the equivalence (2.4).

Note that the equivalence relation (2.4) explicitly depends on ε. In order to avoid such an inconvenience, let us introduce a new chart t straightening the family v.

Define the family of maps
$$\mathbf{t}\colon (\mathbb{R}^1, 0) \times (\mathbb{R}_+^1, 0) \to \mathbb{R}^1 \times (\mathbb{R}_+^1, 0), \qquad \mathbf{t} = (t, \mathrm{id}),$$
where
$$t = t(x, \varepsilon) = \frac{1}{2\sqrt{\varepsilon}} \ln\left|\frac{x - \sqrt{\varepsilon}}{x + \sqrt{\varepsilon}}\right| + \frac{1}{2} a(\varepsilon) \ln|x^2 - \varepsilon|, \qquad \varepsilon > 0. \tag{2.5}$$

Note that this family of maps depends explicitly on the choice of the germ $a(\varepsilon)$: from now on we assume that condition (2.1) is satisfied. The map $(x, \varepsilon) \mapsto \mathbf{t}(x)$ transforms the field $v = (x^2 - \varepsilon)(1 + a(\varepsilon)x)^{-1}\partial/\partial x$ into the constant field $\partial/\partial t$; in the coordinate t, the family $\widetilde{F}$ becomes the unit shift $\mathrm{id} + 1$. Therefore the transition map Φ (when written in this coordinate) commutes with the unit shift: if we denote $(\varphi, \mathrm{id}) = \mathbf{t} \circ \Phi \circ \mathbf{t}^{-1}$, then the preservation of the second coordinate implies that
$$\varphi(t + 1) = \varphi(t) + 1. \tag{2.6}$$

The difference $\varphi - \mathrm{id} = \psi$ is a 1-periodic function: $\psi(t+1) = \psi(t)$. If $\Phi, \widetilde{\Phi}$ are two families of maps equivalent in the sense of (2.4), then the corresponding functions $\psi, \widetilde{\psi}$ differ by a shift in the source space:
$$\widetilde{\psi}(t, \varepsilon) = \psi(t + \theta(\varepsilon), \varepsilon). \tag{2.7}$$

This form is similar to the one used in §2 of paper I.

2.4. The moduli space. The above construction motivates the following definitions modeled after the patterns given in Paper I.

Consider the Cartesian product $\mathscr{M} = \mathbb{R}_+^2 \times \mathscr{P}$, where the half-plane $\mathbb{R}_+^2$ is endowed with the coordinates $\varepsilon > 0$, $a \in \mathbb{R}^1$ and $\mathscr{P}$ denotes the space of smooth 1-periodic functions on the real line.

DEFINITION. Two elements $(\varepsilon, a, \psi), (\widetilde{\varepsilon}, \widetilde{a}, \widetilde{\psi}) \in \mathscr{M}$ are called *equivalent*, if $(\varepsilon, a) = (\widetilde{\varepsilon}, \widetilde{a})$ and there exists a $\theta \in \mathbb{R}$ such that $\psi = \widetilde{\psi} \circ (\mathrm{id} + \theta)$.

Denote the space of all equivalence classes by M. The above construction permits to associate to every saddle-node family $\mathbf{F}$ in preliminary normal form a parametrized curve in the moduli space:

$$\mu = \mu_{\mathbf{F}} \colon (\mathbb{R}^1_+, 0) \to \mathcal{M}, \qquad \varepsilon \mapsto (\varepsilon, a(\varepsilon), \psi(t, \varepsilon)). \tag{2.8}$$

This curve is smooth for $\varepsilon > 0$: this means that $a(\varepsilon)$ is the smooth germ and the last component $(\psi(\,\cdot\,, \varepsilon))$ is a smooth family of 1-periodic functions.

We will use the notion of a smooth parametrized curve also in the quotient space M, meaning that there exists a smooth curve consisting of representatives of equivalence classes.

In the terminology introduced above, the results obtained in 2.2, 2.3 can be formulated in a geometric way.

PROPOSITION. *Two saddle-node families* $\mathbf{F}_1$ *and* $\mathbf{F}_2$ *in preliminary normal form are conjugated only if the corresponding parametrized curves* μ_1, $\mu_2 \colon (\mathbb{R}^1_+, 0) \to \mathcal{M}$ *in the moduli space are pointwise equivalent*:

$$\forall \varepsilon > 0, \quad \mu_1(\varepsilon) \sim \mu_2(\varepsilon).$$

To transform this proposition into a full-scale classification theorem, one needs to investigate possible types of parametrized curves which can be realized as invariants of SN-families. This investigation is based on a detailed description of the normalizing charts $\mathbf{H}_\pm$, provided by the *sectorial normalization theorem* formulated in the next section.

§3. Embedding in a flow: the key to the classification of saddle-node families

Consider a smooth SN-family $\mathbf{F}$ in preliminary normal form along with the associated family v of vector fields. We construct a conjugacy between the family $\mathbf{F}$ and its formal normal form $\widetilde{\mathbf{F}} = g_v^1$ in domains of a special form.

3.1. Embedding in sectors. Let $\Omega_+ \subset (\mathbb{R}^2, 0)$ be a closed domain of the form

$$\{-\delta^2 \leqslant \varepsilon \leqslant 0, \ |x| \leqslant \delta\} \cup \{0 \leqslant \varepsilon \leqslant \delta^2, \ \delta \geqslant x \geqslant -2\varepsilon\},$$

and Ω_- its mirror image in the ε-axis. Here δ is a small positive parameter to be chosen afterwards.

SECTORIAL EMBEDDING THEOREM FOR SN-FAMILIES. *For all sufficiently small $\delta > 0$ in each domain $\Omega_\pm$ there exist maps $\mathbf{H}_\pm$ smooth in the sense of Definition 4, §0,*

$$\mathbf{H}_\pm \colon (x, \varepsilon) \mapsto (H_\pm(x, \varepsilon), \varepsilon),$$

possessing the following properties:

(1) *each $\mathbf{H}_\pm$ preserves the parameter and conjugates $\mathbf{F}$ and its formal normal form $\widetilde{\mathbf{F}}$*;

(2) *for $\varepsilon \leq 0$ the two maps identically coincide*: $H_+(\cdot, \varepsilon) \equiv H_-(\cdot, \varepsilon)$;

(3) *in the connected component $|x| \leq 2\varepsilon$ of the intersection $\Omega_+ \cap \Omega_-$ belonging to the positive half-plane $\{\varepsilon > 0\}$, the two maps differ by a function which is flat at the origin*:

$$\mathbf{H}_+ \circ \mathbf{H}_-^{-1} = (\mathrm{id} + \varphi(x, \varepsilon), \mathrm{id}),$$

where $\varphi(x, \varepsilon)$ along with all its derivatives decreases more rapidly than any power of ε as $\varepsilon \to 0^+$, $|x| \leq 2\varepsilon$.

REMARK. The fact that an individual nonhyperbolic mapping $F(\cdot, 0)$ can be represented as the time 1 map for a vector field on the line was proved by F. Takens [T2]. This Takens generator is uniquely determined. On the other hand, by the corollary to Lemma 2, each hyperbolic singularity $x = \pm\sqrt{\varepsilon}$ uniquely defines the "hyperbolic" generator in the open quadrants $\{\varepsilon > 0, \pm x > 0\}$. The sectorial embedding theorem means that the "hyperbolic generators" can be smoothly extended on the x-axis by the above Takens generator.

Finally note that for $\varepsilon < 0$ there are no fixed points at all, so the existence of the generator becomes a trivial statement, while uniqueness no longer holds. The first assertion of the theorem implies that the smooth continuations of the "hyperbolic generators" across both positive and negative semiaxis can be chosen to coincide with each other in the negative half-plane.

The proof of the theorem is rather technical, although transparent: it is postponed till the second part of the paper. Now we turn to applications.

3.2. Classification of saddle-node families. Let $\mathbf{F}_1$, $\mathbf{F}_2$ be two SN-families already in their preliminary normal forms. Then the coincidence of the corresponding formal normal forms (i.e., the identity $a_1(\varepsilon) \equiv a_2(\varepsilon)$ for $\varepsilon > 0$) is a necessary condition for their equivalence. One can define the corresponding normalizing charts $\mathbf{H}_{\pm, i}$, $i = 1, 2$, and the transition functions $\mathbf{\Phi}_i = (\Phi_i, \mathrm{id})$. This necessary condition being satisfied, the equivalence relation (2.4) makes sense for $\varepsilon > 0$, since the field v is the same for both families. Now we can formulate the following main result.

CLASSIFICATION THEOREM FOR SN-FAMILIES. *Two local SN-families in preliminary normal form are smoothly conjugate if and only if they are formally equivalent (i.e., they have the same formal normal form), and their functional invariants $\mathbf{\Phi}_i$ are equivalent in the sense of* (2.4).

Moreover, any pair $(\widetilde{\mathbf{F}}, \mathbf{\Phi})$ consisting of a standard family (the formal normal form) and a family of functions defined on $\Omega_+ \cap \Omega_-$ and commuting with $\widetilde{\mathbf{F}}$ can be realized in an appropriate SN-family as the invariant of smooth classification, provided that Φ differs from the identity by a function flat at the origin.

REMARK. The necessity part of the theorem is evident as explained above. Note also that we do not require the coincidence of the formal parts for

$\varepsilon < 0$, because Corollary 2 to the theorem in §1 implies that two C^∞-smooth families of fields coinciding for positive ε are automatically C^∞-equivalent.

PROOF. Applying the sectorial embedding theorem, one may assume that the entire neighborhood of the origin in the (x, ε)-plane is covered by two charts $\Omega_{\pm, i}$, $i = 1, 2$, with local coordinates $(x_\pm^i, \varepsilon)$, $x_\pm^i = H_{\pm, i}(x, \varepsilon)$ such that in each chart the corresponding family F_i is precisely the time 1 map for the same vector field v. Moreover, since the transition functions for the two families differ only by a flow map, one can choose another pair of normalizing charts, say, for F_1, so that the transition functions will coincide identically on their domains: $\Phi_1(\cdot, \varepsilon) \equiv \Phi_2(\cdot, \varepsilon)$ for $\varepsilon > 0$. Define the conjugacy $\mathbf{H}$ between the families as the map which is identical in the charts of the same "sign" so that the following diagram is commutative:

$$\begin{array}{ccccc} \mathbb{R}^2 & \xleftarrow{H_{+,1}} & \Omega_{+,1}, \Omega_{-,1} & \xrightarrow{H_{-,1}} & \mathbb{R}^2 \\ \text{id} \downarrow & & \mathbf{H} \downarrow & & \downarrow \text{id} \\ \mathbb{R}^2 & \xleftarrow{H_{+,2}} & \Omega_{+,2}, \Omega_{-,2} & \xrightarrow{H_{-,2}} & \mathbb{R}^2 \end{array}$$

Since all the charts are smooth and the definition of $\mathbf{H}$ on the intersection of domains is self-consistent, the above diagram provides the desired equivalence.

Let us proceed with the proof of the realization part of the theorem. We follow the standard pattern suggested in Paper I. Without loss of generality we may assume that the formal normal form of the family to be realized is the standard one: $\widetilde{F} = g_v^1$, $v = (x^2 - \varepsilon)(1 + a(\varepsilon)x)^{-1}\partial/\partial x$.

Consider the disjoint union of two open sets $\operatorname{int}\Omega_\pm$ belonging to two distinct copies of the real plane $\mathbb{R}^2$ with the coordinates $(x_\pm, \varepsilon_\pm)$. Define the smooth maps $\mathbf{F}_\pm^*$ on these sets as the standard family $\widetilde{F}$ written in these coordinates, see (2.2). Let $\mathbf{\Phi} = (\Phi(x, \varepsilon), \text{id})$ be the given map which is to be realized as the modulus of the classification. By the assumptions, $\mathbf{\Phi}$ maps the sector $\{|x| < 2\varepsilon, 0 < \varepsilon < \varepsilon_0\}$ into a nearby curvilinear one with the same vertex, preserves the ε-coordinate and commutes with the restriction of $\widetilde{\mathbf{F}}$ on the sector. Moreover, this map is ∞-tangent to the identity at the origin.

Define the quotient space Ω of the union $\Omega_+ \cup \Omega_-$ using $\mathbf{\Phi}$ as the identifying map. More precisely, we identify two points $(x_\pm, \varepsilon_\pm) \in \operatorname{int}\Omega_\pm$ if $\varepsilon_+ = \varepsilon_-$ (this common parameter value will be denoted by ε), and one of the following holds: either $\varepsilon < 0$ and $x_+ = x_-$, or $\varepsilon > 0$ and $x_+ = \Phi(x_-, \varepsilon)$.

As usual, we may consider the sets $\operatorname{int}\Omega_\pm$ as charts on Ω with $\mathbf{\Phi}$ extended by the identity to the negative half-plane as the transition function for the corresponding atlas. The quotient space is homeomorphic to a punctured neighborhood of the origin on the plane. We show that Ω can be completed by a point in such a way that the completion $\widetilde{\Omega}$ retains the structure of a smooth manifold. Indeed, define the map $\mathbf{H}: \Omega \to \mathbb{R}^2$ in the coordinates

$(x_\pm, \varepsilon_\pm)$ by the formula

$$\mathbf{H}(x_\pm, \varepsilon) = (\theta_+ x_+ + \theta_- x_-, \varepsilon),$$

where $\{\theta_\pm\}$ is a smooth partition of unity subordinate to the covering $\operatorname{int}\Omega_\pm$ with polynomial growth of derivatives [H]. The polynomial growth condition means that all the derivatives of the truncating functions $\theta_\pm$ can be estimated by certain powers $|\varepsilon|^{-l}|x|^{-d}$ in the sectors $\Omega_\pm$ as $(x, \varepsilon) \xrightarrow[\Omega_\pm]{} (0, 0)$, the exponents depending on the order of the derivative. On the intersection of charts, this map admits the following representation:

$$\theta_+ x_+ + \theta_- x_- = \theta_+ \Phi(x_-, \varepsilon) + \theta_- x_- = x_- + \theta_+ \varphi(x_-, \varepsilon), \qquad (\ddagger)$$

(for simplicity we put φ equal zero for negative values of ε). The map $\mathbf{H}$ defines the embedding of Ω in $\mathbb{R}^2$. We endow Ω with the smooth structure inherited from the plane. Adding to Ω the point corresponding to the origin in the plane ensures that the result $\tilde{\Omega}$ is diffeomorphic to an entire neighborhood of the origin. Denote this neighborhood by U.

Now we define the saddle-node family in U. To do this, recall that there are two maps $\mathbf{F}_\pm^*$ in the charts $\operatorname{int}\Omega_\pm$. Since the identifying map used in the construction of the quotient space commutes with them ($\mathbf{F}_+^* \circ \Phi = \Phi \circ \mathbf{F}_-^*$), we can consider the unified map $\mathbf{F}^*: \Omega \to \Omega$. We pull it back into $U \setminus \{0\}$, using the chart $\mathbf{H}$, by setting $\mathbf{F} = \mathbf{H} \circ \mathbf{F}^* \circ \mathbf{H}^{-1}$. Obviously, the map $\mathbf{F}$ is smooth outside the origin.

Now our problem is to extend $\mathbf{F}$ to the origin while preserving its smoothness. Set $\mathbf{F}(0) = 0$. One needs only to prove that this extension is smooth at the origin. From ($\ddagger$) it follows that the chart H differs from either x_+ or x_- by functions flat at the origin, while the map $\mathbf{F}^*$ is given in both charts $x_\pm$ by the same formula. Hence all derivatives of $\mathbf{F}$ with respect to the chart H have the same limits at the origin as those of $\mathbf{F}_\pm^*$ with respect to $x_\pm$. Since the latter functions are smooth, the conditions of the Whitney continuation theorem [H] hold for them. These same conditions also hold for F at the origin by continuity. This ensures the possibility of smooth extension. This extension is unique for obvious reasons: $\lim_{x,\varepsilon \to 0} \mathbf{F}(x, \varepsilon) = 0$. So we have constructed a smooth local family of line diffeomorphisms.

Let us verify that the family $\mathbf{F}$ is indeed of the SN-type. To prove this, note that this family for $\varepsilon > 0$ has two hyperbolic fixed points with the same multiplicators as in the family $\tilde{\mathbf{F}} = g_\nu^1$. The smoothness of $H_\pm$ at zero (see item (3) of the sectorial normalization theorem for SN-families) implies that the fixed points of the map $F(\cdot, 0)$ have multiplicity 2. Therefore the preliminary normal form condition is satisfied and the formal normal form is as required. The functional invariant for $\mathbf{F}$ is equal to Φ, since the $x_\pm \circ \mathbf{H}^{-1}$ form the normalizing atlas with that same transition function. The realization part is proved.

The sectorial normalization theorem has a geometric corollary for parametrized curves (in the function space) associated with smooth SN-families.

These curves are smooth not only on the interior of the interval $0 < \varepsilon < \varepsilon_0$, but also at the left boundary point $\varepsilon = 0$. This means that the family $\psi(t, \varepsilon)$ of functional invariants constructed in the preceding section possesses certain limits for all its derivatives either in t or in ε as $\varepsilon \to 0^+$. Moreover, the function $\psi(t, \varepsilon)$ tends to zero together with all its derivatives as $\varepsilon \to 0^+$. Indeed, a similar property holds for the differences $\varphi = \Phi - \mathrm{id}$ as asserted in the theorem. The rest follows from the polynomial growth of the straightening chart t given by (2.5).

§4. Applications to bifurcation theory

In order to give an example of the application of the sectorial embedding theorem, let us prove the sufficiency of the Malta-Palis conditions [MP] describing the simultaneous occurrence of multiple saddle connections in a generic two-parameter family of planar vector fields.

4.1. Semistable cycles and multiple saddle connections. Consider a planar vector field having a semistable limit cycle with a monodromy transformation

$$\Delta \colon (\mathbb{R}^1, 0) \to (\mathbb{R}^1, 0), \quad \Delta(x) = x + ax^2 + O(x^3), \quad a \neq 0,$$

such that there are at least two topologically distinct trajectories tending to the cycle from the inside as well as from the outside (for example, these trajectories may be stable and unstable separatrices of hyperbolic saddles, see Figure 2).

In a generic deformation of the field such a cycle disappears and one or more saddle connections (i.e., heteroclinic orbits of the field) may occur.

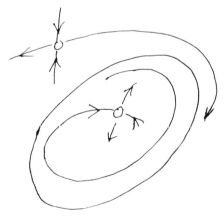

FIGURE 2. Semistable limit cycle with topologically distinguished curves on both sides.

Let us examine whether the simultaneous occurrence of several saddle connections is possible or not. Proceeding in the standard way, we reduce the problem to the investigation of the orbits of the monodromy maps. These maps form a saddle-node family of line diffeomorphisms $x \mapsto F(x, \varepsilon)$.

The topologically distinguished trajectories intersect the transversal to the cycle on which the monodromy is defined. This intersection for $\varepsilon = 0$ consists of semi-infinite orbits of F (either positive orbits for the trajectories converging to the cycle or negative ones corresponding to the trajectories converging to the cycle after time reversal). We consider the case where there are at least two orbits on each side.

Take a unique representative from each orbit in its intersection with the transversal. Denote them by $x_\pm^i$, where x_+^i, $i = 1, 2$, is the pair of points belonging to the trajectories converging to the cycle with time increasing, while x_-^i, $i = 1, 2$, stand for the other pair of orbits.

Without loss of generality we may assume that $x_+^1 < x_+^2 < 0 < x_-^1 < x_-^2$. Note that, by transversality, all the $x_\pm^i$ smoothly depend on the parameter. We are interested in conditions guaranteeing that no pair of these points belongs to the same orbit of F for $\varepsilon < 0$. These conditions were found in [MP]. In order to formulate them, recall that the individual monodromy map $\Delta(\cdot, 0)$ can be embedded in a flow v by the Takens theorem. Define the two numbers $\tau_\pm$ by the following conditions:

$$g_v^{\tau_\pm} x_\pm^1 = x_\pm^2. \tag{4.1}$$

The values $\tau_\pm$ measure the distance between the orbits coming from the same side, in units of time. So they make sense only $\mod \mathbb{Z}$.

THEOREM [MP]. *If we have*

$$\tau_+ \neq \tau_- \mod \mathbb{Z}, \tag{4.2}$$

then the simultaneous occurrence of two saddle connections is impossible.

We shall prove that this necessary condition is actually very close to the sufficient one.

THEOREM ON MULTIPLE SADDLE CONNECTIONS. *If condition* (4.2) *is violated in a generic two-parameter family of planar vector fields, then there are countably many values of the parameters (accumulating to the critical value) for which two saddle connections occur simultaneously.*

PROOF. Without loss of generality one may assume that the violation of (4.1) implies $\tau_+ = \tau_-$ (otherwise another representative from the orbit must be chosen). Let (ε, μ) be the two parameters, so that the monodromy map takes the form $\Delta = F(\cdot, \varepsilon, \mu)$. It follows from the transversality theorem that values of parameters for which the map F has a double fixed point

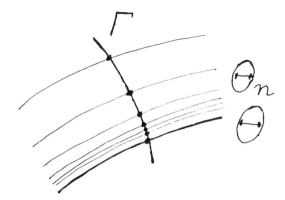

FIGURE 3. Bifurcation diagram on the parameter plane.

belong to a smooth curve Θ passing through the origin on the parameter plane.

One may assume that this curve is $\{\varepsilon = 0\}$ and that for $\varepsilon < 0$ there are no fixed points at all. Apply the sectorial embedding theorem to the family $\mathbf{F}$. This theorem provides a smooth family of vector fields $v = v(x, \varepsilon, \mu) \partial/\partial x$ defined for negative ε and such that $\mathbf{F}$ is the time 1 map for v. So one may define instead of two numbers $\tau_\pm$, two germs of functions $\tau_\pm(\varepsilon, \mu)$, $\varepsilon < 0$ by the following conditions:

$$\tau_\pm(\varepsilon, \mu) = \int_{x_\pm^1(\varepsilon,\mu)}^{x_\pm^2(\varepsilon,\mu)} \frac{dx}{v(x, \varepsilon, \mu)}. \tag{4.3}$$

Since the field v is nowhere vanishing on the segment of integration, these functions are smooth. For a generic family of maps the curve

$$\Gamma = \{(\varepsilon, \mu) \in \mathbb{R}^2 : \tau_+(\varepsilon, \mu) = \tau_-(\varepsilon, \mu)\}$$

is smooth and transversal to Θ.

Note that a saddle connection between the points x_+^1 and x_-^1 occurs if and only if

$$T_1(\varepsilon, \mu) = \int_{x_+^1(\varepsilon,\mu)}^{x_-^1(\varepsilon,\mu)} \frac{dx}{v(x, \varepsilon, \mu)} = n \in \mathbb{N} \tag{4.4}$$

(a similar condition for the other pair has the form $T_2 \in \mathbb{N}$). The corresponding curves $\Theta_n = \{\varepsilon, \mu : T_1(\varepsilon, \mu) = n\}$ in the parameter plane are accumulating to Θ as $n \to \infty$, see Figure 3. This follows from the fact that the integral (4.4) uniformly tends to infinity as $\varepsilon \to 0$.

Now note the following obvious fact: on the curve Γ one has $\tau_+ = \tau_-$, therefore each point of the intersection $\Theta_n \cap \Gamma$ corresponds to the occurence of a double saddle connection. This intersection contains countably many points accumulating to the origin and lying on the smooth curve Γ.

4.2. Asymptotics of the trapping time. Another application of the sectorial embedding theorem is the computation of the asymptotics of trapping time. Consider a planar analytic vector family of vector fields having a semistable limit cycle for a critical value of the parameters and suppose that by small perturbations of parameters this cycle is pushed away from the real into the complex domain. Nevertheless all trajectories for the perturbed system remain trapped near the place where the cycle was for a sufficiently long time. This time was estimated in [DV] in terms of the imaginary parts of the multipliers of the (complex) fixed points of the monodromy.

The analyticity is irrelevant, since the sectorial embedding theorem implies the following estimate of the trapping time for smooth families.

Suppose that a smooth generic one-parameter family of planar vector fields depending on a real parameter ε possesses a semistable limit cycle C_0 for $\varepsilon = 0$, which splits into two hyperbolic cycles $C_{\pm}(\varepsilon)$ when ε becomes positive. Take any transversal γ to the semistable cycle. Then these two hyperbolic cycles intersect γ at two points between which the distance is $d(\varepsilon) \sim \sqrt{\varepsilon}$.

For any sufficiently narrow annulus-like open domain U containing C_0 and any sufficiently small *negative* ε denote by $N = N(U, \varepsilon)$ the number of full turns made by a trajectory confined within U. We are going to estimate the principal term of the asymptotics of $N(U, \varepsilon)$.

THE TRAPPING TIME ESTIMATE. *For any* U

$$N(U, \varepsilon) = \frac{\pi}{2d(|\varepsilon|)} + O(1), \qquad \varepsilon < 0, \ \varepsilon \to 0^-. \tag{4.5}$$

PROOF. First we point out that the number N is defined in terms of orbits rather than of the field itself: multiplication of the field by a positive nonvanishing function changes the trapping time and the "period" of one rotation, preserving the number N. Moreover, we can replace the domain U by the flow box U_0 bounded by two segments of integral curves between subsequent intersections with γ and segments of the transversal γ itself: this will change the number N by a bounded term.

Let $\Delta = \Delta_\gamma(\cdot, \varepsilon)$ be the monodromy map. Then the trapping "time" is equal to

$$N(U_0, \varepsilon) = \inf_n \{n : \Delta^{[n]}(-\delta, \varepsilon) > +\delta\},$$

where δ is a small positive number (corresponding to the boundary of the flow box U_0).

Let us embed the family of monodromy maps in the smooth flow. Then the trapping time is equal to the integral defining the time necessary to get from the outer boundary of the annulus to the inner one. This integral can be explicitly computed provided that the flow is standard, and the principal

term of its asymptotics equals
$$N(U_0, \widetilde{\varepsilon}) \sim \int_{-\delta}^{+\delta} \frac{dx}{x^2 - \widetilde{\varepsilon}} = \frac{\pi}{\sqrt{|\widetilde{\varepsilon}|}},$$
where $\widetilde{\varepsilon}$ is the parameter occurring in the preliminary normal form; this parameter is expressed via the distance d between the singular points of the generator as $\sqrt{-\widetilde{\varepsilon}} = d(-\widetilde{\varepsilon})$, whence the specific form of the formula (4.5). □

Evidently this result can be generalized for the multidimensional case.

§5. Sectorial normalization and smooth classification of flip families

In this section we show that in domains of a certain special form a flip family is conjugated with the superposition of the time $1/2$ map for an odd family of vector fields and with the standard symmetry $x \mapsto -x$ preserving this family.

5.1. Iteration square of flip families: embedding in a flow. Let $\mathbf{F} = (F, \mathrm{id})$ be a flip family in the preliminary normal form. As it was shown in 1.2, this family has the form
$$F(x, \varepsilon) = -x + x(x^2 - \varepsilon)f(x, \varepsilon) \tag{5.1}$$
with a smooth function f such that $f(0, 0) = 1$. The iteration square $\mathbf{F}^{[2]}$ is the family which is a transversal deformation of the line diffeomorphism tangent to the identity with the third order at the origin:
$$\mathbf{F}^{[2]} = (F_2, \mathrm{id}), \qquad F_2(x, \varepsilon) = x + x(x^2 - \varepsilon)f_2(x, \varepsilon). \tag{5.2}$$
The specific trait of the case (5.2) is that $F_2(\cdot, 0)$ has a triple fixed point at the origin instead of a double. The representation (5.2) implies that the fixed points of $\mathbf{F}^{[2]}$ lie on the "trident" $\{x(x^2 - \varepsilon) = 0\}$. Nevertheless, all the fixed points of F_2 for $\varepsilon \neq 0$ are hyperbolic. This circumstance allows us to modify the construction developed in §3 in order to obtain the smooth classification of flip families.

Let $\overline{\Omega}_+, \overline{\Omega}_-$ be two closed domains in the (x, ε)-plane:
$$\begin{aligned}\overline{\Omega}_- &= \{-\delta^2 \leq \varepsilon \leq 0, |x| \leq \delta\} \cup \{0 \leq \varepsilon \leq \delta^2, |x| \geq \varepsilon\}, \\ \overline{\Omega}_+ &= \{0 \leq \varepsilon \leq \delta^2, |x| \leq 4\varepsilon\}.\end{aligned} \tag{5.3}$$
The intersection $\overline{\Omega}_+ \cap \overline{\Omega}_-$ consists of two sectors $\{0 \leq \varepsilon \leq \delta^2, \varepsilon \leq \pm x \leq 4\varepsilon\}$, each of which is wide enough to include some fundamental domain of the action of $\mathbf{F}_2$ in the positive half-plane $\varepsilon > 0$.

Consider a smooth family of vector fields of the form
$$w(x, \varepsilon) = x(x^2 - \varepsilon)(1 + b(\varepsilon)x^2)\partial/\partial x + 0 \cdot \partial/\partial \varepsilon, \tag{5.4}$$
where the smooth function $b = b(\varepsilon)$ is chosen in such a way that g_w^1 has the same multipliers of all fixed points as the family $\mathbf{F}^{[2]}$.

SECTORIAL NORMALIZATION THEOREM FOR "TRIDENT" FAMILIES. *For all sufficiently small values of $\delta > 0$, in each domain $\overline{\Omega}_\pm$ there exist smooth maps $\mathbf{H}_\pm = (H_\pm, \mathrm{id})$ which conjugate $\mathbf{F}^{[2]}$ and g_w^1; these maps differ on the intersection $\overline{\Omega}_+ \cap \overline{\Omega}_-$ by a function which is flat at the origin:*

$$\mathbf{H}_+ \circ \mathbf{H}_-^{-1} = (\mathrm{id} + \varphi, \mathrm{id}), \qquad \varphi = \varphi(x, \varepsilon),$$
$$|D_x^\alpha D_\varepsilon^\beta \varphi(x, \varepsilon)| \to 0 \quad as \;\; x \to 0^+, \qquad (x, \varepsilon) \in \overline{\Omega}_+ \cap \overline{\Omega}_-.$$

We do not give the proof of this theorem, since it reproduces essentially that of the analogous theorem for saddle-node families. Let us turn directly to implications.

5.2. Involutions and flows. We have just asserted the possibility of embedding orientation-preserving families which are the squares of flip-type ones in a flow. Since a true flip is orientation-reversing, it cannot be represented as the time 1 map for any flow. Nevertheless we can split a flip family into the superposition of the standard involution and a flow map.

Consider two fields $w_\pm = (\mathbf{H}_\pm)_*^{-1} w$ on $\overline{\Omega}_\pm$. Both of them are generators for the family $\mathbf{F}^{[2]}$ on the corresponding domains, therefore they are preserved by $\mathbf{F}^{[2]}$:

$$\mathbf{F}_*^{[2]} w_\pm = w_\pm. \tag{5.5}$$

Our immediate goal is to prove that the initial orientation-reversing flip family $\mathbf{F}$ also satisfies this property.

PROPOSITION. *Both fields $w_\pm$ are preserved by $\mathbf{F}$:*

$$\mathbf{F}_* w_\pm = w_\pm. \tag{5.6}$$

COROLLARY. *$\mathbf{F}$ commutes with the time t maps $g_{w_\pm}^t$ for all $t \in \mathbb{R}$.*

PROOF. Consider the fields $\widetilde{w}_\pm = \mathbf{F}_* w_\pm$ on $\overline{\Omega}_\pm$. Let us prove that they also are generators for the family $\mathbf{F}^{[2]}$. In this case the hyperbolicity of fixed points of $\mathbf{F}^{[2]}|_\varepsilon$, where $\varepsilon = \mathrm{const} \neq 0$, implies that $\widetilde{w}_\pm = w_\pm$, since the generator is unique.

Indeed, $\mathbf{F}^{[2]}$ preserves $\widetilde{w}_\pm$: from the commutation of powers of $\mathbf{F}$ and property (5.5), it follows that

$$\mathbf{F}_*^{[2]} \widetilde{w}_\pm = (\mathbf{F}^{[2]} \circ \mathbf{F})_* w_\pm = (\mathbf{F} \circ \mathbf{F}^{[2]})_* w_\pm = \mathbf{F}_* w_\pm = \widetilde{w}_\pm.$$

Now note that any field preserved by $\mathbf{F}^{[2]}(\cdot, \varepsilon)$, $\varepsilon \neq 0$, differs from the generator of $\mathbf{F}^{[2]}$ by a scalar factor, so $\widetilde{w}_\pm = \lambda(\varepsilon) w_\pm$. Computation of 1-jets at the fixed points yields $\lambda(\varepsilon) \equiv 1$ for $\varepsilon \neq 0$. From continuity arguments it follows that $\widetilde{w}_\pm \equiv w_\pm$. □

Define the pair of maps

$$\sigma_\pm = \mathbf{F} \circ g_{w_\pm}^{-1/2}.$$

These maps are involutions: this follows from the previous corollary and the definition of $w_\pm$. Moreover, these involutions preserve the corresponding fields: $(\sigma_\pm)_* w_\pm = w_\pm$.

LEMMA. *In both domains $\overline{\Omega}_\pm$ there exists a smooth map $\mathbf{H}_\pm$ which takes the corresponding field $w_\pm$ to polynomial form* (5.4) *and at the same time transforms $\sigma_\pm$ into the standard symmetry $x \mapsto -x$.*

PROOF. We demonstrate the statements in several steps. For the sake of simplicity the subscripts referring to the choice of the domain will be omitted.

1. *Polynomial normal form*. There exists a smooth transformation taking the family w to polynomial normal form (see Corollary 1 to the theorem in §1). This polynomial is of degree 5; since the zeros of the field lie on the "trident" $\{x(x^2 - \varepsilon) = 0\}$, one may assume that the field is of the form

$$w = x(x^2 - \varepsilon)(b_0 + b_1 x + b_2 x^2) \partial/\partial x, \qquad (5.7)$$

where $b_i = b_i(\varepsilon)$ stand for smooth germs.

2. *Oddity*. Let us show that an odd polynomial in (5.7) may in fact be chosen. This step is implemented somewhat differently for $\overline{\Omega}_+$ and $\overline{\Omega}_-$.

If the case of $\overline{\Omega}_-$ is considered, then one can note that the eigenvalues of w corresponding to the zeros $\partial w/\partial x(\pm\sqrt{\varepsilon}, \varepsilon)$, coincide, since there is an automorphism of the field which swaps these zeros. So for $\varepsilon > 0$ we have $b_1(\varepsilon) \equiv 0$. The theorem of §1 implies that one may put b_1 identically equal to zero: indeed, the function $b_1(\varepsilon) x^2(x^2 - \varepsilon)$ is divisible by $x^2(x^2 - \varepsilon)^2$ in $\overline{\Omega}_-$ provided that b_1 is as above.

In the case of $\overline{\Omega}_+$ the situation is even simpler, since $b_1 x^2(x^2 - \varepsilon)$ is divisible by the same factor in $\overline{\Omega}_+$.

3. *The involution*. It remains only to prove that any orientation-reversing automorphism preserving the odd family

$$w = x(x^2 - \varepsilon)(1 + b(\varepsilon)x^2) \partial/\partial x = xk(x^2, \varepsilon) \partial/\partial x \qquad (5.8)$$

must be the standard symmetry. Indeed, if we write $\sigma: (x, \varepsilon) \mapsto (s(x, \varepsilon), \varepsilon)$, then the graph $y = s(x, \varepsilon)$ must be an integral curve of the following system on the (x, y)-plane:

$$\frac{dy}{dx} = \frac{yk(y^2, \varepsilon)}{xk(x^2, \varepsilon)} \qquad (5.9)$$

(see (5.8)). Note that for $\varepsilon \neq 0$ the singular point $(0, 0)$ of (5.9) is the *dicritical node*: they became nonsingular after blowing up [AI]. So we have uniqueness of smooth continuation of integral curves through dicritical nodes: for any direction there is a *unique* pair of integral curves of the system in a neighborhood of a dicritical node which becomes a smooth curve tangent to the given direction after adding the singular point. If we choose the antidiagonal direction, then the curve $y = -x$ is the integral one for (5.9) in the case

$\varepsilon \neq 0$, hence it must coincide with the graph of s. In the case $\varepsilon = 0$, we have the same assertion: $s = \{y = -x\}$ is true by continuity. So we have proved the last assertion of the lemma. □

5.3. The "sectorial embedding" in the flip case. Let $\mathbf{F} = (F, \mathrm{id})$ be a flip family in preliminary normal form (1.6); denote by $\mathbf{F}_0 = (F_0, \mathrm{id})$ the local flip family

$$\mathbf{F}_0 = \sigma \circ g_w^{1/2},$$

where $w = w_\mathbf{F}$ is the smooth odd family (5.8) constructed using the corollary to Lemma 1B, §1, and σ is the standard symmetry with respect to the ε-axis.

SECTORIAL NORMALIZATION THEOREM FOR THE FLIP CASE. *If the parameter δ determining the size of the domains $\overline{\Omega}_\pm$ is sufficiently small, then there exists a pair of maps $\mathbf{H}_\pm = (H_\pm, \mathrm{id})$, defined and smooth on these domains, such that:*

(1) *in each domain the corresponding map conjugates the restriction $\mathbf{F}|_{\overline{\Omega}_\pm}$ with the normal form $\mathbf{F}_0|_{\overline{\Omega}_\pm}$:*

$$\mathbf{H}_\pm \circ (\mathbf{F}|_{\overline{\Omega}_\pm}) = (bF_0|_{\mathbf{H}_\pm(\overline{\Omega}_\pm)}) \circ \mathbf{H}_\pm;$$

(2) *any other pair $\widetilde{\mathbf{H}}_\pm$ of maps satisfying the above property must differ from $\mathbf{H}_\pm$ by a flow transformation in the target space: there exists a pair of smooth germs $\tau_\pm = \tau_\pm(\varepsilon)$ such that*

$$\widetilde{\mathbf{H}}_\pm = g_w^{\tau_\pm} \circ \mathbf{H}_\pm,$$

where

$$g_w^{\tau_\pm}(x, \varepsilon) = (g_{w(\cdot, \varepsilon)}^{\tau_\pm(\varepsilon)} x, \varepsilon), \qquad (x, \varepsilon) \in \overline{\Omega}_\pm;$$

(3) *the maps $\mathbf{H}_\pm$ normalizing the family $\mathbf{F}$ may be chosen so as to have a common ∞-jet at the origin: in other words, their iteration ratio $\Phi = \mathbf{H}_+ \circ \mathbf{H}_-^{-1} = (\mathrm{id} + \varphi(x, \varepsilon), \mathrm{id})$ differs from the identity by a function φ defined in the union of two sectors with the common vertex at the origin, and flat at this vertex.*

PROOF. The existence of the normalizing charts follows immediately from the above lemma. The third assertion holds, since the embedding theorem for "trident" families asserts an analogous property. It only remains to prove the second assertion which is a kind of uniqueness.

Note that any other normalizing charts must preserve the local generators $w_\pm$ due to the hyperbolicity of the fixed points of $\mathbf{F}^{[2]}$. Hence the normalizing charts must differ by a flow map as it was shown in §2. A minor difficulty arises from the fact that the intersection $\overline{\Omega}_- \cap \{\varepsilon = \mathrm{const} > 0\}$ is not connected, so *a priori* two different values of the shift time, each for its own connected component of the intersection, are possible. In fact, nothing of that kind occurs. Indeed, the iteration ratio Φ of the normalizing charts commutes not only with $\mathbf{F}^{[2]}$, but also with $\mathbf{F}$: this implies that Φ must commute with the standard involution, thus being odd. If we try to find an odd map of the form $g_w^{\tau_1(\varepsilon)} \circ \Phi \circ g_w^{-\tau_2(\varepsilon)}$ defined on $\overline{\Omega}_+ \cap \overline{\Omega}_-$, we necessarily come to the conclusion that $\tau_1 \equiv \tau_2$. $\square$

The classification theorem for flip families follows from the sectorial normalization theorem precisely as it was in the SN-case. We can introduce a straightening chart $\mathbf{t}$ and define an analogous notion of equivalence on the set of transition functions written in this chart.

Note the following fact. Since the intersection $\overline{\Omega}_+ \cap \overline{\Omega}_-$ consists of two sectors, one may expect that the invariant consists of two transition functions. This is not so, since the equivalence of transition functions, say, in the upper sector of the intersection, implies automatically the equivalence of those corresponding to the lower one: indeed, the family $\mathbf{F}$ swaps these sectors.

In order to get rid of repetitions, we omit the realization statement (which is true for sure) and formulate the classification theorem for flip families in the following way.

CLASSIFICATION THEOREM FOR FLIP FAMILIES. *Two flip families* $\mathbf{F}_1$, $\mathbf{F}_2$ *in preliminary normal form are conjugate if and only if the corresponding "trident" families* $\mathbf{F}_1^{[2]}$, $\mathbf{F}_2^{[2]}$ *are conjugate.*

Clearly, this is simply a reformulation of the equivalence of the transition functions.

§6. Smooth orbital classification of Andronov-Hopf families

6.1. Semimonodromy. Let $\mathbf{V} = V(z, \varepsilon) \partial/\partial z$, $z \in (\mathbb{C}, 0) \simeq (\mathbb{R}^2, 0)$ be a smooth local AH-family. As we mentioned in the Introduction, it takes the form

$$\mathbf{V}(z, \varepsilon) = z A(z, \varepsilon) \partial/\partial z,$$
$$\operatorname{Re} A(0, 0) = 0, \quad \operatorname{Im} A(0, 0) = \omega > 0, \quad (\partial \operatorname{Re} A/\partial \varepsilon)(0, 0) \neq 0, \quad (6.1)$$

with a smooth complex-valued function A in a certain complex coordinate system on the plane. Note that for each $\varepsilon \in (\mathbb{R}^1, 0)$ the field $\mathbf{V}(\cdot, \varepsilon)$ has a unique singularity at the origin which is hyperbolic for $\varepsilon \neq 0$.

Our study of AH-families is based on the investigation of the *first return map* defined as follows.

Let γ be a germ of a smooth curve γ passing through the singularity and tangent to a certain direction. Note that the polar angle function $\operatorname{Arg} z$ on the complex plane $\mathbb{C}$ monotonically increases along orbits of the family (6.1), at least in a sufficiently small neighborhood of the singularity. So for any point $b \in \gamma$ close enough to the origin the positive semitrajectory $g_\mathbf{V}^t b$, $t > 0$, intersects γ at least once.

DEFINITION. A *semimonodromy map* (or, more precisely, a local family of maps) for a family **V**, defined on a smooth curve γ, is the map taking a point $b \in \gamma$, $b \neq 0$, to the first intersection of the positive semiorbit starting at b with γ.

We shall denote the semimonodromy map by Δ_γ.

An analogous construction is well known for the case when there is a periodic orbit Γ of the field **V** and γ is a smooth curve transversal to Γ. In this case the general theorems on differentiable dependence of solutions of ODE's on initial conditions imply the differentiability of the corresponding first return map. In our situation these results cannot be applied, since there is no transversality at the singular point. For a generic singularity, the first return map, even if defined, is not smooth. Nevertheless, the AH-case proves to be an exception.

PROPOSITION. *If* **V** *is an AH-family in the form* (6.1), *then for any smooth curve* γ *passing through the origin, the semimonodromy* Δ_γ *is a smooth local family of orientation-reversing maps depending smoothly on the parameter. If* γ_1 *is another smooth curve, then the corresponding semimonodromy* Δ_{γ_1} *is conjugate with* Δ_γ.

PROOF. This assertion becomes evident after passing to polar coordinates on the plane.

More precisely, let C be a cylinder with coordinates $r \in (\mathbb{R}^1, 0)$, $\varphi \in S^1 \simeq \mathbb{R}/2\pi\mathbb{Z}$. Define the map $\Pi : C \to \mathbb{C}$, $\Pi(r, \varphi) = r \exp i\varphi$ (the inverse Π^{-1} is given by $r = |z|$, $\varphi = \operatorname{Arg} z \pmod{2\pi}$). The inverse image $\widetilde{\mathbf{V}} = \Pi_*^{-1} \mathbf{V}$ of the field **V** is smooth on C and admits a unique smooth continuation on the circle $\Gamma = \{r = 0\}$. Indeed, one can write the equation corresponding to the field **V** in the form

$$A(z, \varepsilon) = \dot{z}/z = (\operatorname{Ln} z)\dot{} = (\ln|z| + i \operatorname{Arg} z)\dot{} = \dot{r}/r + i\dot{\varphi},$$

whence

$$\dot{r} = r \operatorname{Re} A(r \exp i\varphi, \varepsilon), \quad \dot{\varphi} = \operatorname{Im} A(r \exp i\varphi, \varepsilon).$$

Since $\operatorname{Im} A(0, 0) \neq 0$, it is clear from these formulas that Γ is an invariant curve for $\widetilde{\mathbf{V}}$. Now note that the inverse image $\Pi^{-1}(\gamma)$ consists of two smooth curves $\gamma_\pm$ and of the circle Γ, and the covering transformation $T: (r, \varphi) \mapsto (-r, \varphi + \pi)$ maps $\gamma_\pm$ one onto the other diffeomorphically. Moreover, the restriction $\Pi_\pm = \Pi|_{\gamma_\pm}$ maps $\gamma_\pm$ diffeomorphically onto γ. Finally note that the map $\Delta : \gamma_+ \to \gamma_-$ taking each point to the first intersection of the corresponding positive semiorbit of $\widetilde{\mathbf{V}}$ with the other curve, is smooth, since both $\gamma_\pm$ are transversal to $\widetilde{\mathbf{V}}$. All these facts imply that the semimonodromy Δ_γ, which can be expressed as the superposition $\Pi_+ \circ \Delta \circ \Pi_-^{-1} = \Pi_- \circ T \circ \Delta \circ \Pi_-^{-1}$, is a smooth map.

The second assertion is no less evident, since for another choice of the curve γ_1 one has another pair of smooth curves $\gamma_{1, \pm}$ on C, and the desired smooth conjugation is realized in the polar coordinates by a map $\gamma_+ \to \gamma_{1,+}$ defined again as the first intersection correspondence.

REMARK. Computation of higher order terms shows that the semimonodromy map associated with an AH-family is indeed of SN-type.

6.2. Semimonodromy and orbital equivalence. If there are two AH-families V_1, V_2, and Δ_{γ_i}, $i = 1, 2$, are their semimonodromies corresponding to the curves γ_i, then orbital equivalence between V_i implies smooth equivalence of the germs Δ_{γ_i}. Indeed, multiplication of a field by a positive function does not change its semimonodromy. On the other hand, if H is a map conjugating the flows of V_i, then its restriction to γ_1 conjugates the semimonodromies Δ_{γ_1} and $\Delta_{H(\gamma_1)}$, the latter being conjugated with Δ_{γ_2} by the above proposition.

We are interested in the inverse statement.

CLASSIFICATION THEOREM FOR ANDRONOV-HOPF FAMILIES. *Two AH-families are orbitally equivalent if and only if the corresponding monodromies, being flip-type local families, are smoothly conjugate.*

Any flip family can be realized as the semimonodromy map for an appropriate AH-family.

The rest of the section is devoted to the proof of this statement.

An idea is evident. If there are two AH-families $zA_i(z, \varepsilon)\partial/\partial z$, then one can divide these families by appropriate real functions so that $\operatorname{Im} A_i \equiv 2\pi$. Choose the germ of the positive semiaxis as a semitransversal γ_i in both cases. Then the semimonodromies associated with this choice are simply the time $1/2$ maps $g_i^{1/2}$ for the corresponding flows $\{g_i^t\}$, $i = 1, 2$. Suppose that these semimonodromies are smoothly conjugate, i.e., there exists a smooth real family h of maps defined on the real line and conjugating $g_i^{1/2}$. Extend h to an entire neighborhood of the origin $(\mathbb{C}, 0)$ by the formula

$$H(z, \varepsilon) = g_2^{(2\pi)^{-1} \operatorname{Arg} z} \circ h \circ g_1^{-(2\pi)^{-1} \operatorname{Arg} z} z, \quad z \neq 0, \qquad (6.2)$$
$$H(0, \varepsilon) = 0.$$

The relation (6.2) is correct, since for any point z one has $\operatorname{Arg} g_1^{-\operatorname{Arg} z/2\pi} z = 0$ by the above convention imposed on the functions A_i. The map H in fact is not multivalued, though the function Arg is: if we choose another value of Arg, differing by $2\pi k$, $k \in \mathbb{Z}$, from the initial one, then the whole expression will not be changed. This is a consequence of the identity

$$g_2^1 \circ h = h \circ g_1^1,$$

that follows from the above conjugacy. The only problem is to prove smoothness of H at the origin. To do this, one needs a certain special coordinate system in which the given AH-family will be almost rotationally symmetrical.

6.3. Preliminary normal form for the Andronov-Hopf case.

LEMMA. *Any local AH-family* $\mathbf{V}$ *is orbitally equivalent to a family* $\mathbf{V}_0$ *of the form*

$$\mathbf{V}_0(z, \varepsilon) = z(R(z, \varepsilon) + 2\pi i + F(z, \varepsilon))\partial/\partial z, \tag{6.3}$$

where:

(1) *both functions* R, F *are real and smooth*;
(2) R *is rotationally symmetrical, i.e.,* $R(z, \varepsilon) = \rho(z\bar{z}, \varepsilon)$ *with some smooth real* $\rho\colon (\mathbb{R}^1, 0) \times (\mathbb{R}^1, 0) \to (\mathbb{R}^1, 0)$;
(3) F *is flat at the origin* $z = 0$ *for all* ε.

We shall refer to the field $z(R + 2\pi i)\partial/\partial z$ as the *symmetrical part* of $\mathbf{V}_0$, while $F\,\partial/\partial z$ will be called the *flat perturbation*.

PROOF. Let us associate with each smooth local family $\mathbf{V}$ its *semiformal Taylor series*

$$\hat{\mathbf{V}} = \left(z \sum_{k,l \geq 0} a_{k,l}(\varepsilon) z^k \bar{z}^l\right)\frac{\partial}{\partial z},$$

with $a_{k,l}$ being smooth functions defined in some neighborhood of zero, *common for all* k, l, in the parameter space.

On the set of such semiformal vector fields a natural action of the group of semiformal transformations of the form

$$(z, \varepsilon) \mapsto \left(z + \sum_{k,l \geq 0,\, k+l \geq 2} h_{k,l}(\varepsilon) z^k \bar{z}^l,\ \varepsilon\right)$$

is defined. Normal forms in the semiformal category are very close to those of Poincaré-Dulac (see [A]). For example, using semiformal transformations, one can put the series $\hat{\mathbf{V}}$ corresponding to the initial AH-family into the form

$$\hat{\mathbf{V}}_1 = z\left(\sum_{k \geq 0} a_{k,k}(\varepsilon)(z\bar{z})^k\right)\frac{\partial}{\partial z}, \tag{6.4}$$

containing only terms corresponding to the resonance $\lambda + \bar\lambda = 0$, which takes place for $\varepsilon = 0$.

Consider the semiformal transformation putting the initial series into the form (6.4) and extend it to a smooth family of plane transformations with the same Taylor series, using the Borel-Whitney continuation theorem. After applying this extension to the family $\mathbf{V}$, one obtains a smooth family of planar vector fields having (6.4) as its semiformal Taylor series. This means that the new family acquires the desired form after division by an appropriate real positive smooth function.

For a more detailed exposition of the semiformal techniques see [IY].

6.4. Proof of the sufficiency assertion.

Consider a family $\mathbf{V}$ in the form (6.3) and choose the positive semiaxis as the semitransversal γ. Then the semimonodromy associated with such choice equals

$$\Delta_\gamma = -g_w^{1/2} + (\text{flat function}), \tag{6.5}$$

where $w = x\rho(x^2, \varepsilon)\partial/\partial x$ is an odd family of vector fields on the real line. For the family w the origin is an isolated singularity, hyperbolic for $\varepsilon \neq 0$.

Suppose that there is a smooth family $h = h(x, \varepsilon)$ conjugating two maps Δ_{γ_1}, Δ_{γ_2}, both of the form (6.5) with different vector fields w: w_1 and w_2 respectively. The above-mentioned hyperbolicity implies, that:

(1) the function h is almost odd in the given coordinate: $h(x, \varepsilon) + h(-x, \varepsilon) = $ (flat function);
(2) h almost conjugates the corresponding fields:

$$h_* w_1 - w_2 = \text{vector field, flat at the origin for all } \varepsilon.$$

The proof of both statements deals with formal Taylor series. It is almost evident when $\varepsilon \neq 0$ and the fields $w - 1$, w_2 are equivalent to linear ones near the singular point $x = 0$. The case $\varepsilon = 0$ comes by continuity.

Let $\mathbf{V}_1$, $\mathbf{V}_2$ be two families in the form (6.3) with semimonodromies as above, conjugated by a smooth map $\mathbf{h} = (h, \text{id})$. Consider the antisymmetrization $\widetilde{h}(x, \varepsilon) = (1/2)(h(x, \varepsilon) - h(-x, \varepsilon))$, which differs from h by a flat function. This antisymmetrization can be extended to a plane transformation $\widetilde{H}(z, \varepsilon) = (z/|z|)\widetilde{h}(|z|, \varepsilon)$. Since $\widetilde{h}$ is odd, $\widetilde{H}$ is smooth. After applying $\widetilde{h}$ to the first family $\mathbf{V}_1$ we obtain another family (denoted again by the same symbol), so that:

(1) $\mathbf{V}_1$ is again in the form (6.3);
(2) its symmetrical part coincides with that of $\mathbf{V}_2$ up to a flat symmetrical field;
(3) there exists a conjugation (denoted again by h) between semimonodromies of $\mathbf{V}_i$, which differs from the identity by a flat function.

This list of properties implies the smoothness of the map H defined by (6.2). Indeed, one can easily check that all maps in the superposition have the form id+(flat function). For h this is explicitly asserted above. The flow maps g_i^t for fields $\mathbf{V}_i$ differ by a flat function uniformly in t, since their ∞-jets at the origin coincide. Finally, the polar angle function $\text{Arg } z$ along with all its derivatives has no more than polynomial growth in $|z|^{-1}$ as $|z| \to 0$. So the whole superposition differs from the identity by a flat function, thus being smooth at the origin.

6.5. Realization. Consider a flip family $\mathbf{F}$. Let us prove that there exists an AH-family of vector fields on the plane which has the semimonodromy conjugated to $\mathbf{F}$.

To do that, we shall use the sectorial embedding theorem for flip families established in §5. Let $\overline{\Omega}_+$, $\overline{\Omega}_-$ be the standard domains in which the given flip family has the form

$$\mathbf{F} = \sigma \circ g_w^{1/2}, \quad w = x(x^2 - \varepsilon)(1 + b(\varepsilon)x^2)\partial/\partial x, \quad \sigma(x, \varepsilon) = (-x, \varepsilon), \quad (6.6)$$

and denote by Φ the odd transition function between the normalizing charts.

Define two domains $S_{\pm}$ in $\mathbb{C} \times (\mathbb{R}^1, 0)$ as the sets of (z, ε) such that $(|z|, \varepsilon)$ belongs to $\overline{\Omega}_{\pm}$. In each domain $S_{\pm}$ consider the symmetrical family of vector fields

$$\mathbf{V}_{\pm} = z[2\pi i + (z\bar{z} - \varepsilon)(1 + b(\varepsilon)z\bar{z})]\partial/\partial z, \qquad z \in S_{\pm}.$$

If we choose the positive semiaxis as the semitransversal γ, then the semimonodromy of $\mathbf{V}_{\pm}$ (where it is defined), has the standard form (6.6). Denote the flow maps of these fields by $g_{\pm}^t$ respectively.

Now consider the intersection $S_+ \cap S_-$. This intersection is the region between two right cones centered around the positive semiaxis. In each plane $\varepsilon = \text{const} > 0$ it is seen as the annulus $\{\varepsilon < |z| < 4\varepsilon\}$. The transition function Φ restricted to the plane $\varepsilon = \text{const}$ can be identified with a map of this annulus onto a nearby annulus, odd and commuting with the time 1 map for the flow of the field w.

Define a map $\mathbf{H}$ of the intersection $S_+ \cap S_-$ into a nearby intersection by the formula

$$\mathbf{H}(z, \varepsilon) = g_-^{(2\pi)^{-1} \operatorname{Arg} z} \circ \Phi \circ g_+^{-(2\pi)^{-1} \operatorname{Arg} z} z.$$

One can see that $\mathbf{H}$ is well defined because Φ and $g_{\pm}^{1/2}$ commute. Moreover, since Φ differs from the identity by a function flat at the origin, the same is true for $\mathbf{H}$.

Using $\mathbf{H}$ as the identifying map, one obtains from the disjoint union $S_+ \cup S_-$ an abstract 3-dimensional manifold with a foliation originating from $\varepsilon = \text{const}$ and with a vector field tangent to this foliation.

The general argument described at least twice (in 3.2 above and in Paper I) is now applicable: we endow the quotient space with a smooth structure using a partition of unity and then show that the field obtained is indeed of the type required. By construction, this field has the given family $\mathbf{F}$ as the semimonodromy.

The proof of the classification theorem for Andronov-Hopf families is complete.

Part II. Proof of the sectorial embedding theorem

§7. Jets and operations on them

From now on our exposition will be entirely devoted to the proof of the sectorial embedding theorem, see 3.1. The embedding is implemented in two steps. In both steps our main tools will be jets and flat functions, defined below. A jet is, roughly speaking, a collection of derivatives with prescribed values on a given fixed subset $K \subseteq \mathbb{R}^n$.

The first step is to solve the embedding problem in jets on an appropriate set. After this task being completed, we obtain a smooth field whose time 1 map coincides with the given SN-family up to a correction which is flat on

K. This means that all derivatives of the correction are rapidly decreasing if their argument tends to K.

The second step is to show that two SN-families differing by a flat correction are smoothly conjugate. In fact this is not true, but the conjugation exists in the sector-like domains $\Omega_{\pm}$ introduced above. The main reason that this step can be carried out is that the set K contains not only fixed points of $\mathbf{F}$, but also the entire fiber corresponding to nonhyperbolic map.

REMARK. The above approach can be implemented not only in the case of SN-families. For example, the Sternberg linearization theorem for hyperbolic maps can be proved by using it. In this case the set K is simply the origin, the corresponding jets are identified with formal Taylor series. The established Poincaré-Dulac theory of formal normal forms provides us with the means to implement the first step. The second one is more or less standard. This proof is given in detail in [IY].

7.1. Jets. The notion of formal series is a particular case of the more general construction known as smooth function in the sense of Whitney, or integrable jet.

DEFINITION. Let K be a compact subset of the space $\mathbb{R}^n$ with fixed coordinates $X = (X_1, \ldots, X_n)$. By a *k-jet of functions on K* for $k \leqslant \infty$ we shall mean a family of functions $f_\alpha : K \to \mathbb{R}^n$ continuous on K and indexed by a multi-index $\alpha \in \mathbb{Z}_+^n$, $|\alpha| \leqslant k$ (where, as usual, $|\alpha| = \alpha_1 + \cdots + \alpha_n$).

We shall denote jets by hats over symbols: if not specified otherwise, only ∞-jets will be considered. If necessary, the set on which a jet is defined, will be indicated by the subscript, e.g., $\widehat{f}_K$. The set of all k-jets on the set K is denoted by $J^k(K)$, that of ∞-jets, by $J(K)$.

A jet of order k can be truncated to a jet of order k' if $k' < k \leqslant \infty$. This truncation endows the set of ∞-jets with the structure of projective limit of spaces of finite order jets.

DEFINITION. A k-jet $\widehat{f} = \{f_\alpha\}|_{|\alpha| \leqslant k}$, $k \leqslant \infty$, is called *integrable*, if there exists a function $f : \mathbb{R}^n \to \mathbb{R}$ of class C^k such that for all α, $|\alpha| \leqslant k$, its derivative of order α, when restricted to K, coincides with f_α:

$$D^\alpha f|_K = f_\alpha.$$

Any function f satisfying this condition will be called a *continuation* of the jet $\widehat{f}$.

EXAMPLES. 1. If $K = \{0\} \in \mathbb{R}^n$, then an ∞-jet is a collection of reals $\{c_\alpha\}$. From the Borel-Whitney theorem it follows that all such jets are integrable.

2. If $K \simeq \mathbb{R}^{n-k}$ is the coordinate plane $\{X_1 = \cdots = X_k = 0\}$, then all integrable jets on K can be identified with semiformal series of the form

$$\sum_{\beta = (\beta_1, \ldots, \beta_k, 0, \ldots, 0), \beta_i \geqslant 0} f_\beta(X_{k+1}, \ldots, X_n) X^\beta$$

with f_β being smooth functions on K.

A continuation of a given jet, if it exists, is not unique in general. But any two continuations f, $\tilde{f}$ of a given jet $\widehat{f}_K$ differ by a function $\varphi = f - \tilde{f}$ which is *flat* on K in the following sense.

DEFINITION. A smooth function $\varphi \colon \mathbb{R}^n \to \mathbb{R}$ is said to be *flat* on a compact subset $K \subseteq \mathbb{R}^n$, if

$$\forall \alpha \in \mathbb{Z}_+^n, \ \forall N \in \mathbb{N} \ \exists C = C_{\alpha, N} \in \mathbb{R}: \quad |D^\alpha \varphi(X)| \leq C \operatorname{dist}(X, K)^N. \quad (7.1)$$

REMARK. The notion of jet as it was introduced above is very close to that defined algebraically. Let K be an irreducible analytic subset with I_K denoting its ideal, i.e., the set of functions vanishing on K. An algebraic k-jet is by definition an equivalence class modulo $(I_K)^k$. Clearly, only integrable jets can fit this definition, since equivalent functions have the same derivatives.

7.2. Continuation principles. Now we state the fundamental principle of continuation of jets. The criterion of integrability for jets looks like a consequence of the Taylor formula. If a jet $\widehat{f} = \{f_\alpha\}$ is integrable, then by the Taylor formula in the form of Peano one has the following relations between its components: for any finite k not exceeding the order of the jet,

$$f_\alpha(X) = \sum_{\beta \in \mathbb{Z}_+^n, |\beta| \leq k - |\alpha|} f_{\alpha+\beta}(Y) \frac{(X-Y)^\beta}{\beta!} + o(|X-Y|^{k-|\alpha|}). \quad (7.2)$$

It turns out that the condition (7.2) is not only necessary, but also sufficient for the integrability of a jet.

WHITNEY CONTINUATION THEOREM [M]. 1. *For a given k-jet $\widehat{f}$ of finite order to be integrable it is necessary and sufficient that the conditions* (9.1) *hold for all $|\alpha| \leq k$ uniformly over $X, Y \in K$.*

2. *An infinite jet $\widehat{f}$ (with the order $k = \infty$) is integrable if and only if all its finite truncations are integrable.*

REMARK. The complete formulation is given only for aesthetic reasons: in fact, we shall use only the second assertion of the theorem.

The last general point to be discussed is integrability of jets defined on the union of two compact sets $K, L \subseteq \mathbb{R}^n$. Suppose that two integrable jets $\widehat{f}_K, \widehat{f}_L$ are given, coinciding on the intersection: $\widehat{f}_K = \widehat{f}_L|_{K \cap L}$. One can formulate an explicit condition on the mutual position of the two sets, guaranteeing that the "combined" jet

$$\widehat{f}_{K \cup L}(X) = \begin{cases} \widehat{f}_K(X) & \text{if } X \in K, \\ \widehat{f}_L(X) & \text{if } X \in L, \end{cases}$$

is also integrable. This condition, called the Lojasiewicz condition [M], is automatically satisfied if both K and L are analytic subsets. This will be sufficient for our purposes.

7.3. Operations with jets. The notion of the jet of functions leads to definitions of other objects of analogous nature, such as jets of vector fields and multidimensional maps. We will deal mostly with fibered maps of the form $\widehat{\mathbf{F}} = (\widehat{F}, \text{id})$ and vector fields $\widehat{v}\, \partial/\partial x$ parallel to the x-axis (we assume that $X = (x, \varepsilon) \in \mathbb{R}^2$). The corresponding "coordinate" jets will be considered on one of the following sets: the origin $O \in \mathbb{R}^2$, the parabola Γ of fixed points of SN-families in preliminary normal form, the x-axis L and the total critical set $K = \Gamma \cup L$.

The problem is to define geometrical operations on such objects. For example, what is the time 1 map for a jet of a vector field? The answer, at least for integrable jets, is the following. Take any continuation of the jet and consider the result of the corresponding geometrical operation applied to a smooth object (e.g., the field), then take the jet of the result and check that the procedure is independent of the choice of the continuation. Only the last step can be nontrivial, although in what follows we will omit its examination. Let us finish this section with examples.

EXAMPLES. 1. Let $\widehat{F} = (\widehat{f}_1, \dots, \widehat{f}_n)$, $\widehat{f}_i \in J(O)$, be a jet "preserving the origin", that is, $\widehat{f}_{i,0}(0) = 0$. Then for any $\widehat{g} \in J(O)$ the superposition $\widehat{g} \circ \widehat{F}$ is defined. Indeed, independence of the continuation of g is evident. As for the choice of continuation of $\widehat{F}$, it is clear that the difference between $g \circ F$ and $g \circ (F + \varphi)$ is flat at the origin, provided that φ is.

Another way to check that the superposition is well defined is to compute all formal derivatives of $g \circ F$ at the origin and express them through the derivatives of g and f_i constituting their jets.

2. If $\widehat{v}_L\, \partial/\partial x$ is a jet of a vector field on the x-axis L in the plane, then its time 1 map is well defined in jets $J(L)$. This follows from the fact that L consists of the entire orbits of fields parallel to it: the difference between flow maps for fields $v\, \partial/\partial x$ and $(v + \varphi)\, \partial/\partial x$ is flat on L, provided that φ is; this is a consequence of the equation in variations along the given trajectory of the vector field.

§8. Embedding problem in the space of jets on the x-axis

In this section we solve the standard embedding problem in jets on the x-axis, that is, modulo a correction flat on this set.

To find a solution, we replace the equation $g_{\widehat{v}}^1 = \widehat{\mathbf{F}}$, which is nonlinear in $\widehat{v}$, by another, called the *transfer equation*. The latter is linear.

As a preliminary step we find a solution to the embedding problem in jets at the origin. The following general result is due to F. Takens [T1].

8.1. Embedding problem in jets at the origin. Recall that a jet at the origin O can be identified with a formal Taylor series. Let

$$\widehat{F}_O = (\widehat{f}_1, \dots, \widehat{f}_n), \qquad \widehat{f}_k = \{f_{k,\alpha} : \alpha \in \mathbb{Z}_+^n\} \in J(O),$$

be an integrable jet of a map $F : \mathbb{R}^n \to \mathbb{R}^n$ with free terms equal to zero. The embedding problem in jets at the origin consists in finding a jet of vector fields $\hat{v} = \hat{v}_1 \partial/\partial X_1 + \cdots + \hat{v}_n \partial/\partial X_n$, $\hat{v}_j \in J(O)$, such that $g_{\hat{v}}^1 = \hat{F}_O$. Such a jet is called the *formal generator* of $\hat{F}$.

THEOREM [T1]. *Suppose that the linear terms* $\|f_{k,\alpha}\|_{k=1,\ldots,n,|\alpha|=1}$ *of the jet* $\hat{F}$ *are of the form* id$+N$, *where* N *is a nilpotent linear operator. Then there exists a unique jet of vector fields* $\hat{v}$ *which is the formal generator of* $\hat{F}$.

COROLLARY 1. *If* $\mathbf{F} = (F, \mathrm{id})$ *is a smooth saddle-node family, then there exists a smooth local family* $v(x, \varepsilon) \partial/\partial x$ *of vector fields on the line such that*

$$\mathbf{F} = g_v^1 + (\varphi, 0),$$

where $\varphi = \varphi(x, \varepsilon)$ *is a smooth germ flat at the origin:* $j_{(0,0)}^\infty \varphi = 0$.

Indeed, find a formal generator $\hat{v}$ and extend it to a smooth fibered field v. The (smooth) time 1 map g_v^1 will have the same Taylor expansion as $\mathbf{F}$. □

Another result by Takens asserts the *exact* solvability of the embedding problem for *one-dimensional smooth maps tangent to the identity with finite order*.

THEOREM [T2]. *Let*

$$f: (\mathbb{R}^1, 0) \to (\mathbb{R}^1, 0), \quad f(x) = x + cx^k + \cdots, \quad c \neq 0,$$

be a smooth map. Then there exists a unique smooth vector field

$$v_f = (cx^k + \cdots) \partial/\partial x$$

such that $f = g_{v_f}^1$.

COROLLARY 2. *For any SN-family* $\mathbf{F} = (F, \mathrm{id})$ *the restriction* $F|_{\varepsilon=0}$ *admits an embedding in a smooth flow; the ∞-jet of the corresponding generator at the origin* $x = 0$ *is unique.*

8.2. The transfer equation. From now on we shall consider jets on the x-axis L, if not specified otherwise.

If $F: (\mathbb{R}^n, 0) \to (\mathbb{R}^n, 0)$, $F = (f_1, \ldots, f_n)$ is a smooth flow map,

$$F = g_v^1, \quad v = \sum v_j \partial/\partial X_j,$$

then the field v is preserved by F: $F_* v = v$, or, in coordinate form,

$$\sum_j \frac{\partial f_i}{\partial X_j} v_j(X) = v_i(f_1(X), \ldots, f_n(X)). \tag{8.1}$$

In general, the inverse statement is not true: not every field preserved by F is a generator for the map. But in the case of lowest dimension, under some reasonable conditions imposed on F any nonzero field preserved by F generates F after multiplication by an appropriate function.

The corresponding assertion for the space of jets on L is the following.

LEMMA 1. *Let $\widehat{\mathbf{F}}_L = (\widehat{F}_L, \mathrm{id})$ be a jet of a saddle-node family in the preliminary normal form. Suppose that a jet $\widehat{v}_L \partial/\partial x$ of a vector field satisfies the formal counterpart of the equation (8.1), called the transfer equation:*

$$(\partial \widehat{F}/\partial x) \cdot \widehat{v} = \widehat{v} \circ \widehat{\mathbf{F}}, \tag{8.2}$$

where the subscript L is omitted for simplicity. If $v_0(x) = x^2 + \cdots$, then there exists a formal series $\widehat{\tau} = 1 + \sum \tau_j \varepsilon^j$ such that $g^{\widehat{\tau}}_{\widehat{v}\partial/\partial x} = \widehat{\mathbf{F}}$.

PROOF. Let v_{F_0} be the generator of F_0 given by Corollary 2 of 8.1. We shall prove that

$$v_0(x) = \partial/\partial x = v_{F_0}(x).$$

Indeed, all the vector fields in Lemma 1 are parallel to the x-axis, and therefore they differ by a factor at any nonsingular point. Moreover, $v_0 = x^2(1 + o(1))$ by assumption, and $v_{F_0} = (x^2 + \cdots)\partial/\partial x$. Thus

$$v_0^{(x)}\partial/\partial x = \alpha(x)v_{F_0}^{(x)}$$

for some smooth function α: $\alpha(0) = 1$; the equality holds in some interval containing zero. Both of the fields $v_0 \partial/\partial x$ and v_{F_0} are preserved by the map F_0; therefore, so is the function α. Hence, by continuity, $\alpha(x) \equiv \alpha(0) = 1$.

The key idea of the next step is the following. Let $v\partial/\partial x$ be a vector field in $(\mathbb{R}^2, 0) \cap \{\varepsilon < 0\}$ having no singular points, and let $\mathbb{F}$ be a saddle-node family. Define the function T:

$$T(x, \varepsilon) = \int_x^{F(x,\varepsilon)} \frac{d\xi}{v(\xi, \varepsilon)}, \qquad \varepsilon < 0.$$

Then

$$g^{T(x,\varepsilon)}_{v\partial/\partial x}(x, \varepsilon) = \mathbb{F}(x, \varepsilon), \qquad \varepsilon < 0.$$

Now let v and $\mathbb{F}$ be the extensions of $\widehat{v}$ and $\widehat{\mathbb{F}}$ in equation (8.2). Suppose that the function T is defined by the preceding formula precisely for these v and $\mathbb{F}$. Then we shall prove the following two statements:

(1)
$$|v(x, \varepsilon)| > \varepsilon/2$$

for $\varepsilon > 0$, x and ε small enough.

(2)
$$T(x, \varepsilon) = \tau(\varepsilon) + \flat_L^-,$$

where $\tau(\varepsilon) = T(a, \varepsilon)$ for some small a (the choice of a is of no importance); $\flat_L^-$ denotes a function that is flat on L and defined for $\varepsilon \leq 0$ (respectively, in a neighborhood of L). Note that the function τ has a smooth extension into an entire neighborhood of zero.

Assertion (2) implies
$$g^1_{\tau v \partial/\partial x} - F = \flat_L$$
and therefore proves the lemma. It remains to prove (2); for this we need (1). But (1) is an immediate consequence of the decomposition
$$\hat{v} = x^2 - \varepsilon + O(x^3) + O(x\varepsilon) + O(\varepsilon^2),$$
which follows directly from the assumption of the lemma. Now let us prove (2). We have
$$T(x, \varepsilon) - \tau(\varepsilon) = T(x, \varepsilon) - T(a, \varepsilon) = \int_a^x \frac{d\xi}{v(\xi, \varepsilon)} - \int_{F(a,\varepsilon)}^{F(x,\varepsilon)} \frac{d\eta}{v(\eta, \varepsilon)}.$$
The coordinate change $\eta = F(\xi, \varepsilon)$ together with (8.2) implies
$$\int_{F(a,\varepsilon)}^{F(x,\varepsilon)} \frac{d\eta}{v(\eta, \varepsilon)} = \int_a^x \frac{d\xi}{v(\xi, \varepsilon) + \flat_L^-} = \left(\int_a^x \frac{d\xi}{v(\xi, \varepsilon)}\right)(1 + \flat_L^-).$$
The last equality follows from (1). This implies (2). □

8.3. Solution of the transfer equation in jets.

LEMMA 2. *The transfer equation admits a solution $\hat{v} \simeq \sum v_j(x)\varepsilon^j$ in $J(L)$ such that $v_0(x)\partial/\partial x$ is the generator for $F(\cdot, 0)$.*

PROOF. Let $\hat{v}_0$ be the formal generator for $\widehat{F}$ constructed by using the first Takens theorem. Take any continuation of $\hat{v}_0$ and consider the jet $\hat{v}_L$ of this continuation on L. Since the transfer equation is linear, we shall seek its solution in the form $\hat{v}_L + \hat{w}_L$. For simplicity, we write $\hat{v}$, $\hat{w}$ instead of $\hat{v}_L$, $\hat{w}_L$. The jet $\hat{w}$ must satisfy a nonhomogeneous transfer equation
$$\frac{\partial \widehat{F}}{\partial x}\hat{w} - \hat{w} \circ \widehat{F} = \hat{h} \stackrel{\text{def}}{=} \frac{\partial \widehat{F}}{\partial x}\hat{v} - \hat{v} \circ \widehat{F}. \qquad (8.3)$$
By choice of $\hat{v}$, the jet $\hat{h} \simeq \sum h_j(x)\varepsilon^j$ has flat coefficients h_j at $x = 0$.

Let us prove, using induction, that (8.3) admits a solution $\hat{w} \simeq \sum w_j(x)\varepsilon^j$ with all w_j flat at $x = 0$. The equation in jets is equivalent to a series of functional relations between coefficients w_j, h_j, f_j. The first one corresponds to the zero order terms:
$$\frac{\partial f_0}{\partial x} w_0 - w_0 \circ f_0 = h_0. \qquad (8.4_0)$$
The other equalities possess an analogous form:
$$\frac{\partial f_0}{\partial x} w_j - w_j \circ f_0 = h_j + \begin{pmatrix} \text{polynomial combination of functions} \\ w_k \text{ with } k < j \text{ and their derivatives} \\ \text{with smooth coefficients} \end{pmatrix}. \qquad (8.4_j)$$

This is easily proved by using the Taylor formula for the superpositions $w_k \circ \widehat{F}$ when $k \leqslant j$.

Note that all equations (8.4_j) for all $j \geq 0$ also have the form of a transfer equation, though nonhomogeneous: if we interpret them in invariant terms, then the right-hand sides as well as the unknown functions w_j will be coordinates of vector fields on the real line, which must satisfy certain conditions with respect to the map f_0 of the line into itself. So we may choose another chart on $\mathbb{R}$ in order to prove that all equations (8.4_j) admit smooth solutions flat at $x = 0$.

The second theorem of Takens implies that the map f_0 can be embedded in the flow of the field $(x^2 + \cdots)\partial/\partial x$. Consider first the negative semineighborhood $0 > x > -\delta$ endowed with the chart

$$t(x) = \int_{-\delta}^{x} \frac{ds}{(s^2 + \cdots)},$$

taking the generator to the constant field $\partial/\partial t$ and its time 1 map f_0 to the unit shift $\mathrm{id} + 1$, both defined on the semiaxis $t > 0$. Note that the function t has no more than polynomial growth at $x = 0$, and the same is true for its derivatives and the inverse function $x = x(t)$. In this chart the transfer equations (8.4_j) take the form

$$\widetilde{w}_j(t+1) - \widetilde{w}_j(t) = \widetilde{h}_j(t) + \begin{pmatrix} \text{polynomial combination of } \widetilde{w}_k \\ \text{and their derivatives for } k < j \\ \text{with smooth coefficients of no} \\ \text{more than polynomial growth} \end{pmatrix} \quad (8.5_j)$$

(for $j = 0$ the last term disappears). Since all h_j are flat at the origin, the functions $\widetilde{h}_j$ are flat at infinity, that is, they decrease together with all their derivatives more rapidly than any finite power t^{-N} as $t \to +\infty$.

Denote the right-hand side of (8.5_j) by r_j. By summing, we obtain the formal solution

$$\widetilde{w}_j(t) = -\sum_{k \geq 0} r_j(t+k). \tag{8.6}$$

In fact, the series (8.6) converges to a smooth function flat at infinity. If, using induction, we have proved that all the equations for $k < j$ admit flat solutions, then the function r_j will also be flat at infinity. For such functions the above assertion on convergence of the series becomes evident: indeed, we can differentiate (8.6) as many times as we wish; the right-hand side will be majorized by the convergent series $\sum_k k^{-N}$ with N as large as we wish. So by induction we deduce that the transfer equation corresponding to the unit shift on the semiaxis admits a unique smooth solution flat at infinity, provided that the right-hand side is flat. After returning to the initial chart x, we conclude that the initial transfer equations (8.4_j) admit solutions on $0 > x > -\delta$, flat at $x = 0$, since flatness overcomes the polynomial growth of the transition function between the charts x and t. The same is evidently true for the positive semineighborhood, whence we get the solvability of the transfer equations in jets on L. The desired property of $v_0(x)\partial/\partial x$ follows from Lemma 1. □

§9. Solution of the embedding problem in jets on the parabola Γ and on the critical set $K = L \cup \Gamma$

We start with some facts concerning the embeddings of automorphisms of commutative algebras in a one-parameter group of such automorphisms. The algebraic formulation is as follows.

Let $F = (F_1, \ldots, F_n) \colon \mathbb{R}^n \to \mathbb{R}^n$ be a smooth map and K be the set of its fixed points. Suppose K is compact. Consider the commutative associative algebra $\mathscr{E}_K$ of germs on K of smooth functions defined in a neighborhood of K.

For any germ $f \in \mathscr{E}_K$ the germ $f \circ F$ of the same nature is defined. The mapping

$$F^* \colon \mathscr{E}_K \to \mathscr{E}_K, \qquad F^* f = f \circ F$$

is an automorphism of the algebra. Conversely, any automorphism A of the analytic subalgebra $\Omega_k \subset \mathscr{E}_K$ is generated by some smooth map: let $F_j \stackrel{\text{def}}{=} AX_j$ be the images by A of the coordinate functions X_j. It is easy to prove that $A = F^*$, where $F = (F_1, \ldots, F_n)$.

If there is a one-parameter group of automorphisms A^t, $t \in \mathbb{R}$, then the corresponding germs F^t also form a group: $F^t \circ F^s = F^{t+s}$. So we can define the germ of the generator $v = (d/dt)|_{t=0} F^t$. Thus the embedding problem at the algebraic level consists in embedding a given automorphism into a one-parameter group of automorphisms. We will call the corresponding problem the *algebraic embedding problem*.

Now let us pass to jets. If K consists of fixed points of F, then we can define the corresponding action F_k^* on the set $J^k(K)$ of k-jets. For any smooth function f the derivatives of the superposition $f \circ F$ of order no greater than k can be expressed via those of f and F of order $\leqslant k$. So we may use the corresponding formulas to define the action of the jet $\widehat{F}_K \in (J^k(K))^n$ on $J^k(K)$.

Integrable jets constitute not only a set, but a commutative algebra. The corresponding action of the *integrable* jet $\widehat{F}$ is an automorphism of the algebra. The embedding problem in jets thus acquires its algebraic formulation.

If A is a finite-dimensional automorphism, sufficiently close to the identity, then the corresponding one-parameter subgroup can be written explicitly in the form of a convergent power series. The solution of the algebraic embedding problem is much more difficult if the corresponding algebra is infinite-dimensional. We are interested in the case of jets on the parabola Γ. Even finite jets on Γ constitute an infinite-dimensional algebra. The key point is to use the fibered structure of the map $\widehat{F}$ in order to endow this algebra with the structure of a *smooth family of finite-dimensional algebras over* $\mathbb{R}$.

9.1. Local families of finite-dimensional algebras and the binomial expansion.

DEFINITION. A family $Q(\varepsilon)$, $\varepsilon \in (\mathbb{R}^1, 0)$, of finite-dimensional commutative algebras of the same dimension d over $\mathbb{R}$ is called *smooth*, if there exists a basis $e_1(\varepsilon), \ldots, e_d(\varepsilon) \in Q(\varepsilon)$ depending on the parameter such that the structural constants are smooth functions of the parameter:

$$e_i(\varepsilon)e_j(\varepsilon) = \sum_{k=1}^{d} c_{ij}^k(\varepsilon) e_k(\varepsilon), \qquad c_{ij}^k(\cdot) \in C^\infty(\mathbb{R}^1, 0).$$

A family $A(\varepsilon): Q(\varepsilon) \to Q(\varepsilon)$ of automorphisms of the algebras is called *smooth*, if in the above basis all its matrix elements are smooth functions of the parameter.

With evident modifications this definition can be extended to the case of a parameter ranging over a semineighborhood $(\mathbb{R}^+, 0)$ with smoothness in the sense of Whitney.

We show that under certain reasonable conditions, a finite dimensional automorphism can be embedded in a group, this embedding being smooth in additional parameters, if any.

LEMMA 1. *Let* $A = A(\varepsilon)$ *be a smooth family of automorphisms of a smooth local family* $Q(\varepsilon)$ *of finite-dimensional commutative algebras over* $\mathbb{R}$. *Suppose that* $A(0) = \mathrm{id} + N$, *where* N *is a nilpotent operator. Then in some neighborhood* $\varepsilon \in (\mathbb{R}^1, 0)$ *there exists a family* $A^t(\varepsilon)$ *of one-parameter groups, smooth both in* t *and* ε *simultaneously, such that*

$$\forall \varepsilon \in (\mathbb{R}^1, 0), \quad A^1(\varepsilon) \equiv A(\varepsilon).$$

This family can be given by the binomial series

$$A^t = (\mathrm{id} + (A - \mathrm{id}))^t \stackrel{\mathrm{def}}{=} \mathrm{id} + t(A - \mathrm{id}) + (1/2!)t(t-1)(A - \mathrm{id})^2 + \cdots \\ + (1/k!)t(t-1)\cdots(t-k+1)(A - \mathrm{id})^k + \cdots \quad (9.1)$$

(*dependence on the parameter is omitted for simplicity*).

REMARK. For $\varepsilon = 0$ the series (9.1) is in fact a finite sum, since $A(0) - \mathrm{id}$ is nilpotent by assumption.

PROOF. 1. First we show that the series (9.1) converges for all sufficiently small ε. Indeed, by choosing a suitable norm on $Q(0)$, one can get $\|A(0) - \mathrm{id}\| < 1/2$, because the differences are nilpotent. This norm can be continuously extended to $Q(\varepsilon)$ for ε sufficiently small, while preserving the above inequality for $A(\varepsilon)$ instead of $A(0)$.

Since the radius of convergence of the (scalar) binomial series

$$(1+z)^t = 1 + tz + (1/2!)t(t-1)z^2 + \cdots$$

is equal to 1 for all t, the series (9.1) can be majorized, hence it converges for all $t \in \mathbb{R}$ and $\|A^t - \mathrm{id}\| < 1$. The smoothness of its sum is evident. Note also that it depends analytically on the matrix elements of the automorphism A.

2. The group identity
$$A^t A^s = A^{t+s} \tag{9.2}$$
follows from the formal identity between scalar binomial series
$$\sum_{k=0}^{\infty} \binom{t}{k} z^k \cdot \sum_{k=0}^{\infty} \binom{s}{k} z^k = (1+z)^t (1+z)^s = (1+z)^{t+s} = \sum_{k=0}^{\infty} \binom{t+s}{k} z^k,$$
in which $A(\varepsilon) - \mathrm{id}$ should be substituted for z: since the powers of A commute and the series converge, this argument justifies property (9.2).

3. It remains only to verify that the sum of the series (9.1) for all t defines an automorphism of the corresponding algebra. This fact can be deduced from the general formula relating differentiations and automorphisms [P], but the proof given in [P] uses additional geometric structures, so we give an independent one.

Denote by $\mathrm{GL}(Q(\varepsilon))$ the finite-dimensional Lie group of linear transformations of the algebra $Q(\varepsilon)$. Automorphisms of $Q(\varepsilon)$ form a subgroup $\mathrm{Aut}(Q(\varepsilon))$ which itself is a Lie group. Denote by $B_\rho \subset \mathrm{GL}(Q(\varepsilon))$ the ball of radius ρ around the identity in the given norm.

Since the exponential map is a local diffeomorphism for any $A \in B_\rho$, there exists a unique one-parameter subgroup $\{A^t\}$ of $\mathrm{GL}(Q(\varepsilon))$ such that $A^1 = A$ and $\forall t \in [0, 1]$, $A^t \in B_\rho$, provided that ρ is sufficiently small. In particular, this holds for all $A \in \mathrm{Aut}(Q(\varepsilon)) \cap B_\rho$. But, since $\mathrm{Aut}(Q(\varepsilon))$ itself is a Lie group, there must exist a one-parameter subgroup of automorphisms with the same property. Therefore the uniqueness of the subgroup implies that the sum of the series (9.1), which gives the formula for the subgroup in $\mathrm{GL}(Q(\varepsilon))$, is in fact an automorphism, at least for A sufficiently close to the identity operator and for $t \in [0, 1]$. The rest follows from the analyticity of formula (9.1) in A and t: it defines an automorphism for all t, A such that the series converges. In order to complete the proof, note that A belongs to the circle of convergence, provided that $A(0) - \mathrm{id}$ is nilpotent.

The arguments given here also prove the first Takens theorem, see 8.1. □

9.2. Space of jets on the parabola. Consider a local SN-family **F** in preliminary normal form. Denote by $\mathscr{E}_{x,\varepsilon}$ the set of smooth germs of functions of two variables x, ε. By $\mathscr{E}_\varepsilon$ we denote the set of germs of smooth functions of the parameter $\varepsilon \in (\mathbb{R}^1, 0)$, and $\mathscr{E}_{\varepsilon, +}$ will stand for the set of germs in $(\mathbb{R}^1_+, 0)$, smooth in the sense of Whitney.

Consider the ideal $\mathfrak{n} \in \mathscr{E}_{x,\varepsilon}$, generated by a germ $x^2 - \varepsilon \in \mathscr{E}_{x,\varepsilon}$. The symbol $\mathfrak{Q}_N$ denotes the quotient algebra: $\mathfrak{Q}_N = \mathscr{E}_{x,\varepsilon}/\mathfrak{n}^N$. Clearly, this algebra can be identified with the subset of all integrable N-jets on the parabola, because any two functions differing by an element from $\mathfrak{n}^N$, have the same derivatives on it.

Define an automorphism $A: \mathscr{E}_{x,\varepsilon} \to \mathscr{E}_{x,\varepsilon}$ corresponding to the map $\mathbf{F}$:

$$\forall f \in \mathscr{E}_{x,\varepsilon}, \quad Af \stackrel{\text{def}}{=} f \circ \mathbf{F}. \tag{9.3}$$

Since $A\mathfrak{n} \subset \mathfrak{n}$, the quotient map $A_N : \mathfrak{Q}_N \to \mathfrak{Q}_N$

$$A_N(f \bmod \mathfrak{n}^N) = (Af) \bmod \mathfrak{n}^N$$

can be defined.

Suppose that we have found a one-parameter subgroup A_N^t of automorphisms of $\mathscr{E}_{x,\varepsilon}$ preserving $\varepsilon \in \mathscr{E}_{x,\varepsilon}$, in which A_N can be embedded. Consider their action on the element $x \in \mathscr{E}_{x,\varepsilon}$ of the algebra. More precisely, let $a_N^t = a_N^t(x, \varepsilon)$ be any representative of the germ $A_N^t x \in \mathfrak{Q}_N$. Denote

$$v_N = v_N(x, \varepsilon) = d/dt|_{t=0} a_N^t(x, \varepsilon).$$

Then, as one can easily check, the N-jet of v_N on the parabola (i.e., the equivalence class modulo $\mathfrak{n}^N$) is well defined, and

$$\mathbf{F} = g_v^1 \bmod \mathfrak{n}^N, \quad v = v_N(x, \varepsilon) \partial/\partial x,$$

so the embedding problem is solved in the space of jets on the parabola.

Unfortunately, $\mathfrak{Q}_N$ is not finite-dimensional, so we cannot apply the results of the preceding section directly. Instead we interpret $\mathfrak{Q}_N$ as a smooth family of algebras depending on the parameter ε. To do this, we need endow it with the appropriate structure.

Recall that by the Weierstrass division theorem any smooth function $f \in \mathscr{E}_{x,\varepsilon}$ can be expressed in the form

$$f(x, \varepsilon) = P_{2N-1}(x, \varepsilon) + (x^2 - \varepsilon)^N \varphi(x, \varepsilon), \tag{9.4}$$

where P is a polynomial in x with coefficients from $\mathscr{E}_\varepsilon$, of degree $\leqslant 2N-1$. The representation (9.4) is in general unique only in the analytical category. Nevertheless, all coefficients $p_j = p_j(\varepsilon)$, $j = 0, \ldots, 2N-1$, of P are uniquely defined for $\varepsilon > 0$: this follows from the fact that all roots of the polynomial $(x^2 - \varepsilon)^N$ are real for such values of the parameter.

Define the smooth family $Q_N = Q_N(\varepsilon)$, $\varepsilon \in (\mathbb{R}^1_+, 0)$ of finite-dimensional commutative algebras over $\mathbb{R}$ as follows. Choose the monomials $e_j = x^j$, $j = 0, \ldots, 2N-1$, as the basis and define the multiplication on the set of linear combinations $a_0(\varepsilon)e_0 + \cdots + a_{2N-1}(\varepsilon)e_{2N-1}$ by the formula

$$e_i \cdot e_j = \begin{cases} e_{i+j} & \text{if } i+j \leqslant 2N-1, \\ \sum_{k=0}^{\infty} c_k^{i+j}(\varepsilon) e_k & \text{otherwise.} \end{cases}$$

Here the c_{ij}^k are the coefficients of the Weierstrass representation for x^l with $l \geqslant 2N$:

$$x^l = \sum c_k^l(\varepsilon) x^k + (x^2 - \varepsilon)^N \varphi_l(x, \varepsilon). \tag{9.5}$$

There exists a natural projection from $\mathscr{E}_{x,\varepsilon}$ to $Q_N(\varepsilon)$: the image of any function $f \in \mathscr{E}_{x,\varepsilon}$ is interpreted as a multijet at the points $\pm\sqrt{\varepsilon}$ for $\varepsilon > 0$. Now define a smooth family of automorphisms $A_N(\varepsilon) \colon Q_N(\varepsilon) \to Q_N(\varepsilon)$: if

$$F(x, \varepsilon) = \sum a_k(\varepsilon) x^k + (x^2 - \varepsilon)^N \varphi_F(x, \varepsilon)$$

is the Weierstrass decomposition for the function $F(x, \varepsilon) = Ax$, then put

$$A_N(\varepsilon) e_1 = \sum a_k(\varepsilon) e_k, \qquad A_N e_0 = e_0,$$

and extend the morphism $A_N(\varepsilon)$ to the set of combinations $\sum f_k e_k$ using multiplicativity and additivity: $A_N e_k = (A_N e_1)^k$, $k \geq 2$.

The families of algebras and automorphisms constructed in this way possess the following properties:

(1) there exists a family of natural projections $\pi_N(\varepsilon) \colon \mathfrak{Q}_N \to Q_N(\varepsilon)$, taking each function $f \in \mathscr{E}_{x,\varepsilon}$ to the naturally ordered set of coefficients of its Weierstrass polynomial P defined by (9.4);

(2) the family $A_N(\varepsilon)$ of automorphisms form a "section" of A_N, such that the diagram

$$\begin{array}{ccc} \mathfrak{Q}_N & \xrightarrow{A_N} & \mathfrak{Q}_N \\ \pi_N(\varepsilon) \downarrow & & \downarrow \pi_N(\varepsilon) \\ Q_N(\varepsilon) & \xrightarrow{A_N(\varepsilon)} & Q_N(\varepsilon) \end{array}$$

is commutative;

(3) both families $Q_N(\varepsilon)$ and $A_N(\varepsilon)$ are smooth in the sense of the above definition.

PROPOSITION. *If $\mathbf{F}$ is an SN-family, then the automorphism $A_N(0)$ differs from the identity by a nilpotent term.*

PROOF. By definition, the action of $A_N(0)$ is given by the formula

$$\forall f \in \mathscr{E}_x, \qquad A_N(f \bmod x^{2N}) = f \circ F(\cdot, 0) \bmod x^{2N}.$$

Since $F(x, 0) = x + x^2 + \cdots$, we have

$$A_N(0) e_j = (x + x^2 + \cdots)^j \bmod x^{2N} = x^j + \cdots \bmod x^{2N} = e_j + \sum_{k > j} \alpha_k e_k,$$

and the matrix corresponding to it is upper triangular with units on the diagonal. □

So we have found a generator for $\mathbf{F}$ in N-jets on the small segment $\Gamma \cap \{|\varepsilon| < \delta_N\}$. Taking any continuation (in fact, it can be polynomial in x), we obtain the following

COROLLARY 1. *For any natural N there exists a polynomial vector field*

$$\mathbb{V}_N = v_N \frac{\partial}{\partial x}, \quad v_N = \sum_{j=0}^{2N-1} v_{N,j} x^j, \quad v_{N,j} = v_{N,j}(\varepsilon), \quad \varepsilon \in (\mathbb{R}_+^1, 0)_N,$$

such that

$$\mathbf{F} - g_{v_N}^1 = 0 \mod (x^2 - \varepsilon)^N. \tag{9.6}$$

REMARK. If we extend the coefficients $v_{N,j}$ to the negative semiaxis $\varepsilon \in (\mathbb{R}_-^1, 0)$, then the above equality still holds independently of the choice of the continuation.

Now we construct the entire ∞-jet of the generator on the entire Γ.

LEMMA 2. *If $\mathbf{F}$ is a saddle-node family in preliminary normal form, then there exists an integrable ∞-jet $\widehat{v}$ on the parabola Γ such that its formal time 1 map has the same ∞-jet on Γ as $\mathbf{F}$ does. This means that for any vector field $\mathbf{v} = v\partial/\partial x$, where v is a smooth continuation of $\widehat{v}$, the difference $\mathbf{F} - g_{\mathbf{v}}^1$ is flat on Γ.*

PROOF. From the hyperbolicity mentioned above, it follows that there exist open subsets $U_\pm \subset \mathbb{R}^n$ containing both "horns" $\Gamma_\pm = \{x = \pm\sqrt{\varepsilon}, \varepsilon > 0\}$ of the parabola Γ, such that in each of them there is a smooth generator $\mathbf{v}_\pm = v_\pm \partial/\partial x$, $v_\pm : U_\pm \to \mathbb{R}$. Define the ∞-jet $\widehat{v}$ on $\dot{\Gamma} = \Gamma \setminus \{0\} = \Gamma_+ \cup \Gamma_-$ as the restriction of the functions $v_\pm$ and its derivatives. By definition, the jet $\widehat{v}$ is integrable on every compact subset of $\dot{\Gamma}$. Moreover, the hyperbolicity arguments imply that the jet of the generator is uniquely defined on $\dot{\Gamma}$: if any other vector field $\widetilde{\mathbf{v}} = \widetilde{v}\partial/\partial x$ satisfies the property

$$g_{\widetilde{\mathbf{v}}}^1 = 0 \mod (x^2 - \varepsilon)^N, \tag{9.7}$$

then the $(2N-1)$-jet of the corresponding function $\widetilde{v}$ on $\dot{\Gamma}$ is the same as that of $v_\pm$. We assert that all the functions v_α constituting the jet $\widehat{v}$ admit a (unique) continuous extension to the vertex of the parabola, forming an integrable ∞-jet on Γ.

Indeed, Corollary 1 provides us with the sequence $\mathbf{v}_N = v_N \partial/\partial x$ of vector fields, each of them smooth in its own neighborhood U_N of the origin, such that condition (9.7) holds. Since all the U_N have nonempty intersections with $\dot{\Gamma}$, the derivatives $D_{x,\varepsilon}^\alpha v_N$ are uniquely defined independently of N, provided that $|\alpha| \leqslant N$, continuous on their domains and coincide with v_α on $\dot{\Gamma}$. Therefore each v_α admits a unique continuous extension to the origin as the restriction of $D^\alpha v_N$ for any $N > |\alpha|$. Denote this extended ∞-jet by the same symbol $\widehat{v}$.

Let us prove the integrability of $\widehat{v} = \{v_\alpha\}$. In fact, it is evident. Fix any $N < \infty$. The entire parabola is covered by three open sets U_+, U_-, U_N. By construction, $\widehat{v}_N = \{v_\alpha\}|_{|\alpha| \leqslant N}$ is integrable in these sets: there exist smooth continuations v_+, v_-, v_N of the jet $\widehat{v}_N$ in each of them. Consider a partition of unity $1 = \varphi_+ + \varphi_- + \varphi_N$ subordinate to this covering and such that the sets $\operatorname{supp} \varphi_\pm$ do not contain the origin. It is evident that the combination $\varphi_+ v_+ + \varphi_- v_- + \varphi_N v_N$ is the continuation of the jet $\hat{v}_N$ which is therefore proved to be integrable. □

Our next step is to show that the integrable ∞-jet $\hat{v}$ on Γ, together with the ∞-jet on the x-axis constructed in the preceding section, constitutes an integrable jet on the union of these sets. This is a consequence of a well-known result by Lojasiewicz [M].

The formal generator (i.e., the solution of the embedding problem in ∞-jets at the origin) is unique by the Takens theorem, see 7.1. Therefore the restriction of the jet $\hat{v}$ to the origin must coincide with it as well as the restriction of the semiformal generator. Since both Γ and the x-axis are analytic subsets of the plane, they are regularly positioned in the sense of Malgrange [M], so any two integrable ∞-jets coinciding on their intersection constitute an integrable jet on the union of these subsets ([M], Theorem 5.5).

COROLLARY 2. *For a given saddle-node family* **F** *in preliminary normal form, there exists a smooth vector field* $\mathbf{v} = v \partial/\partial x$ *such that the difference* $\mathbf{F} - g_{\mathbf{v}}^1$ *is flat on the set* $\varepsilon\{x^2 - \varepsilon\} = 0$.

PROOF. Take any continuation v of the "combined" jet. □

§10. Sectorial embedding

10.1. Reduction to the conjugation problem. In the preceding sections we have solved the embedding problem in jets on the critical set $K = \{\varepsilon(x^2 - \varepsilon) = 0\}$. Take any continuation of the corresponding jet $\hat{v}_K \partial/\partial x$ of vector fields and denote by $\widetilde{\mathbf{F}}$ its time 1 map. By definition of jets, the difference between $\mathbf{F}$ and $\widetilde{\mathbf{F}}$ is a map $(\varphi, 0)$ with first coordinate φ flat on K. Our task will be finished if we prove that any two such maps are conjugated by a (fibered) transformation defined and smooth in the sector-like domains $\Omega_\pm$. As before, the reasoning is most conveniently carried out in appropriate charts.

The parabola Γ divides both domains, each into two parts. We prove that in each part there exists a smooth fibered transformation $\mathbf{H} = (H, \mathrm{id})$ conjugating the two maps and flat on K. All four cases are treated similarly, so we restrict ourselves to the domain

$$\Omega = \{-\delta^2 < \varepsilon < 0, \ |x| < \delta\} \cup \{0 \leqslant \varepsilon < \delta^2, \ -\delta < x < -\sqrt{\varepsilon}\}.$$

CONJUGATION THEOREM. *For any function φ flat on the set K there exists a smooth map (H, id) conjugating $\widetilde{\mathbf{F}} = g^1_{v \partial/\partial x}$ and $\widetilde{\mathbf{F}} + (\varphi, 0)$ in Ω; the map H differs from the identity map by a term which is flat on K.*

Clearly, the positive solution of the sectorial embedding problem follows immediately from this theorem. The rest of this section is the proof of the conjugation theorem.

10.2. An explicit formula for the conjugating map.
Introduce in Ω another chart straightening the field $v\,\partial/\partial x$. Let

$$t(x,\varepsilon) = \int_{-\delta}^{x} \frac{ds}{v(s,\varepsilon)}, \qquad (x,\varepsilon) \in \Omega,$$

be the time function. It is smooth on the interior of Ω, and grows at most polynomially on the boundary: this means that the function t as well as any of its derivatives can be majorized by an appropriate (negative) power of the distance from K:

$$\forall \alpha, \ \exists C = C_\alpha, \ N = N_\alpha: \quad |D^\alpha t(x,\varepsilon)| \leqslant C|\varepsilon(x^2 - \varepsilon)|^{-N}. \tag{10.1}$$

The map $T: (x,\varepsilon) \mapsto (t(x,\varepsilon), \varepsilon)$ takes the domain Ω into a subset of the semistrip $\Pi = \{|\varepsilon| < \delta^2, t > 0\}$. That same map takes the field $v\,\partial/\partial x$ into the constant one $\partial/\partial t$, the map $\widetilde{F}$ into the unit shift $\mathrm{id}+1$ and the perturbed map into a map of the form

$$\Phi: (t,\varepsilon) \mapsto (t + 1 + R(t,\varepsilon), \varepsilon),$$

R being flat at infinity as well as on the center line of the semistrip Π:

$$\forall \alpha \in \mathbb{Z}_+^2, \ N \in \mathbb{Z}_+^1, \ \exists C = C_{\alpha,N}: \quad |D^\alpha R(t,\varepsilon)| \leqslant C t^{-N} |\varepsilon|^N. \tag{10.2}$$

We prove the conjugation theorem in this chart: we shall establish the existence of a smooth map

$$H: (t,\varepsilon) \mapsto (t + h(t,\varepsilon), \varepsilon)$$

with h defined on $T(\Omega)$ and flat in the same sense (10.2) as R is, which conjugates Φ with the unit shift $\mathrm{id}+1$.

If H is such a map, then $H \circ \Phi = (\mathrm{id}+1) \circ H$, that is,

$$\mathrm{id}+1 + R + h \circ \Phi = \mathrm{id} + h + 1,$$

whence

$$h \circ \Phi = h - R. \tag{10.3}$$

As this was done when proving the solvability of the transfer equation in §8, we add to (10.3) the series of relations

$$h \circ \Phi^{[j+1]} = h \circ \Phi^{[j]} - R \circ \Phi^{[j]}, \qquad j = 1, 2, \ldots, \tag{10.3$_j$}$$

and obtain a formal solution to (10.3) as the series

$$h = \sum_{j=0}^{\infty} R \circ \Phi^{[j]}. \tag{10.4}$$

Thus we have reduced the continuation theorem to the following

MAIN LEMMA. *The series* (10.4) *converges to a smooth function on the semistrip, which is flat both at infinity and on the center line, provided that R is flat in the same sense.*

REMARK. Strictly speaking, the formulas (10.3$_j$) make no sense if the corresponding orbit leaves the domain $T(\Omega)$. In order to overcome this inconvenience, we extend the function R to the entire semistrip subject only to the flatness condition (clearly, this is possible), and after this procedure we extend the map Φ using the same formula. After such globalizations are made, we can write any sums over orbits of Φ without further comments.

The proof of the Main Lemma is based on the same ideas that were exploited in the proof of the solvability of the transfer equation. If Φ were exactly the unit shift, the above arguments *verbatim* would prove the assertion. In our case we have some additional work to do.

10.3. Estimates. For the sake of simplicity, we denote

$$\Phi^{[j]} = (F_j, \text{id}), \quad F_j = \text{id} + j + R_j, \quad j = 0, 1, \ldots.$$

We shall prove that the series (10.4) together with all its formal derivatives can be majorized by an appropriate scalar series $\sum_j j^{-N}$ with N as large as we wish. The estimates given below are based on the fact that any orbit of Φ is close enough to an arithmetical progression.

The functions R_j satisfy the following set of recursive equations:

$$R_{j+1} = R_j + R_j \circ \Phi^{[j]}. \tag{10.5}$$

DEFINITION. A family $\{G_j\}_{j=1,\ldots}$ of smooth functions bounded on the semistrip is called *shrinking*, if

$$\forall N \in \mathbb{N}, \ \exists C = C_N: \quad |G_j|_\Pi \leqslant C j^{-N}.$$

We intend to prove that the family $R \circ \Phi^{[j]}$ is shrinking, as well as all families $D^\alpha(R \circ \Phi^{[j]})$ obtained from it by differentiation.

PROPOSITION 1. 1. *If the semistrip Π is sufficiently narrow, then $F_j(t, \varepsilon) > j/2$.*

2. *All the R_j are uniformly bounded in Π.*

PROOF. 1. Take a strip in which $|R| < 1/2$; this is always possible because R is flat, see (10.2). In this case for the distance we have

$$F_{j+1}(t, \varepsilon) - F_j(t, \varepsilon) = 1 - R \circ \Phi^{[j]} > 1/2,$$

and the statement is obtained by adding these inequalities.

2. The uniform boundedness of R_j comes from the identity (10.5). Indeed, from (10.2) we deduce that

$$|R \circ F_j| < C|\varepsilon|^N (j/2)^{-N}$$

uniformly in Π. Taking $N = 2$ and adding the equalities, we majorize R_j by partial sums of the convergent series:

$$|R_j|_\Pi \leq C|\varepsilon|^2 \sum_{k=0}^{j} k^{-2}. \quad \square$$

As a corollary, we obtain the following statement.

PROPOSITION 2. *For any α the family $(D^\alpha R) \circ \Phi^{[j]}$ is shrinking.*

PROOF. The function R is flat at infinity, that is, it decreases more rapidly than any negative power of its argument t. But the t-coordinate of the function $\Phi^{[j]}$ in the semistrip is asymptotically equivalent to $j + \text{const}$. Hence we obtain the desired estimate of the superposition in Π.

Now note that the derivatives $D^\alpha(R \circ \Phi^{[j]})$ can be expressed as linear combinations of the derivatives $(D^\beta R) \circ \Phi^{[j]}$ computed at the corresponding point of the orbit, the coefficients being polynomial expressions in the derivatives of $D^\beta F_j$ with $|\beta| \leq |\alpha|$. The latter coincide with the derivatives of the functions R_j up to addition of the derivatives of the identity function (by definition of R_j).

Hence, taking into account Corollary 2, we conclude that the family $R \circ \Phi^{[j]}$ is shrinking provided that the family R_j is uniformly bounded on Π along with all its derivatives (the bound may depend on the order of the derivative).

It remains to prove the latter assertion. Differentiating the recursive equations (10.5), we obtain

$$D^\alpha R_{j+1} = D^\alpha F_j + \begin{pmatrix} \text{linear combination of derivatives of } R \\ \text{computed at the point } \Phi^{[j]} \text{ whose co-} \\ \text{efficients are polynomials in derivatives} \\ \text{of } R_j \text{ of order no greater than } |\alpha| \end{pmatrix}. \quad (10.6)$$

The family R_j is evidently bounded: this follows from (10.5) and Corollary 2. Indeed,

$$|R_k|_\Pi \leq |R|_\Pi + \sum_{j=1}^{k} C_2 j^{-2} < \infty.$$

Suppose that all the families of derivatives of R_j of order less than m are proved to be bounded. Let us prove the boundedness of all derivatives of order m. Write equations (10.6) for all derivatives of order m and note that the higher order derivatives appear in the right-hand sides *linearly*: the coefficients before them, as well as the free terms not containing those derivatives, form shrinking families by the induction assumption, as the products of families shrinking by Corollary 2 and certain bounded families. Denote $b_j = \max_{|\alpha|=m} |D^\alpha R_j|_\Pi$.

For these bounds we have

$$b_{j+1} \leq b_j + b_j C j^{-N} + C j^{-N} \quad (10.7)$$

for any choice of N (with a certain constant $C = C_N$ depending on N). All that remains to do is to prove that any scalar sequence b_j satisfying (10.7) is bounded. This fact can be easily demonstrated by induction: in fact, if we set $N = 3$, then we have the estimate

$$b_k \leqslant K \sum_{j=0}^{k} j^{-2}$$

valid for all sufficiently large K. Since the inverse square series converges, we have proved boundedness of all derivatives of R_j. Therefore the series (10.4) and all its termwise derivatives have terms forming shrinking families, so the sum (10.4) represents a function smooth in the semistrip.

As to the flatness of this function, we note that all the above arguments can be applied with some evident modifications to sums of the form

$$\sum_{j \geqslant 0} \varepsilon^{-p} t^p (R \circ \Phi^{[j]})(t, \varepsilon)$$

with an arbitrary $p \in \mathbb{N}$, proving that the function $\varepsilon^{-p} t^p h(t, \varepsilon)$ is bounded. Therefore we conclude that h is flat in the required sense.

The proof of the Main Lemma is complete. □

REFERENCES

[A] V. I. Arnold, *Supplementary chapters in the theory of ordinary differential equations*, "Nauka", Moscow, 1978; English transl., *Geometrical methods in the theory of ordinary differential equations*, Springer-Verlag, Berlin and New York, 1988.

[AAIS] V. I. Arnold, V. S. Afraimovich, Yu. S. Il'yashenko, and L. P. Shil'nikov, *Bifurcation theory*, Itogi Nauki i Tekhniki: Sovremennye Problemy Mat.: Fundamental'nye Napravleniya, vol. 5, VINITI, Moscow, 1986, pp. 5–218; English transl. in Encyclopaedia of Math. Sci., vol. 5, Springer-Verlag, Berlin and New York, 1989.

[AI] V. I. Arnold and Yu. S. Il'yashenko, *Ordinary differential equations*, Itogi Nauki i Tekhniki: Sovremennye Problemy Mat.: Fundamental'nye Napravleniya, vol. 1, VINITI, Moscow, 1985, pp. 7–150; English transl. in Encyclopaedia of Math. Sci., vol. 1 [Dynamical Systems, I], Springer-Verlag, Berlin and New York, 1988.

[DV] R. van Damme and T. P. Valkering, *Transient periodic behavior related to a saddle-node bifurcation*, J. Phys. A **20** (1987), 4161–4167.

[H] L. Hörmander, *The analysis of linear partial differential operators* I: *Distribution theory and Fourier analysis*, Springer-Verlag, Berlin, Heidelberg, New-York, and Tokyo, 1983.

[IY] Yu. S. Il'yashenko and S. Yu. Yakovenko, *Finite differentiable normal forms for local families of diffeomorphisms and vector fields*, Uspekhi Mat. Nauk **46** (1991), no. 1, 3–39; English transl. in Russian Math. Surveys **46** (1991).

[K] V. P. Kostov, *Versal deformations of differential forms of degree α on the line*, Funktsional. Anal. i Prilozhen. **18** (1984), no. 4, 81–82; English transl. in Functional Anal. Appl. **18** (1984).

[M] B. Malgrange, *Ideals of differentiable functions*, Oxford University Press, Oxford, 1966.

[MP] I. P. Malta and J. Palis, *Families of vector fields with finite modulus of stability*, Lecture Notes in Math., vol. 898, Springer-Verlag, New York and Berlin, 1981, pp. 212–229.

[P] M. M. Postnikov, *Lie groups and algebras*, "Nauka", Moscow, 1982. (Russian)

[T1] F. Takens, *Partially hyperbolic fixed points*, Topology **10** (1971), 133–147.
[T2] _____, *Normal forms for certain singularities of vector fields*, Ann. Inst. Fourier (Grenoble) **23** (1973), no. 2, 163–195.

CHAIR OF DIFFERENTIAL EQUATIONS, DEPARTMENT OF MECHANICS AND MATHEMATICS, MOSCOW STATE UNIVERSITY, MOSCOW 119899, RUSSIA

Translated by S. YAKOVENKO